BRITISH AND IRISH PUG MOTHS

BRITISH AND IRISH PUG MOTHS

(Lepidoptera: Geometridae, Larentiinae, Eupitheciini)

A guide to their identification and biology

Adrian M. Riley
and the late Gaston Prior

Illustrated by Adrian M. Riley
with photographic plates of set specimens by
David Wilson

Harley Books (B.H. & A. Harley Ltd.)
Martins, Great Horkesley,
Colchester, Essex CO6 4AH, England

All rights reserved. No part of this publication
may be reproduced, stored in a retrieval system,
or transmitted, in any form or by any means,
electronic, mechanical, photocopying, recording
or otherwise without the prior permission of
the publisher.

Text set in Plantin by
Rowland Phototypesetting Ltd.
and printed by St Edmundsbury Press Ltd.,
Bury St Edmunds, Suffolk

Line drawings and maps scanned by
Swaingrove Ltd., Rougham, Suffolk

Colour reproduced and printed by
Hilo Colour Printers Ltd., Colchester, Essex

Bound by Woolnoughs Bookbinding Ltd.,
Irthlingborough, Northants

British and Irish Pug Moths –
a guide to their identification and biology
© Adrian Riley, 2003,
and the estate of Gaston Prior

British Library Catalogue-in-Publication Data
applied for

ISBN 0 946589 51 8

Contents

	Preface	7
	Acknowledgements	9
	Introduction	11
1:	How to use this book	13

The distribution maps; Adult flight periods; Synonymy; Colour plates; Genitalia and the male abdominal plate; Identification of the adult moths; Larval foodplants and larval identification; Infraspecific terminology and checklist of infraspecific taxa; Systematic checklist of the British and Irish pug moths

2:	Historical review of the species	25
3:	Breeding and rearing pugs	29
4:	Descriptions of the British and Irish pug species	33
5:	Genitalia	146
6:	Illustrations of the larvae	164
7:	Vice-county distribution maps	169
	Colour Plates	195
Appendix I:	Glossary	221
Appendix II:	Table of phenology	224
Appendix III:	Foodplants and associated larvae	226
	References and bibliography	236
	Synonymic index of genera, species, subspecies, forms and aberrations	257

*This book is dedicated to Betty Prior
and the staff of the Rothamsted Insect Survey,
past and present,
and to the memory of Gaston Prior
(1914–1994)*

Preface

This book aims to provide as comprehensive an account as possible of the British and Irish pug moths (Geometridae, Larentiinae, Eupitheciini). Extensive data obtained through the Rothamsted Insect Survey (RIS) national light-trap network have allowed a thorough revision of the phenology and national distribution of the taxa comprising this tribe. The authors believe that here, for the first time, are provided clear diagnostic drawings of the wing patterns and genitalia of both sexes, and also distribution maps showing the status of each species according to Watsonian and Praeger vice-counties of Britain and Ireland; also diagnostic drawings of all the larvae; a phenological chart of the known British species; a list of foodplants and the associated larvae; relevant synonymy; and a fully comprehensive bibliography.

Under the species headings in Chapter 4, synonyms included are those which have been encountered in course of consulting the British literature, relating to description, variation and life histories. However, we do not claim this synonymy to be comprehensive. As our title indicates, this work is essentially a guide to the identification and biology of pug species. Any errors of omission or commission are the responsibility of the authors.

Sadly, my co-author, Gaston Prior, who made a major contribution to that part of the text dealing with synonymy and life histories, died in April 1994 following a protracted illness. It therefore became my responsibility to complete and later revise the whole work for publication. It is with much pleasure and no little relief that I now present it to my fellow lepidopterists.

September, 2003

Adrian Riley
Sculthorpe, Norfolk

Acknowledgements

The authors are greatly indebted to the following for their diverse advice and information generously given during the preparation of this book: J. D. Bradley, D. Carter; J. Fenn; D. S. Fletcher; B. Goater; G. M. Haggett; B. H. Harley; T. Karisch; B. R. Kerry; H. D. Loxdale; J. Reid; D. K. Riley; M. R. Shaw; L. R. Taylor; W. G. Tremewan; P. M. Waring; H.-J. Weigt and I. P. Woiwod.

We are particularly grateful to David Carter and the Trustees of the Natural History Museum, London, for granting access to the national collection, and to Bernard Skinner, the late Laurie Christie, J. Reid and H.-J. Weigt for providing specimens. We thank also Jim Porter and R. Austin for the loan of photographs for reference.

In researching literature sources, our task was made easier by the efforts of Joan Johnson, formerly of the Rothamsted Library and David Carter of the Natural History Museum. Translations of German and French literature were kindly made by N. Loxdale and J. Simpkins, formerly of Rothamsted.

Photographs for the colour plates of set specimens were taken by David Wilson. Many specimens were borrowed from the collections of J. Reid and B. Skinner. Colour transparencies of live moths were kindly loaned by Jim Porter, Ulrich Ratzel of Karlsruhe, Germany, Robert Thompson and Paul Waring. The photograph of *Eupithecia massiliata* was provided by Brian Goodey and skilfully incorporated in Plate 3 by Alan Rowland.

Working photographs used during the formulation of the species groups were taken by A. Martin of Rothamsted. Although not used in the published plates, his work was invaluable and very much appreciated.

In compiling details of the national distributions of each species the authors are only too aware of their dependence on the generosity and patience of a host of local recorders throughout the British Isles. Special acknowledgement is due to R. M. Palmer, G. Hancock, M. R. Young, the late R. I. Lorimer, M. R. Shaw and K. P. Bland for their help in collating Scottish records, and to K. G. M. Bond, T. A. Lavery and I. Rippey for their work in Ireland. We also take this opportunity to thank past and present operators of the Rothamsted Insect Survey (R.I.S.) light-traps for their support. Records were also gratefully received from the following individuals, some sadly no longer with us: D. J. L. Agassiz; M. Albertini; D. Allen; W. Angell; M. Anthony; V. W. Arnold; R. Austin; B. R. Baker; D. A. Barbour; R. J. Barnett; N. L. Birkett; K. Bland; I. Bolt; K. G. M. Bond; D. Boyce; A. Bromby; A.M. Broome; D. C. G. Brown; P. W. Brown; C. J. Cadbury, J. M. Campbell; J. M. Chalmers-Hunt; I. Christie; J. Clayton; A. Coates; J. Coates; R. V. Collier; G. A. Collins; J. Cooter; G. D. Craine; P. Cramp; T. J. Crawford; A. Creaser; J. N. D'Arcy; K. Darwin; P. Davey; M. Dempsey; D. Dey; B. Dickerson; E. Donaldson; D. W. Duncan; T. C. Dunn; R. Elliott; D. Evans; D. Eyre; G. T. Foggitt; J. B. Formstone; A. P. Fowles; R. G. Gabb; L. S. Garrad; C. Garrett-Jones; S. Gauld; R. Gaunt; C. Gibson; B. Goater; B. Goodey; J. P. Guest; G. M. Haggett; G. Hancock; L. W. Hardwick; M. Harper; M. Harvey; W. Henderson; S. Hewitt; S. Hind; D. Hipperson; B. Holmes; M. Holmes; R. Holmes; G. A. N. Horton; J. S. A. Hunter; P. P. Jennings; R. Johnson; A. Kearsley; S. Keiller; A. Kennard; A. King; R. Knight; F. R. La Forte; J. R. Langmaid; A. Law; K. Leaver; R. Leverton; A. G. Long; R. Long; N. R. Lowe; S. Lucas; J. McCleary; R. McCormick; V. McLoughlin; B. J. MacNulty; J. McPhail; S. A. Moran; I. K. Morgan; M. J. Morgan; R. Morris; C. Moscrop; A. A. Myers; S. Nash; G. Neil; G. H. Oldham; M. O'Meara; R. M. Palmer; S. M. Palmer; J. D. Parrack; M. L. Passant; E. C. Pelham-Clinton; M. Pennington; E. G. Philp; V. Philpott; A. Pickles; C. W. Plant; J. Potts; C. R. Pratt; A. Prichard; J. Radford; J. W. Rayner; N. D. Redgate; J. Reid; R. Revel; O. W. Richards; A. Richardson; M. Robinson; A. Russell; C. I. Rutherford; K. Saul; A. Scott; E. A. Seddon; J. Seddon; A. N. B. Simpson; I. F. Smith; F. N. Smith; A.-M. Smout; A. Spalding; R. Sheppard; N. J. Steeden; P. Sterling; P. J. Stevenson; B. Stewart; R. Thompson; M. C. Townsend; J. Trevor; A. Tyner; I. Viles; J. W. Ward; R. G. Warren; J. Warwick; S. Warwick; A. Watchman; P. Q. Winter; P. Wormell; S. Wright, and C. Yates.

We thank Paul Waring for his assistance in tracing county recorders in Britain and for allowing access to the database compiled during his ongoing survey of the rarer British Macrolepidoptera (JNCC, Species Conservation Branch). We also thank Tony Davis, Moth Conservation Officer, for extracting further British records from the Butterfly Conservation database.

Finally, but not least, we are greatly indebted to Jackie Fountain who undertook the mammoth task of word-processing the manuscript and to Basil and Annette Harley for their painstaking editing and all the sterling work that went into the making of this book.

Introduction

The British and Irish pug moths comprise the four genera *Eupithecia* Curtis, *Chloroclystis* Hübner, *Pasiphila* Meyrick, and *Gymnoscelis* Mabille, totalling 52 species. Ten of these species are further divided into 16 geographical forms or subspecies and the British status of a further three of them is at present unclear. The total of 71 recognized taxa constitutes the largest single group of closely related macrolepidoptera in the British Isles.

The group is characterized by a distinctive sclerotized plate on the ventral surface of the eighth abdominal segment in the male, which provides a useful diagnostic character for identification. The Dentated Pug (*Anticollix sparsata* (Treitschke)) lacks this plate, and the form of the male genitalia and differences in the wing venation also preclude its inclusion in the Eupitheciini. It has therefore been omitted from this work.

Identification of the adults is often a formidable problem in the study of this group. Many of the species are so similar in appearance that they are often confused or misidentified. Consequently, knowledge of the group's taxonomy, biology, phenology and distribution has always been uncertain. This book attempts to provide information which will enable the adults and larvae to be identified accurately and methodically. The colour plates depict the adults first in systematic order and then again in groups of superficially similar species. Diagnostic notes and line drawings of the distinctive characters that separate each species are given in the text. Special attention has been paid to those diagnostic features which remain when the moth becomes worn. Where identification is not possible by superficial examination, the illustrations of genitalia, which concentrate on the main diagnostic features, should be used.

An aim of this book is to encourage the enthusiast to rear the pugs. Notes on their biology, and suggestions for collecting and rearing the early stages are given. These are to help enthusiasts and local recorders find new localities and those rearing them to gain first-hand knowledge of their biology. It is hoped they will also stimulate a more intensive study of pug moths and thereby add to our knowledge of this rather neglected group.

1: How to use this book

The Distribution Maps

The distribution maps show for the first time the occurrence and status of each species in the Watsonian and Praeger vice-counties of Britain and Ireland respectively. Data for these maps have been collected from the Rothamsted Insect Survey (RIS) light-trap network (Taylor, 1986), the Biological Records Centre, Institute of Terrestrial Ecology, Monks Wood (Harding, 1992), county recorders, museums and regional lists, an exhaustive search of the entomological literature, and from many private collectors and recorders.

The symbols on the maps represent the following categories:

Category A – Generally distributed •

Species in this category are usually frequently recorded in most habitat types and are not restricted to one or a few specific foodplants.

Category B – Not generally distributed •

The range of these species is restricted to a greater or lesser degree by habitat requirement (e.g. heathland; oak woodland), or by a dependence on one or a few specific foodplants which are not themselves generally distributed (e.g. heather; thyme). The species may be locally very common and frequently recorded. Conversely, they may be widespread but only infrequently recorded. The relative abundance of species within this category may be anomalous as the recording of some may be more thorough than of others. Therefore it seems inadvisable to attempt subdivision of this category into 'locally scarce' or 'locally common'. It is more accurate to regard this category as an overall indication of distribution rather than localized abundance.

Category C – Uncertain status ○

The presence of a species in this category rests on nineteenth-century or subsequent occasional records there being no clear evidence of any extant established colonies.

It must be stressed that the absence of a symbol within a vice-county does not necessarily indicate the absence of a species but merely a lack of records. Assignment to the various category levels is based on available data. Despite the absence of data from certain vice-counties, it is generally possible to derive an overview of national patterns of distribution.

Adult flight periods

The flight periods for each species are based on records from the RIS light-trap network. Where insufficient data were available from this source, those given in Skinner (1984) have been cited.

The synonymy

The vernacular and scientific names of most species of Lepidoptera have changed over the years – sometimes several times. This is especially the case with the pug moths as problems with identification, particularly in the nineteenth century, often led to some species being known under several different names. Although the majority of these problems of identity have now been resolved, the discarded names remain in the literature, which can be confusing when the history of a species is being traced or when old records are evaluated. To clarify this we have included a complete synonymic index of the British species. It comprises vernacular/scientific and scientific/vernacular cross references to synonyms of all the known British pugs, their subspecies, forms and aberrations. It is integrated into the species index so that the reader is automatically referred to the appropriate page in the text.

The colour plates

These depict all the species, subspecies, common forms and some aberrations occurring in the British Isles. Most previous works have illustrated the moths only in their generally accepted phylogenetic order but, in this book, we felt that identification would be made easier if groups of superficially similar taxa were also illustrated together. Worn specimens of some species are also included. However, comparison with the illustrations should not be the sole method of identification, and a recommended procedure is given under the heading 'Identification of the adult moths' (p. 15). Colour photographs of live adults, wherever possible showing their natural resting positions, are reproduced on Plates A to D as a further aid to their identification.

Genitalia and the male abdominal plates

In order to ascertain the distribution of each species, it is obviously essential that wild-caught moths are correctly identified. However, due to wear and the close similarity of some species, this is often impossible to achieve by means of superficial characters alone. In such cases, the only recourse is to examine certain key diagnostic structures. The most useful of these for identification purposes are the sclerotized plate on the eighth abdominal tergite in the male and the bursa copulatrix (corpus bursae) in the female.

Illustrations of the internal genitalia in previous works (Petersen, 1909; Pierce, 1914; Bleszynski, 1965; and Agassiz et al., 1981) have lacked the required detail of the male aedeagus and female bursa copulatrix. These structures, and the shape and arrangement within them of the male aedeagal cornuti and of the female bursa and signa respectively, are unique to each species and are therefore the most important diagnostic characters. Various authors (e.g. Mikkola

(1992a,b) have suggested they form part of a species-specific 'lock and key' arrangement. Chapter 5 (p. 146) contains detailed drawings of the male aedeagus and the female bursa copulatrix (Figs 13–26) from slide preparations made by Riley.

The present authors agree with Mikkola (1992a) that, in most species, the male valva is unreliable as a diagnostic feature. Consequently only those species with a distinctive form of valva are figured (Fig.12). For difficult groups, the precise diagnostic features are shown.

The diagnostic importance of the male abdominal plate, situated on the eighth abdominal sternite, was first recognized by Buchanan White (1891). These plates can be examined by brushing the scales from the ventral surface of the segment. They are illustrated (Figs 27–29), together with the internal genitalia, in Chapter 5 (pp. 161–163).

Identification of the adult moths

Dichotomous keys of the external features of pugs have been published in Meyrick (1928), Agassiz *et al.* (1981) and Johnson (2002) but the present authors consider that such keys are not wholly satisfactory for the identification of the pugs to species level, due to their overdependence on colour or shades of colour. At infraspecific level this dependence is even greater. Although in freshly emerged specimens subtle differences in colour can sometimes be diagnostic, the description and personal interpretation of these is somewhat subjective. Moreover, with age or wear, the original colours usually change and are sometimes lost altogether, making reference to them misleading. In our opinion, based on many years' experience, the most accurate, efficient and indeed easiest means of identifying Pugs is by comparison with good quality illustrations of both fresh and also old and worn specimens, supported by a succinct diagnostic text.

Identification procedure:

Stage 1. Compare the specimen with those illustrated in the colour plates and find a species or group of species which it fits.

Stage 2. Refer to the text in Chapter 4 for the species which it most resembles and consult the section 'Characteristic features'. It should be noted that the measurement of the wingspan is twice the distance from the centre of the thorax to the apex of the forewing. An average figure is cited but there is often variation.

Stage 3. If the majority of features do not correspond, return to Stage 1.

Stage 4. If the majority of features do correspond, refer to the section 'Similar species' and its accompanying line drawings and compare the features with those of all the species listed.

Stage 5. If an identification cannot be made at Stage 4, the genitalia or the abdominal plate should be examined and compared with the figures in Chapter 5.

Stage 6. Having made an identification, the following final checks should be made:

i. that the specimen was caught within the appropriate geographical range;
ii. that the flight period corresponds with that given in the text, at least approximately;
iii. that the specimen was caught in the stated preferred habitat or that this occurs within a reasonable distance of the place of capture;
iv. that at least one of the listed larval foodplants is present nearby.

If any doubt remains, the genitalia should be re-examined.

Larval foodplants and larval identification

Under the species accounts in Chapter 4, only the main foodplants are given. However, Appendix III (p. 226) lists all the known foodplants on which the larvae of the pugs feed. Where a foodplant has been recorded in the literature but has not been confirmed by the present authors, the relevant reference is given.

For each foodplant, the relevant pug species are listed. For example, a larva found feeding on the flowers of blackthorn can probably be narrowed down to two possibilities: *Pasiphila rectangulata* or *P. chloerata*. The drawings of these species in Chapter 4 should then be consulted.

The larval figures in Chapter 6 (Figs 30–33) show the typical form of each species, and variation is discussed in the accompanying text. They are based on the authors' notes and sketches of living larvae, with occasional reference to the works of Buckler (1899), Dietze (1910; 1913), Juul (1948), and Haggett (1981a). Black-and-white drawings of the larvae are preferred to colour photographs because pug larvae vary greatly in colour; it is therefore more important to compare the structure and markings.

Infraspecific terminology

In the older literature, in particular, many of the terms describing infraspecific forms of polymorphic taxa have been used ambiguously. Unfortunately, some ill-defined terminology has persisted. Therefore, for the purpose of this book, those terms used to describe such taxa in the Eupitheciini have been revised using the following definitions:

Subspecies

A subspecies is a taxon which is predominant, distinct in appearance and allochronic (separated by different adult flight periods) or allopatric (geographically isolated) from the nominotypical or other subspecies. Genetic isolation is an essential prerequisite for the use of this term. It has often been used erroneously to describe clinal geographic forms (see below) which are sympatric (occupying the same locality) over part of their respective distributional ranges. Island subspecies, which appear to be part of a discontinuous cline, must be distinct in appearance from the subspecies, or the predominant form, occurring on the nearest adjacent island or the mainland.

The name given to the subspecies can either be used trinomially, e.g. *Eupithecia denotata jasioneata* Crewe, or be preceded by the abbreviation 'subsp.', e.g. *Eupithecia denotata* (Hübner) subsp. *jasioneata* Crewe.

Form

A form is a taxon distinct in appearance and of infrasubspecific rank, but sympatric or synchronic with the nominal taxon, which can recur or predominate at a given geographical location. Although a form may predominate in some populations, it can occur elsewhere at very low frequency (e.g. *E. vulgata* f. *scotica* Cockayne, which is to be found occasionally in southern England). Provided these individuals conform to the original description of the form, they should be referred to as such and not as aberrations. A form conforming to the original description of a subspecies, but found outside its isolated geographical range (e.g. some of the dark *fumosae*-like individuals of *E. venosata* which occur in Orkney), should be referred to as a form approaching that subspecies. Under such circumstances it would be unwise and unnecessary to describe a further formal taxon for these individuals. Forms occur in response to various environmental factors such as climate, pollution and larval foodplant as well as to selective pressures such as predation. They may be controlled genetically (e.g. the melanic form *atropicta* of *E. vulgata*), or by external influences (e.g. the *Calluna*-feeding form *goossensiata* of *E. absinthiata*). The name given to a form is preceded by the abbreviation 'f.', e.g. *Eupithecia vulgata* (Haworth) f. *scotica*.

The continuous spatial gradation of superficial characters from one form to another is termed a 'cline'. Such geographic forms are sometimes referred to as races, e.g. the Scottish 'race' of *E. vulgata* f. *scotica*.

Aberration

An aberration is an individual variant, resulting from genetic mutation or extreme environmental anomalies, which can occur erratically in any population. The degree of mutation may be extreme or slight; superficial or structural. Theoretically, it would be possible for the frequency of an aberration to increase over a period of time if it were subjected to neutral or positive selection. In such circumstances, the taxon would then be regarded as a form. In the past, many authors applied the term 'aberration' to any varietal taxon, regardless of how it had been caused or its frequency. Where these are now known to satisfy the criteria for a higher taxonomic ranking, their status has herein been altered accordingly. Where their status is uncertain, the original rank is retained. The name given to an aberration is preceded by the abbreviation 'ab.', e.g. *Gymnoscelis rufifasciata* (Haworth) ab. *nigrofasciata* Dietze.

Variety

A largely outdated blanket term denoting any individual showing morphological or superficial deviation from the nominotypical form.

Race

A synonymic term used to describe geographic forms or hostplant-related populations of a given species, e.g. the sea buckthorn-feeding 'race' of *E. fraxinata* or the Scottish 'race' of *E. vulgata*. (See *Form* above.)

Checklist of infraspecific taxa

The infraspecific taxa of the eupitheciine subspecies and common forms are re-assessed below according to the preceding criteria. The status expressed in previous works is given in brackets.

Eupithecia tenuiata (Hübner)   Slender Pug
  f. (subsp.) *cinerae* Gregson (1888)
Occurs regularly in some Scottish populations, along with individuals of the nominotypical form.
  f. (ab.) *johnsoni* Harrison (1931)
Occurs regularly in N.E. England and the Midlands.

Eupithecia pulchellata Stephens   Foxglove Pug
  f. (subsp.) *hebudium* Sheldon (1899)
Of erratic distribution and apparently sympatric with the nominotypical form over parts of its geographical range.

Eupithecia exiguata (Hübner)   Mottled Pug
  f. (subsp.) *muricolor* Prout (1938)
There is no evidence for spatial or temporal isolation of this form.

Eupithecia venosata (Fabricius)   Netted Pug
  subsp. (subsp.) *ochracae* Gregson (1886)
Restricted to Orkney where it is allopatric from morphologically distinct taxa in Shetland and the Scottish mainland.
  subsp. (subsp.) *fumosae* Gregson (1887)
Predominates in Shetland where it is allopatric from superficially similar taxa in Orkney. Individuals approaching subsp. *fumosae* are also known from the Inner Hebrides, Rannoch and Forres.
  f. (subsp.) *hebridensis* Parkinson Curtis (1944a)
Stated to occur as a form slightly darker than the nominotypical form in Canna, Islay and Skye. However, the species seems extremely variable in these localities and no one form appears to predominate.
  subsp. (subsp.) *plumbea* Huggins (1962a)
Restricted to the Blasket Islands (Ireland) and parts of the adjacent mainland. Isolated from populations containing individuals which are similar but morphologically distinct.

Eupithecia trisignaria Herrich-Schäffer   Triple-spotted Pug
  f. (ab.) *angelicata* Prout (1938)
Melanic form which occurs commonly in some populations.

Eupithecia intricata (Zetterstedt).
  subsp. (subsp.) *arceuthata* Freyer (1841)   Freyer's Pug
Morphologically distinct and isolated from subsp. *intricata* of mainland Europe.
  subsp. (subsp.) *millieraria* Wnukowsky (1929)   Edinburgh Pug
Appears to be allopatric from subsp. *arceuthata* and morphologically distinct.
  subsp. (subsp.) *hibernica* Mere (1964)   Mere's Pug
Restricted to the Burren, Co. Clare, and apparently isolated from populations of subsp. *arceuthata*.

Eupithecia satyrata (Hübner)   Satyr Pug
  f. (subsp.) *curzoni* Gregson (1884a,b)
Found mainly in Shetland but also occurs in Orkney and parts of the mainland.
  f. (subsp.) *callunaria* Doubleday (1850b)
The northern English and Scottish form. There appears to be a north-south cline with no apparent geographical or temporal division.
Eupithecia absinthiata (Clerck)   Wormwood Pug
  f. (f.) *goossensiata* Mabille (1869a) (Ling Pug)
The morphology of this form appears to be dependent on the larval foodplant (*Calluna vulgaris*) and it is consequently restricted to heathland. Genetic isolation from peripheral populations of the nominotypical form suggests subspecific status but further research is required to clarify this point. Some authorities consider *goossensiata* to be specifically distinct.
Eupithecia vulgata (Haworth)   Common Pug
  f. (ab.) *atropicta* Dietze (1910)
Melanic form which occurs commonly in some populations.
  f. (subsp.) *scotica* Cockayne (1951b)
The northern English and Scottish form. There is a north-south cline with no geographical or temporal division.
  f. (subsp.) *clarensis* Huggins (1962b)
Although this form is found mainly on the Burren, Co. Clare, intermediates occur between it and the *scotica*-like form found in surrounding localities. As it is sympatric with these intermediate forms, subspecific status is not appropriate.
  f. (ab.) *unicolor* Lempke (1951)
Incorrectly called a melanic form by previous authors. Lempke (1947) describes the ground colour as normal with the usual markings absent.
Eupithecia tripunctaria Herrich-Schäffer   White-spotted Pug
  f. (ab.) *angelicata* Barrett (1877)
Melanic form which occurs regularly throughout the range of the species.
Eupithecia denotata (Hübner)   Campanula Pug
  subsp. (subsp.) *jasioneata* Crewe (1881)   Jasione Pug
Probably allopatric and therefore unable to cross-breed with the nominotypical subspecies under natural conditions. If colonies of subsp. *denotata* (Hübner, 1813) are found to occur on cultivated *Campanula* species in gardens adjacent to or amongst coastal populations of subsp. *jasioneata*, the status of the latter taxon will require reassessment.
Eupithecia subfuscata (Haworth)   Grey Pug
  f. (ab.) *obscura* Dietze (1910)
Melanochroic form which occurs commonly in some areas.
  f. (ab.) *obscurissima* Prout (1914)
Melanic form which occurs commonly in some areas.
Eupithecia icterata (Villers)   Tawny-speckled Pug
  f. (subsp.) *subfulvata* Haworth (1809)
This form and f. *cognata* Stephens (q.v.) were previously given subspecific status. According to the present criteria, this rank is inappropriate.

f. (subsp.) *cognata* Stephens (1831); f. (ab.) *oxydata* Treitschke (1928); f. (ab.) *grisescens* Lempke (1951).
Varying degrees of reduction in fulvous coloration. The extreme is reached in f. *cognata* which has previously been described as a subspecies. This is not appropriate as f. *cognata* is sympatric with f. *subfulvata* and with the other forms listed here.

Eupithecia succenturiata (Linnaeus)   Bordered Pug
  f. (ab.) *obscurata* Lempke (1951)
Form with white of forewings greatly reduced. Occurs regularly throughout the range of the species. This form has been referred to erroneously as *disparata* Hübner (1799), but Hübner describes f. *disparata* merely as having a rust-coloured suffusion along the hind-margin of the forewing. There is no reference to a reduction in the white ground colour.

Eupithecia nanata (Hübner)   Narrow-winged Pug
  subsp. (subsp.) *angusta* Prout (1914)
Morphologically distinct and isolated from the Continental nominate subsp. *nanata*.
  f. (ab.) *oliveri* Prout (1915)
Melanic form which occurs frequently in some populations; occasionally dominant.

Eupithecia fraxinata Crewe   Ash Pug
  f. (ab.) *unicolor* Prout (1915)
Melanic form which occurs regularly in some populations.

Eupithecia virgaureata Doubleday   Golden-rod Pug
  f. (ab.) *nigra* Lempke (1951)
Melanic form which dominates some populations.

Eupithecia abbreviata Stephens   Brindled Pug
  f. (ab.) *hirschkei* Bastelberger (1908)
Dark form which occurs regularly in some populations.
  f. (ab.) *nigra* Cockayne (1953)
Melanic form which occurs frequently in some populations; occasionally dominant.

Eupithecia pusillata ([Denis & Schiffermüller])   Juniper Pug
  subsp. (subsp.) *anglicata* Herrich-Schäffer (1863)   Kentish Tamarisk Pug
Possibly a form but as this taxon is believed to be extinct it is not possible to re-examine its status accurately. Its original status has therefore been retained.
  f. (subsp.) *scotica* Dietze (1913)
A pale form which predominates in some Scottish localities but is sympatric with the nominotypical form.

Eupithecia lariciata (Freyer)   Larch Pug
  f. (ab.) *nigra* Prout (1915)
Melanic form which occurs frequently in some populations.

Pasiphila rectangulata (Linnaeus)   Green Pug
  f. (ab.) *nigrosericeata* Haworth (1809); f. (ab.) *anthrax* Dietze (1913)
Melanic forms which occur frequently in some populations; either form can be dominant.

Systematic checklist of the British and Irish pug moths

In previous systematic lists of the British Lepidoptera (cf. Kloet & Hincks, 1972; Bradley & Fletcher, 1987; Bradley, 1998, 2000) *E. absinthiata* is followed by *E. assimilata* and then by *E. expallidata*. However, the structure of the genitalia of *E. absinthiata* and *E. expallidata* is sufficiently similar to suggest much closer affinity between these two species than has previously been recognized. It is therefore proposed that, as adopted below, *E. expallidata* should follow *E. absinthiata*, and *E. assimilata* be displaced to follow *E. expallidata*. The Log Book numbering system, devised by Bradley (1979) and adopted by the national Lepidoptera recording scheme (Biological Records Centre, Monks Wood), is given below after the species names in square brackets.

Eupithecia tenuiata (Hübner, 1813)   Slender Pug [1811]
  = *tenuiaria* Doubleday, 1849

E. inturbata (Hübner, 1817)   Maple Pug [1812]
  = *subciliata* Doubleday, 1856
  = *subciliaria* Morris, 1861

E. haworthiata Doubleday, 1856   Haworth's Pug [1813]
  = *isogrammaria* sensu Herrich-Schäffer, 1848
  = *haworthiaria* Morris, 1861

E. plumbeolata (Haworth, 1809)   Lead-coloured Pug [1814]
  = *begrandaria* Boisduval, 1840
  = *plumbeolaria* Doubleday, 1849

E. abietaria (Goeze, 1781)   Cloaked Pug [1815]
  = *pini* (Retzius, 1783), nec (Linnaeus, 1758)
  = *strobilata* (Borkhausen, 1794)
  = *togata* (Hübner, 1814–17)
  = *togaria* Boisduval, 1840

E. linariata ([Denis & Schiffermüller], 1775)   Toadflax Pug [1816]
  = *linariaria* (Borkhausen, 1794)
  = *linaria* Boisduval, 1840

E. pulchellata Stephens, 1831   Foxglove Pug [1817]
  = *pulchellaria* Doubleday, 1849

E. irriguata (Hübner, 1813)   Marbled Pug [1818]
  = *variegata* (Haworth, 1809), nec (Scopoli, 1763)
  = *irriguaria* Boisduval, 1840

E. exiguata (Hübner, 1813)   Mottled Pug [1819]
  = *trimaculata* (Haworth, 1809), nec (Villers, 1789)
  = *ochreata* Stephens, 1831
  = *exiguaria* Boisduval, 1840
  = *lanceolaria* Wood, 1854

E. insigniata (Hübner, 1790)   Pinion-spotted Pug [1820]
  = *consignata* (Borkhausen, 1794)
  = *consignaria* Boisduval, 1840

E. *valerianata* (Hübner, 1813)   Valerian Pug [1821]
  = *viminata* Doubleday, 1858
  = *viminaria* Morris, 1861
E. *pygmaeata* (Hübner, 1799)   Marsh Pug [1822]
  = *pygmaearia* Boisduval, 1840
  = *palustraria* Doubleday, 1850
E. *venosata* subsp. *venosata* (Fabricius, 1787)   Netted Pug [1823]
  = *decussata* (Donovan, 1799)
  = *venosaria* Boisduval, 1840
E. *venosata* subsp. *ochracae* Gregson, 1886
E. *venosata* subsp. *fumosae* Gregson, 1887
E. *venosata* subsp. *plumbea* Huggins, 1962
E. *egenaria* Herrich-Schäffer, 1848   Pauper Pug, Fletcher's Pug [1824]
  = *undosata* Dietze, 1875
E. *centaureata* ([Denis & Schiffermüller], 1775)   Lime-speck Pug [1825]
  = *oblongata* (Thunberg, 1784)
  = *centaurearia* Boisduval, 1840
E. *trisignaria* Herrich-Schäffer, 1848   Triple-spotted Pug [1826]
E. *intricata* (Zetterstedt, 1839) subsp. *arceuthata* (Freyer, 1842)
                                                             Freyer's Pug [1827]
E. *intricata* subsp. *millieraria* Wnukowsky, 1929   Edinburgh Pug
  = *anglicata* Millière, 1869, nec Herrich-Schäffer, 1863
  = *helveticaria* sensu auctt.
E. *intricata* subsp. *hibernica* Mere, 1964   Mere's Pug
E. *satyrata* (Hübner, 1813)   Satyr Pug [1828]
  = *satyraria* Boisduval, 1840
  = *curzoni* Gregson, 1884
E. *cauchiata* (Duponchel, 1831)   Doubleday's Pug [1829]
  = *pernotata* sensu Doubleday, 1858
  = *pernotaria* sensu Morris, 1861
E. *absinthiata* (Clerck, 1759)   Wormwood Pug [1830]
  = *minutata* ([Denis & Schiffermüller], 1775)
  = *elongata* (Haworth, 1809)
  = *notata* Stephens, 1831
  = *innotata* sensu Wood, 1835
  = *minutaria* Boisduval, 1840
  = *elongaria* Doubleday, 1849
  = *goossensiata* Mabille, 1869   Ling Pug
  = *knautiata* Gregson, 1874   Scabious Pug (1831)
E. *expallidata* Doubleday, 1856   Bleached Pug [1833]
  = *expallidaria* Morris, 1861
E. *assimilata* Doubleday, 1856   Currant Pug [1832]
  = *assimilaria* Morris, 1861

E. vulgata (Haworth, 1809)   Common Pug [1834]
= *clusterata* (Hübner, 1809)
= *austerata* (Hübner, 1825)
= *austeraria* Boisduval, 1840
= *vulgaria* Morris, 1861
E. tripunctaria Herrich-Schäffer, 1852   White-spotted Pug [1835]
= *albipunctata* (Haworth, 1809), nec (Hufnagel, 1767)
E. denotata subsp. *denotata* (Hübner, 1813)   Campanula Pug [1836]
= *campanulata* Herrich-Schäffer, 1861
E. denotata subsp. *jasioneata* Crewe, 1881   Jasione Pug
E. subfuscata (Haworth, 1809)   Grey Pug [1837]
= *singulariata* sensu Haworth, 1809
= *castigata* (Hübner, 1813)
= *castigaria* Boisduval, 1940
= *blancheata* Cooke, 1881
E. icterata (Villers, 1789)   Tawny-speckled Pug [1838]
E. succenturiata (Linnaeus, 1758)   Bordered Pug [1839]
= *disparata* (Hübner, 1799)
= *succenturiaria* Boisduval, 1840
E. subumbrata ([Denis & Schiffermüller], 1775)   Shaded Pug [1840]
= *scabiosata* (Borkhausen, 1794)
= *piperitata* Stephens, 1829
= *piperata* Stephens, 1831
= *subumbraria* Boisduval, 1840
= *piperaria* Doubleday, 1849
E. millefoliata Rössler, 1866   Yarrow Pug [1841]
= *achilleata* Mabille, 1869
E. simpliciata (Haworth, 1809)   Plain Pug [1842]
= *subnotata* (Hübner, 1813)
= *subnotaria* Boisduval, 1840
E. sinuosaria Eversmann, 1848   Goosefoot Pug [1842a]
E. distinctaria Herrich-Schäffer, 1848, subsp. *constrictata* Guenée, 1857
                                                         Thyme Pug [1843]
= *constrictaria* Morris, 1861
E. indigata (Hübner, 1813)   Ochreous Pug [1844]
= *indigaria* Boisduval, 1840
E. pimpinellata (Hübner, 1813)   Pimpinel Pug [1845]
= *pimpinellaria* Boisduval, 1840
= *denotata* sensu Doubleday, 1856
E. nanata (Hübner, 1813) subsp. *angusta* Prout, 1914   Narrow-winged Pug [1846]
= *angustata* (Haworth, 1809), nec (Gmelin, 1790)
E. extensaria (Freyer, 1844) subsp. *occidua* Prout, 1914   Scarce Pug [1847]
= *prolongata* sensu Dietze, 1910
E. fraxinata Crewe, 1863   Ash Pug [1848–50]

E. virgaureata Doubleday, 1861   Golden-rod Pug [1851]
= *pimpinellata* sensu Doubleday, 1856
= *virgaurearia* Morris, 1861
E. abbreviata Stephens, 1831   Brindled Pug [1852]
= *nebulata* (Haworth, 1809) nec (Scopoli, 1763)
= *subfasciata* Stephens, 1829
= *abbreviaria* Doubleday, 1849
E. dodoneata Guenée, 1857   Oak-tree Pug [1853]
= *dodonearia* Morris, 1861
E. massiliata Millière, 1865   Epping Pug [1853a]
= *peyerimhoffata* Millière, 1870
E. pusillata subsp. *pusillata* ([Denis & Schiffermüller], 1775   Juniper Pug [1854]
= *laevigata* sensu Haworth, 1809
= *sobrinata* (Hübner, 1817)
= *sobrinaria* Boisduval, 1840
= *scotica* Dietze, 1910
E. pusillata subsp. *anglicata* Herrich-Schäffer, 1863   Kentish Tamarisk Pug
= *ultimaria* sensu Westwood, 1854
= *stevensata* Webb, 1896
E. phoeniceata (Rambur, 1834)   Cypress Pug [1855]
E. ultimaria Boisduval, 1840   Channel Islands Pug [1855a]
E. lariciata (Freyer, 1842)   Larch Pug [1856]
E. tantillaria Boisduval, 1840   Dwarf Pug [1857]
= *subumbrata* sensu Hübner, 1799
= *pusillata* sensu auctt.
= *piceata* Prout, 1914
Chloroclystis v-ata (Haworth, 1809)   V-Pug [1858]
= *coronata* (Hübner, 1813)
= *coronaria* (Doubleday, 1849)
Pasiphila chloerata (Mabille, 1870)   Sloe Pug [1859]
P. rectangulata (Linnaeus, 1758)   Green Pug [1860]
= *viridulata* (Hufnagel, 1767)
= *nigrosericeata* (Haworth, 1809)
= *sericeata* (Haworth, 1809)
= *subaerata* (Hübner, 1817)
= *rectangularia* (Boisduval, 1840)
P. debiliata (Hübner, 1817)   Bilberry Pug [1861]
= *nigropunctata* (Chant & Bentley, 1833)
= *debiliaria* (Boisduval, 1840)
Gymnoscelis rufifasciata (Haworth, 1809)   Double-striped Pug [1862]
= *bistrigata* (Haworth, 1809) nec (Borkhausen, 1790)
= *pumilata* (Hübner, 1813)
= *strobilata* sensu Stephens, 1829
= *recictaria* (Boisduval, 1840)
= *globulariata* Millière, 1861

2: An Historical Review of the British and Irish Pugs

Most pug moths are small, indistinctly marked and easily overlooked: this has led, in the past, to problems with their identification, as is obvious in many early works. Some species were often known under several different names, erroneously attributing specific status to mere forms of the same species (see Stephens, 1829–31). Indeed, even today, confusion still surrounds the specific status within two groups – *fraxinata*, *innotata* and *tamarisciata* (see Haggett, 1963), and *goossensiata* and *absinthiata* (see Riley, 1986a). However, over the years the following works have contributed greatly to our understanding of this large and complex group of moths.

The first British entomological book to include a pug moth was *The Aurelian* by Moses Harris (1766) with The Lime-Speck (*Eupithecia centaureata*). Edward Donovan (1799), in his *Natural History of British Insects*, also included only one species, The Netted Pug (*E. venosata*), which he called The Pretty Widow. Adrian Haworth's *Prodromus Lepidopterorum Britannicorum* (1802) catalogues just four species, without descriptions, but in the second part of his *Lepidoptera Britannica* (1803–1828), published in 1809, he gave original descriptions of six valid species in addition to those of a further eleven species since synonymized.

John Curtis in his *British Entomology* (1824–39) and *A Guide to the Arrangement of British Insects* (1829), J. F. Stephens in his *Systematic Catalogue of British Insects* (1829) and Volume 3 of his *Illustrations of British Entomology, Haustellata* (1831), and James Rennie in his *Conspectus of the Butterflies and Moths found in Britain* (1832) each increased the inventory of known pug species, but a notable milestone was reached by W. Wood, whose work, *Index Entomologicus, or a complete illustrated Catalogue of the Lepidopterous Insects of Great Britain* (1839), with 24 species of pugs, was the first to include accurate hand-coloured illustrations.

Henry Doubleday's *A Synonymic List of British Lepidoptera* (1850a) was reissued in a revised edition in 1859. This included 40 species. Henry Tibbats Stainton, in Volume 2 of *A Manual of British Butterflies and Moths* (1859), gave concise descriptions of the adults of 39 species, a few with delicately engraved illustrations, and accounts of the larvae of 23. It was the first standard work covering British moths. This was followed in 1869 by Edward Newman's *An Illustrated Natural History of British Moths*. He included good descriptions of the adults of 47 species (some now relegated to subspecific status), of which 22 were illustrated by engravings, and also of the larvae of 41 species. The larval descriptions were supplied by the Revd H. Harpur Crewe who was one of the greatest authorities on British pugs. Over a period of more than twenty years from 1859 until shortly before his death in 1881, Crewe published excellent detailed descriptions of the larvae of 43 species in no fewer than six journals: *The Entomologist*, *The Zoologist*, *The Entomologist's Annual*, *The Entomologist's Weekly Intelligencer*, *The Weekly Entomologist*, and *The Entomologist's Monthly Magazine*. These have formed the basis of many subsequent works including O. W. Wilson's *The Larvae of the British Lepidoptera and their Foodplants* (1880) which has coloured illustrations of the larvae of 26 British pugs and descriptions in the text of an additional 20. The larvae of 46 species, showing a wide range of variation, were much more accurately and skilfully illustrated in colour by William Buckler (1899), although even Buckler's work contained some errors. Eighty years later, G. M. Haggett (1981a) sought to remedy these faults and omissions in his *Larvae of the British Lepidoptera not figured by Buckler* in which he illustrated nine pug species and subspecies on three colour plates.

Between 1893 and 1907, C. G. Barrett published the 11 volumes of *The Lepidoptera of the British Islands*. In volume 9 (1904), there were descriptions of 47 pugs with useful hand-coloured illustrations of both adults and larvae, the latter mostly after Buckler, but larval variation was not depicted.

Probably the most widely used work on British moths ever published, *The Moths of the British Isles*, was written by Richard South (1907–08). It was the first to illustrate the adults by means of photographs of set specimens (except for three plates specially drawn by Horace Knight), all of which were reproduced by process engraving. Over the next forty years, in the course of many impressions and two further editions, the printing plates became worn and the illustrations were consequently of less and less help in identification. In the new, fourth edition (1961), drawings by H. D. Swain replaced photographed specimens. Although the photographic plates had been unclear and of limited use in the identification of critical species, they were the forerunners of subsequent works which used improved standards of both photography and photo-lithographic reproduction techniques.

The first of these for the identification of pugs was the British Entomological and Natural History Society's *An Identification Guide to the British Pugs* (Agassiz *et al.*, 1981) with photographic colour reproductions of the adults of all the British species, subspecies and common forms recorded up to that date. Of better quality, Bernard Skinner's *Colour Identification Guide to Moths of the British Isles* (1984), illustrated by colour plates from photographs by David Wilson, is now in a second revised edition (1998) and includes all species recorded up to 1996. It has become a standard work for the identification of the adults of the British macrolepidoptera. The larvae have also been comprehensively illustrated photographically in Jim Porter's *Colour Identification Guide to Caterpillars of the British Isles* (1997). A companion volume to Skinner's work, 51 pug larvae are individually depicted on its excellent colour plates with an informative accompanying text covering life histories.

Progressive phylogenetic or systematic classifications, with synonymy, have been published over recent decades: the revised Lepidoptera part of Kloet & Hinck's *A checklist of British Insects* (Kloet & Hincks, 1972); *A recorder's log book*

(Bradley & Fletcher, 1979), in which a numbering system for species was introduced for the National Recording Scheme data bank at ITE, Monks Wood, and is now widely used; and more recently, *Checklist of Lepidoptera recorded from the British Isles* (Bradley, 1998, revised 2000). For reasons explained in the text, the present work does not always adopt this classification and nomenclature.

The entomological journals provide the principal medium for the exchange of ideas and the presentation of newly-found knowledge. Many hundreds of articles and papers concerning the pugs have appeared in these over the years and most of the aforementioned authors relied heavily on information from such publications to make their own works more complete. It is not proposed to discuss such publications here as they are so numerous. However, an extensive list of these papers, supplemented by brief notes on each, is contained in the bibliography at the end of this book.

For ease of reference, a chronological list of additions to the British list of species, subspecies and geographical forms, sometimes under a synonym, is printed below. The authors and dates of publication of these first British records are given under each species' heading in Chapter 4 with a full synonymy and brief resumé of their history.

1766	*Eupithecia centaureata* ([D.& S.])	(Harris, 1766)	Lime-speck Pug
1799	*E. venosata* (Fabr.)	(Donovan, 1799)	Netted Pug
1802	*E. absinthiata* (Cl.)	(Haworth, 1802)	Wormwood Pug
1809	*E. icterata* (Vill.) f. *subfulvata* (Haw.)	(Haworth, 1809)	Tawny-speckled Pug
	E. plumbeolata (Haw.)	(Haworth, 1809)	Lead-coloured Pug
	E. irriguata (Hb.)	(Haworth, 1809)	Marbled Pug
	E. linariata ([D.& S.])	(Haworth, 1809)	Toadflax Pug
	E. insigniata (Hb.)	(Haworth, 1809)	Pinion-spotted Pug
	E. vulgata (Haw.)	(Haworth, 1809)	Common Pug
	E. subfuscata (Haw.)	(Haworth, 1809)	Grey Pug
	E. succenturiata (L.)	(Haworth, 1809)	Bordered Pug
	E. simpliciata (Haw.)	(Haworth, 1809)	Plain Pug
	E. nanata (Hb.)	(Haworth, 1809)	Narrow-winged Pug
	E. pusillata ([D.& S.])	(Haworth, 1809)	Juniper Pug
	Chloroclystis v-ata (Haw.)	(Haworth, 1809)	V-Pug
	Pasiphila rectangulata (L.)	(Haworth, 1809)	Green Pug
	Gymnoscelis rufifasciata (Haw.)	(Haworth, 1809)	Double-striped Pug
1831	*Eupithecia pulchellata* Steph.	(Stephens, 1831)	Foxglove Pug
	E. exiguata (Hb.)	(Stephens, 1831)	Mottled Pug
	E. subumbrata ([D.& S.])	(Stephens, 1831)	Shaded Pug
	E. abbreviata Steph.	(Stephens, 1831)	Brindled Pug
	E. fraxinata (Crewe) (as *E. innotata* (Hufn.))	(Stephens, 1831)	Ash Pug
	E. absinthiata (Cl.) f. *goossensiata* Mab.	(Stephens, 1831)	Ling Pug – f. of Wormwood Pug
	E. icterata (Vill.) f. *cognata* Steph.	(Stephens, 1831)	f. of Tawny-speckled Pug
1833	*Pasiphila debiliata* (Hb.)	(Chant & Bentley, 1833)	Bilberry Pug
1845	*Eupithecia abietaria* (Goeze)	(Stevens, 1845; Newman, 1845)	Cloaked Pug

1850	E. tenuiata (Hb.)	(Doubleday, 1850a)	Slender Pug
	E. pygmaeata (Hb.)	(Doubleday, 1850b)	Marsh Pug
	E. satyrata (Hb.) f. callunaria Doubl.	(Doubleday, 1850b)	f. of Satyr Pug
1851	E. indigata (Hb.)	(Douglas, 1851)	Ochreous Pug
	E. pusillata ([D. & S.]) subsp. anglicata H.-S.	(Westwood, 1851)	Kentish Tamarisk Pug – subsp. of Juniper Pug
1852	E. pimpinellata (Hb.)	(Henslow, 1852)	Pimpinel Pug
1854	E. satyrata f. satyrata (Hb.)	(Greene, 1854)	Satyr Pug
1856	E. haworthiata Doubl.	(Doubleday, 1856)	Haworth's Pug
	E. inturbata (Hb.)	(Doubleday, 1856)	Maple Pug
	E. dodoneata Guen.	(Doubleday, 1856)	Oak-tree Pug
	E. assimilata Doubl.	(Doubleday, 1856)	Currant Pug
	E. tantillaria Boisd.	(Doubleday, 1856)	Dwarf Pug
	E. expallidata Doubl.	(Doubleday, 1856)	Bleached Pug
	E. distinctaria H.-S. subsp. constrictata Guen.	(Doubleday, 1856)	Thyme Pug
1857	E. intricata (Zett.) subsp. millieraria Wnuk.	(Doubleday, 1857)	Edinburgh Pug
c.1858	E. cauchiata (Dup.)	(Doubleday, 1858a)	Doubleday's Pug
1858	E. valerianata (Hb.)	(Doubleday, 1858b)	Valerian Pug
1860	E. intricata (Zett.) subsp. arceuthata (Freyer)	(Crewe, 1860n)	Freyer's Pug
1861	E. trisignaria H.-S.	(Doubleday, 1861)	Triple-spotted Pug
	E. virgaureata Doubl.	(Doubleday, 1861)	Golden-rod Pug
	E. tripunctaria H.-S.	(Crewe, 1861c)	White-spotted Pug
1862	E. fraxinata Crewe (as distinct from E. innotata (Hufn.))	(Crewe, 1862o)	Ash Pug
1864	E. denotata subsp. denotata (Hb.)	(Crewe, 1864c) Doubleday, 1864)	Campanula Pug
	E. lariciata (Freyer)	(Hopley, 1864;	Larch Pug
1875	E. extensaria (Freyer) subsp. occidua Prout	(Doubleday, 1875)	Scarce Pug
1881	E. denotata (Hb.) subsp. jasioneata Crewe	(Crewe, 1881)	Jasione Pug
1884	E. satyrata (Hb.) f. curzoni Gregs.	(Gregson, 1884,a,b)	f. of Satyr Pug
1886	E. venosata (Fabr.) subsp. ochracae Gregs.	(Gregson, 1886)	subsp. of Netted Pug
1887	E. venosata (Fabr.) subsp. fumosae Gregs.	(Gregson, 1887)	subsp. of Netted Pug
1899	E. pulchellata (Steph.) f. hebudium Sheld.	(Sheldon, 1899)	f. of Foxglove Pug
1933	E. millefoliata Rössl.	(de Worms, 1952)	Yarrow Pug
1938	E. exiguata (Hb.) f. muricolor Prout	(Prout, 1938)	f. of Mottled Pug
1944	E. venosata (Fabr.) f. hebridensis W. P. Curt.	(Parkinson Curtis, 1944a)	f. of Netted Pug
1951	E. vulgata (Haw.) f. scotica Cock.	(Cockayne, 1951b)	f. of Common Pug
1953	E. egenaria H.-S.	(Agassiz, 1981)	Pauper Pug
1960	E. phoeniceata (Ramb.)	(de Worms & Messenger, 1960)	Cypress Pug
1962	E. venosata (Fabr.) subsp. plumbea Huggins	(Huggins, 1962a)	subsp. of Netted Pug
	E. vulgata (Haw.) f. clarensis Huggins	(Huggins, 1962b)	f. of Common Pug
1964	E. intricata (Zett.) subsp. hibernica Mere	(Mere, 1964)	Mere's Pug
1972	Pasiphila chloerata (Mab.)	(Pelham-Clinton, 1972)	Sloe Pug
1985	Eupithecia ultimaria Boisd.	(Riley, 1985)	Channel Islands Pug
1991	E. sinuosaria (Eversm.)	(Slade & Agassiz, 1991)	Goosefoot Pug
2002	E. massiliata Mill. & Dard.	(Goodey, 2003)	Epping Pug

3: Breeding and Rearing Pugs

Collecting adults
Live adults can be collected easily using traditional moth-hunting techniques, most popular of which is light-trapping. Data collected from the RIS light-traps show that the sex ratios attracted to light sources vary from species to species. For example, in *Eupithecia vulgata* males are predominantly attracted, whereas in *E. tripunctaria* nearly always the females. This is a problem which must be considered if gravid females are required from light-trap catches.

At night, many species are attracted to flowers, e.g. *E. simpliciata* to ragwort and *Gymnoscelis rufifasciata* to sallow, so searching flower-heads at that time is often a productive method of finding adult females from which to obtain eggs.

During the day, adults can be beaten from clumps of the larval foodplant or, in the case of tree-feeding species, from the boughs of the trees.

Sugaring does not appear to be an effective method of collecting adult pug moths.

Obtaining ova from adult females
Wild-caught females have often already mated. They should be kept in cages or small boxes (e.g. chip boxes of the sort used to package some cheeses, etc., pill boxes or plastic containers lined with netting) and supplied with food by means of tissue or cotton wool soaked in 10 per cent sucrose solution. This can be placed in a shallow tray or glass phial with a 'wick' of the wet tissue or cotton wool protruding from the top. The tissue should be re-wetted with fresh solution each day. Many species will lay freely on pieces of netting, such as net curtain, draped in the cage, or even on the sides of the container itself (particularly if the surface is rough), but, when all else fails, a sprig of the larval foodplant in water placed with the female will often induce oviposition. Mating of adult moths usually occurs within 24 hours of emergence from the pupa.

Eggs can sometimes be obtained by following a female in the wild while she is ovipositing. The best way of doing this is to sit by the foodplant with a torch at night and watch for the arrival of the moths. However, very close observation of the females is necessary, due to the small size of the eggs. A random search without watching a female laying is not usually worth the effort involved.

Care of the ova
Once laid, the ova should be placed in small clear plastic boxes. It is good practice to stand the boxes on a sheet of paper so that the newly-emerged larvae can be seen easily against the white background. Some species, e.g. *E. tenuiata* and *P. rectangulata*, overwinter as ova. It is very difficult to induce oviposition in these species and the breeder should try to provide conditions which are as

natural as possible. If ova are obtained, they should be kept outdoors in plastic boxes until they are due to hatch. Protection from the cold frequently leads to premature hatching and the foodplants are then often unavailable. Ova should be handled only with a moist, camel-hair paintbrush to avoid damage or, preferably, should not be handled at all but left on the surface upon which they were laid.

Obtaining larvae from the wild

The four main methods of collecting larvae are searching, collecting the foodplant, beating and sweeping. Searching for larvae can be a tedious business and is probably best restricted to those species which exhibit characteristic evidence of feeding. For example, the larvae of *E. assimilata* make tiny holes in the leaves of wild hop (*Humulus lupulus*), which are commonly called 'pepper pot' holes. If such evidence is found, the larvae are often present on the underside of the leaf. Likewise, the closed corolla of a foxglove flower suggests (though it does not always guarantee) the larval presence of *E. pulchellata*.

Beating the foodplants is the most common method used by lepidopterists for obtaining larvae. Quite simply it involves shaking, jarring or tapping branches or bunches of the foodplant over a tray, sheet or net and collecting any larvae that fall out. For most situations, the commercially available beating tray is suitable. However, in some circumstances, such as when collecting *E. inturbata* from tall maples, it is more efficient to beat over a large sheet spread on the ground beneath the tree. Occasionally neither of these methods is suitable. For example, when beating wild thyme growing on quarry faces for larvae of *E. distinctaria*, a beating tray blowing in the wind like a kite is unwieldy and unsafe to use. In this case a biscuit-tin lid or something similar is more appropriate.

Sweeping for larvae involves the use of a rigid-framed net or bag (also available commercially) which is vigorously 'swept' through low herbage. The larvae are dislodged and gathered in the bag. This method is very effective for moorland species on heather, such as *E. nanata*.

Some pug larvae feed internally, e.g. *E. abietaria* in spruce cones and *E. plumbeolata* in the corollas of cow-wheat. For these species the most effective method is to collect the particular parts of the plant on which the larvae feed. These can then be kept in boxes or cages to await the later emergence of adults. However, some lepidopterists prefer to keep the bunches of flowers within which the larvae are feeding in pots of water and then to search for the larvae at frequent intervals. There are arguments for and against both methods and the decision on which to use lies with the individual. More detailed notes for these specialized feeders are given under the relevant species in Chapter 4.

General care of larvae

Most species can be reared on foodplants grown in pots or on cut plants placed in water. However, if close observation of the larvae is required they should, where possible, be reared in clear plastic boxes and supplied with freshly-cut food each day. Hygiene is most important. To avoid the diseases which can occur as a result of high humidity and condensation when using this technique, absorbent tissue should be placed on the bottom of the boxes and adequate ventilation should be provided. The boxes should also be cleaned out every day and the tissue replaced. Just prior to pupation, the colour of the larvae of most species changes and often becomes darker or intensifies. All larvae stop feeding and become restless at this time as they search for a suitable place to pupate. They should then be placed on a layer of peat or a mixture of peat and sand in small plastic boxes. If these changes of behaviour are not noticed the larvae will start to spin a cocoon in the tissue. When this happens it is advisable to remove any larvae which are still feeding as they may eat the pupae. Cannibalism can result from overcrowding; some species, e.g. *E. pusillata*, are irritated by the presence of other larvae, including their own kind. Such species quickly turn and bite whatever touches them and these bites are usually fatal. Overcrowding should therefore be avoided. For example an 8 × 5cm box should contain no more than three or four larvae.

Care of pupae

Pupae should be kept in boxes or cages which are large enough to accommodate the emerging adults. They should be placed on a layer of peat or of peat and sand, and covered with moss or a further shallow layer of peat. During warm weather, spraying the moss daily prevents desiccation; for those species which overwinter as pupae, a light spraying with tepid water about once a week is recommended. Mould is often a bigger threat to overwintering pupae than desiccation, so care must be taken to avoid constant dampness. To prevent premature emergence, overwintering pupae must also be kept cold. They should be stored outdoors in boxes to protect them from predators or in an unheated shed or garage. It should be noted, however, that the air temperature inside even an unheated shed can become very warm if it is subjected to direct sunlight. This can cause problems of desiccation and premature emergence. Some breeders (including the present authors) have overwintered pupae in small plastic boxes in the cold tray of a domestic refrigerator (around $5-7^{\circ}$C). If the latter method is used the pupae can be moved to emergence boxes or cages at the appropriate time.

Parasitism

It was our original intention to include lists of all known hymenopterous and dipterous parasitoid species associated with the Eupitheciini. In consultation with M. R. Shaw, Royal Museums of Scotland, who is an authority of parasitic Hymenoptera, we were advised that the current taxonomic uncertainties in these groups would not facilitate a reliable account of host-parasitoid relationships at a specific level. As with the eupitheciines themselves, many of the earlier published parasitoid lists contain errors of identification, the promulgation of which would be unwise. However, the Eupitheciini are a fairly prominent host resource for parasitoids and a few general observations are appropriate.

The hymenopterous genera *Aleiodes* Wesmael, *Zele* Curtis and *Cotesia*

Cameron (Braconidae); *Casinaria* Holmgren, *Dusona* Cameron and *Platylabus* Wesmael (Ichneumonidae); and *Copidosoma* Ratzeburg (Encyrtidae) include species which appear to be specialized on eupitheciine hosts, though none is known to be associated solely with a single pug species. In addition to these, immature stages of pug moths are attacked by a wide range of less specialized parasitoids for which feeding-site and host-size seem to be the dominant preference factor.

Before more specific relationships can be documented, a great deal of reared material must be identified. It is therefore of the utmost importance that all parasitoids emerging from eupitheciine hosts be sent to the appropriate authority. Readers are referred to Shaw & Fitton (1989) for details of how this should be done.

4: The British and Irish Pug Species

Introduction

Full descriptions of each life stage are given, and the feeding and pupation habits of the larvae are described. The flight period, behaviour and habitat of the adult moths, together with descriptions of morphological variation, are noted. There is also advice on where to find each species and how to rear them.

There are three species (*Eupithecia sinuosaria*, *E. cauchiata* and *E. massiliata*) whose life history in the British Isles is unknown. Moreover, their status is uncertain and only a limited amount of information can be given. However, in each case, reference is made to Continental works which describe the biology as known in mainland Europe.

Eupithecia tenuiata (Hübner, 1813)   Slender Pug
= *tenuiaria* Doubleday, 1849

Pl.1, figs 1–3; Pl.7, figs 5–7; Pl.A, fig.1

HISTORY First recorded in Britain under the name *tenuiaria* by Doubleday ([1849],1850a). The larva was first fully described by Crewe (1860c;1861k).

IMAGO *Characteristic features* (Fig.1b). Wingspan *c*.18mm; forewing broad, costa and termen rounded, discal spot ovoid and conspicuous, striae fine and numerous, postmedian fascia biangulate; hindwing discal spot minute or absent, postmedian fascia curved.

Genitalia (Figs 13a,27a ♂; 20a ♀). The male abdominal plate in this species, *E. haworthiata* and *E. plumbeolata* is in the form of two fine, lightly sclerotized prongs. In *E. tenuiata* these are sharply broadened to about half their length. In *E. plumbeolata* they are more or less straight whereas in *haworthiata* they are distinctly curved inwards medially.

Variation. F. *cinerae* (Pl.1, fig.2; Pl.7, fig.7) was considered by Gregson (1888) to be a subspecies but there is no evidence to suggest that it is allopatric from the nominotypical form and therefore it does not warrant subspecific status. The type-specimens were collected by Curzon in Morayshire in 1887 or 1888. The ground colour is a clearer ash-grey than the nominotypical form and there is an ochreous costal streak; the striae are generally weak. Even paler than f. *cinerae* is ab. *niveipicta* Bastelberger (1907) which has a white ground colour. Ab. *fuscosparsata* and ab. *coaequata* Dannehl (1927) are darker, the former being deep brownish grey with very fine striae and dark strigulae, the latter glossy grey a strong brownish wash and no striae, leaving only conspicuous forewing discal spots, terminal dashes, subterminal line and rudimentary costal dots. These darker varieties reach their extreme in f. *johnsoni* Harrison (1931) (Pl.1, fig.3; Pl.7, fig.5), a melanochroic form occurring mainly in the Midlands and northeast England.

Similar species (See Fig.1, p. 35; Plate 7)

E. inturbata (Fig.1a): forewing narrower, costa less arched, apex more pointed; forewing browner, less uniform, submargins with distinctive dark patches; hindwing less uniform; fringes of both wings often chequered.

E. haworthiata (Fig.1e): base of abdomen with distinctive orange patch; forewing discal spot obsolete, postmedian fascia curved, subterminal line more conspicuous, geniculate median line usually present; hindwing discal spot absent; postmedian fascia curved.

E. plumbeolata (Fig.1d): forewing costa straighter; forewing lacking conspicuous discal spot; hindwing discal spot absent; both wings more uniform in shade.

E. valerianata (Fig.1c): forewing costa straighter; forewing lacking conspicuous discal spot, pale dentate or incomplete subterminal line and prominent tornal spot present, postmedian fascia geniculate; general appearance more uniform, ochreous; hindwing usually with pale tornal spot.

E. pygmaeata (Fig.1g): similar in size but forewing costa straight; fore- and hindwings usually uniform dark brown, striae and fasciae often absent, discal spot absent, tornal spot conspicuous.

E. distinctaria (Fig.1h): forewing costa less arched, apex more pointed; forewing discal spot larger and much more prominent, costal spots conspicuous; hindwing less uniform, marginal band usually more noticeably darker, discal spot more prominent.

E. indigata (Figs 1f,i): forewing more elongate, costa straight, apex more pointed; forewing more uniform, lacking numerous striae, discal spot larger, elongate and prominent; hindwing much paler.

E. ultimaria (Fig.1l): forewing narrower with prominent pale postmedian fascia; discal spot on fore- and hindwings distinctively linear.

E. dodoneata (Fig.1m): uncommon melanochroic form similar but hindwing termen concave, apex pointed.

LIFE HISTORY

OVUM Cream when first laid, turning to dark red before overwintering; 0.6mm. In captivity, laid mainly in the bud axils (J. Reid, pers.comm.). June to early March.

LARVA (Fig.30a) *First instar.* Head and prothoracic plate black; body pale greenish white, without markings. *Second instar.* Head and prothoracic plate black, thoracic legs dark brown; body pale greenish white with fine black transverse lines forming an interrupted dorsal stripe; subdorsal line a series of black dashes on a paler ground colour than the rest of the body; a row of black spots along

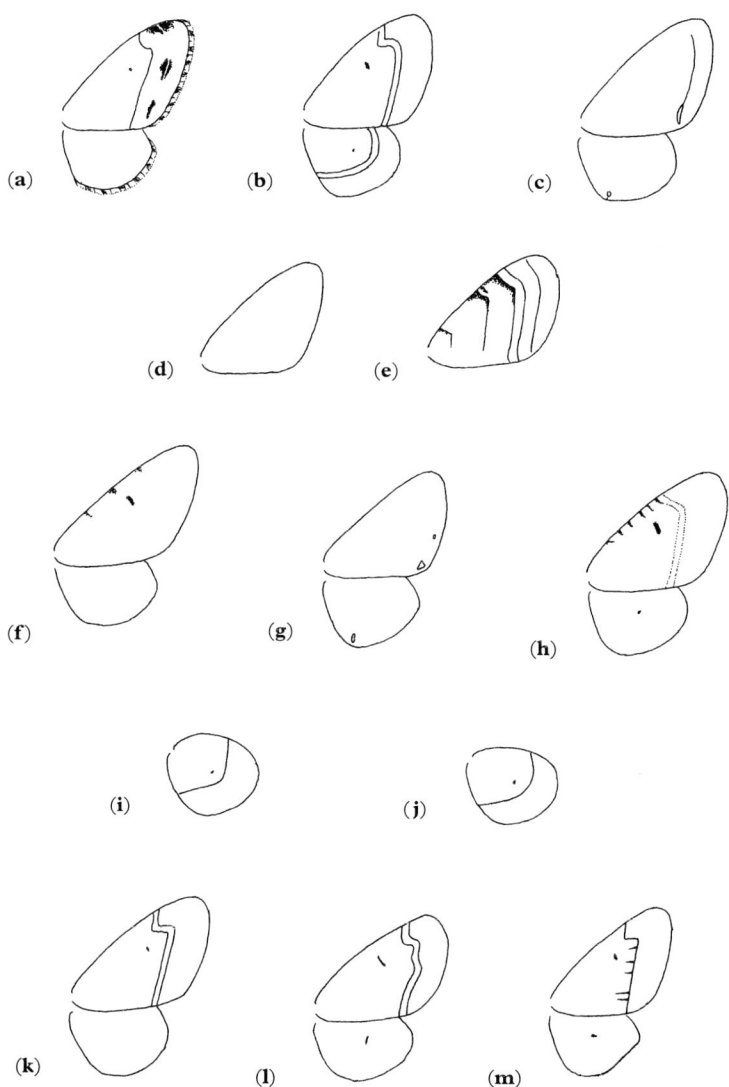

Text figure 1 Wing shapes and patterns (Similar species)

(a) *Eupithecia inturbata* (p. 37) (b) *E. tenuiata* (p. 33) (c) *E. valerianata* (p. 53)
(d) *E. plumbeolata* (p. 41) (e) *E. haworthiata* (p. 39) (f) *E. indigata* (p. 105)
(g) *E. pygmaeata* (p. 55) (h) *E. distinctaria* (p. 103)
(i) *E. indigata*, hindwing underside (p. 105)
(j) *E. distinctaria*, hindwing underside (p. 103)
(k) *E. vulgata* (p. 83) (l) *E. ultimaria* (p. 126) (m) *E. dodoneata* (p. 126)

the spiracular line. *Third and fourth instars.* Short and grub-like, length 10mm when fully grown; head and prothoracic plate black, latter sometimes reduced to two spots, thoracic legs dark brown; integument waxy, wrinkled, covered with sparse fine hairs; pale yellowish or greyish green; dorsal markings variable, slightly heavier than those of the second instar, or reduced to a fine black line which broadens posteriorly; subdorsal and spiracular lines similar to those in second instar.

Feeds internally on male, and exposed on female catkins mainly of goat willow (*Salix caprea*); also grey willow (*S. cinerea*) and eared willow (*S. aurita*). February to mid-April.

PUPA Length 5–6mm; amber yellow. In a tough cocoon on the soil surface or amongst fallen catkins. May and June (Crewe, 1860p).

FLIGHT PERIOD AND HABITAT Univoltine, late June to late August with a peak at the end of July; occurring in the vicinity of sallows. During the day it rests on tree-trunks and at dusk flies around sallows; at night both sexes come readily to light.

DISTRIBUTION (Map 1) Widespread and locally common throughout the British Isles including the Isle of Man and the Channel Islands. It is likely to be found wherever the foodplants grow. The absence of records from a few counties almost certainly reflects lack of observation.

COLLECTING AND REARING Adults are attracted to light in small numbers but are easier to catch at dusk when they fly freely around sallow bushes. Captive females should be fed and supplied with bark on which to oviposit; however, it is very difficult to obtain eggs in captivity. Larvae are easy to collect from the wild but it is necessary to keep the sallow bushes under close observation as larval development is strongly linked to that of the sallow catkins; both male and female catkins are eaten and this leads to two methods of rearing. Male catkins can be beaten from the tree on to a sheet and then placed on a layer of dry peat or sand in a suitable container such as an aquarium covered with netting. The catkins should be sprayed with water every day to prevent desiccation, though not so heavily as to cause constant wetness which results in mould. It is very difficult to gauge the moisture-level accurately and this, along with the frequent presence of the carnivorous larvae of species of *Xanthia* Ochsenheimer (Noctuidae), can make successful rearing difficult. As the larvae feed internally it is also difficult to observe their development. The second method is much more convenient and involves beating the larvae from female catkins, on which they feed externally (Prior, 1978); they can then be reared and observed in the normal way (see Chapter 2).

Eupithecia inturbata (Hübner, 1813)   Maple Pug
= *subciliata* Doubleday, 1856
= *subciliaria* Morris, 1861

Pl.1, fig.4; Pl.7, fig.10; Pl.A, fig.2

HISTORY Discovered in Britain in 1856 by Henry Doubleday and H. G. Knaggs in Cambridgeshire and Kent respectively, and first recorded in the literature by Doubleday (1856). The larva was first described by Crewe (1872a).

IMAGO *Characteristic features* (Fig.1a). Wingspan c.18mm; forewing apex pointed, discal spot minute, postmedian fascia biangulate, submarginal band broken into dark patches; hindwing discal spot minute or absent, postmedian fascia curved; fringes of both wings often chequered.

Genitalia. Figs 13b,27b ♂; 20e ♀

Variation. Restricted mainly to a general darkening of the ground colour. Dark brown specimens have been recorded in Surrey, Kent and Lincolnshire (Agassiz *et al.*, 1981) and Hertfordshire. Black-banded forms (unnamed) have been reared from larvae collected at Bardney Forest, Lincolnshire (G. M. Haggett, pers.comm.).

Similar species (See Fig.1, p. 35; Plate 7)

E. tenuiata (Fig.1b): forewing broader, costa more arched; forewing with numerous striae but more uniform in general appearance, dark submarginal patches absent; hindwing more uniform, postmedian fascia less conspicuous; fringes of both wings not conspicuously chequered.

E. haworthiata (Fig.1e): base of abdomen with orange patch; forewing discal spot obsolete, pale subterminal line usually conspicuous, postmedian fascia curved, general appearance more uniform, dark submarginal patches absent; hindwing discal spot absent; fringes of both wings uniform.

E. plumbeolata (Fig.1d): forewing discal spot obsolete, generally more uniform, lacking conspicuous fasciae, striae and dark submarginal patches; hindwing more uniform, discal spot and postmedian line obsolete; fringes of both wings less noticeably chequered.

E. valerianata (Fig.1c): forewing costa straighter; forewing generally more uniform, lacking conspicuous fasciae, striae and dark submarginal patches, discal spot usually obsolete, dentate or incomplete subterminal line conspicuous, tornal spot prominent; hindwing with pale tornal spot, other markings obsolete or absent; in fresh individuals, overall appearance more glossy.

E. pygmaeata (Fig.1g): similar in size but fore- and hindwings almost uniform dark brown; forewing costa straight; fore- and hindwings with conspicuous tornal spot.

E. indigata (Figs 1f,i): forewing more elongate, costa straighter, apex pointed; fore- and hindwings more uniform; forewing discal spot significantly more conspicuous; fringes of both wings not chequered.

E. dodoneata (Fig.1m) uncommon melanochroic form similar but hindwing termen concave, apex pointed.

E. ultimaria (Fig.1l): forewing discal spot conspicuous and distinctively linear, postmedian fascia more conspicuous, dark submarginal patches absent; hindwing discal spot conspicuous and linear.

LIFE HISTORY

OVUM Pale yellow; 0.6mm long, ovoid. July to early May, probably in the bark crevices of the food plant. The adult female oviposits both on small bushes and on large old trees and is not restricted to the latter as is often stated.

LARVA (Fig.30b) *First instar*. Pale yellowish green; dorsal and subdorsal lines faint dark green. *Second instar*. Pale yellowish green; fine dorsal line dark green or black; subdorsal and spiracular lines yellow; segmental divisions pale green. *Third instar*. Similar to second but with yellow segmental divisions. *Fourth instar*. Short and stumpy, tapering sharply at both ends, length 12mm full-grown; head small, yellowish green or brown; thoracic legs yellow; integument wrinkled or folded and covered sparsely with hairs; pale yellowish green, ventrally pale pink; dorsal line strong, dark green or black surrounded by broad area of purplish pink extending to spiracular region; subdorsal lines cream; segmental divisions pale red.

Feeds on the flowers of field maple (*Acer campestris*). May.

PUPA Length 6mm; abdomen mid-brown, thorax and wings olive-green. In a frail silken cocoon just under the soil surface or in ground-litter. Late May to July.

FLIGHT PERIOD AND HABITAT Univoltine, early July to the end of August, peaking at the end of July. It rests during the day amongst the leaves of maple trees. When disturbed it usually flies quickly to the ground (Tutt, 1906d). Both sexes come to light.

DISTRIBUTION (Map 2) Widely distributed and locally common throughout England, northwards to Lancashire and Durham, and Wales. Although there are many counties within this range for which there are no records, this is probably due to under-recording. The species is likely to be found wherever there are mature maples. Apparently absent from Scotland and Ireland, the Isle of Man, and the Channel Islands.

COLLECTING AND REARING As with other species which overwinter as an ovum, it is very difficult to induce a captive female to lay. Success is most likely if bark is supplied on which the female may oviposit. However, the larvae are easy to collect from the wild. A mature flowering maple is usually a tall tree with flowers mostly near the top. Beating with a tray can be difficult and it is often better to use a sheet spread on the ground beneath the tree. Alternatively, one can often find smaller maples in hedges which also bear flowers; these can be beaten in the usual way. The larvae can easily be missed as their growth is closely linked to the development of the blossom. They are easy to rear in plastic boxes or similar containers but an adequate supply of fresh flowers must be available. Large flowering sprays will keep well in water.

Eupithecia haworthiata (Doubleday, 1856)   Haworth's Pug
= *isogrammaria* sensu Herrich-Schäffer, 1848
= *haworthiaria* Morris, 1861

Pl.1, fig.5; Pl.7, fig.11; Pl.A, fig.3

HISTORY First described and recorded in Britain by Doubleday in 1856. The larva was first described by Crewe (1859b; 1861k).

IMAGO *Characteristic features* (Fig.1e). Wingspan *c*.18mm; base of abdomen with patch of orange scales; forewing broad, costa and termen rounded, discal spot obsolete, dark geniculate median line usually present, postmedian fascia curved, pale subterminal line usually prominent; hindwing discal and tornal spots absent.

Genitalia. Figs 13c,27c ♂; 20c ♀; see *E. tenuiata.*

Variation. Varies little in size and colour. The only named aberration is ab. *coriolutea* Mobius (Goodson, unpubl.) which is 'leather yellow with distinct markings'.

Similar species (See Fig.1, p. 35; Plate 7)

Eupithecia tenuiata (Fig.1b): base of abdomen without orange patch; forewing discal spot conspicuous, subterminal line inconspicuous or absent, dark geniculate median line absent.

E. inturbata (Fig.1a): base of abdomen without orange patch; forewing with small but distinct black discal spot, generally well marked and more blotched overall, submargins with dark patches; hindwing with discal spot and pale postmedian fascia; fringes of both wings usually chequered.

E. plumbeolata (Fig.1d): base of abdomen without orange patch; forewing less brownish-tinged, dark median line absent, postmedian and subterminal lines usually obsolete; generally more uniform overall.

E. valerianata (Fig.1e): base of abdomen without orange patch; forewing almost uniform, most fasciae and striae usually obsolete, median line absent, tornal spot prominent; hindwing uniform, striae obsolete, tornal spot usually conspicuous; fresh specimens appear more glossy.

E. pygmaeata (Fig.1g): base of abdomen without orange patch; forewing costa straighter; fore- and hindwings almost uniform dark brown.

E. indigata (Fig.1f): base of abdomen without orange patch; forewing more elongate, costa straighter, apex more pointed; forewing with prominent discal spot; hindwing discal spot faint but usually present.

LIFE HISTORY

OVUM Cream; 0.5mm long, ovoid. On flower-buds in June and July.

LARVA (Fig.30c) *First instar.* Pale greenish cream; dorsal line faint green. The larva bores into unopened flower-buds of the foodplant, eating the contents. The resulting accumulation of frass produces slightly darker buds which remain unopened. A small exit hole about 0.5mm in diameter is evident after the larva has vacated the bud. Because of the dark frass within, these holes appear black and are often conspicuous. *Second and subsequent instars.* Short and grub like, tapering sharply at both ends; length 10mm when full-grown; head and thoracic legs yellowish brown; integument smooth, greenish cream at first with faint green dorsal and subdorsal lines, later pinkish green and eventually purplish red with darker dorsal line. Feeds on the stamens and pistil of the opened flower. The sepals turn back as the flower opens and then change colour as they dry. As the larva passes through the last three instars it mimics the colour transformations of these sepals perfectly, becoming purplish red just prior to pupation.

Feeds on the flowers of traveller's-joy (*Clematis vitalba*). Allan (1949) states that captive larvae will accept cultivated varieties of *Clematis*. July and August.

PUPA Length 5–6mm; abdomen red, thorax and wings green. In a cocoon just below the surface of the soil or amongst ground debris. Overwinters in this stage. August to June. Cattermole (1986) cites an instance of a pupa of *E. haworthiata* from Dorking, Surrey, passing two winters (1983–5) before emergence.

FLIGHT PERIOD AND HABITAT Univoltine, June and July, with individuals occasionally remaining until early August. The peak emergence appears to be early July. The moth can be found flying near its foodplant at dusk and comes readily to light. Newman (1869) records that it flies in hot sunshine. Tutt (1906c) states that it is easily disturbed during the daytime but is a swift flyer and is difficult to catch.

DISTRIBUTION (Map 3) Widespread and locally common throughout much of England and Wales, northwards to Lincolnshire and Westmorland; apparently absent from Scotland but several records from Ireland; Channel Islands. This small, easily-overlooked species is probably present within its range wherever the larval foodplant is established.

COLLECTING AND REARING The female oviposits freely in captivity. It is also easy to beat the larvae from its widespread, and often common foodplant. It is less time-consuming to beat the trailing branches when the flowers are fully out rather than search for tenanted buds. It is important to wait for some minutes after beating and watch the debris on the beating tray for movement as the larvae greatly resemble loose dried sepals. The moth is often found in urban areas but is easily overlooked and is probably more common than the records suggest.

Eupithecia plumbeolata (Haworth, 1809)   Lead-coloured Pug
= *begrandaria* Boisduval, 1840
= *plumbeolaria* Doubleday, 1849

Pl.1, fig.6; Pl.7, fig.12; Pl.A, fig.4

HISTORY First described and recorded in Britain by Haworth (1809) as *Phalaena plumbeolata*. Larvae were first described by Crewe (1865b).

IMAGO *Characteristic features* (Fig.1d) Wingspan *c*.18mm; forewing costa straight, apex pointed; generally uniform, fasciae and striae obscure; forewing discal spot obsolete; hindwing discal spot absent.

Genitalia. Figs 13e,27d ♂; 20b ♀; see *E. tenuiata*.

Variation. Varies very little in size. Regionally there seems to be only slight variation in colour, specimens from northern Scotland being similar to those from southern England.

Several forms have been named and their salient features are summarized as follows:

Conspicuous discal spot: ab. *singularia* Herrich-Schäffer (1848).

Median area darker and strongly defined: ab. *explicata* Dannehl (1927).

Densely irrorate throughout with dark brown or blackish brown: ab. *lividata* Dannehl (1927).

Large, coarsely scaled and greyish rather than brownish: ab. *enuncleata* Dietze (1910;1913).

Small with numerous light spots 'recalling *E. spissilineata* Metz.': ab. *uralensis* Dietze (1910;1913).

Uniform light leather yellow with the grey markings obsolete or almost so: ab. *flaveolata* Dannehl (1925).

Dead whitish grey with obsolete markings; finely scaled and silky: ab. *plumbalbeolata* Dannehl (1927).

Similar species (See Fig.1, p. 35; Plate 7)

Eupithecia tenuiata (Fig.1b): forewing costa arched, discal spot distinct; overall appearance less uniform.

E. inturbata (Fig.1a): forewing discal spot usually present, submargin with dark patches, generally less uniform; hindwing less uniform, striae and postmedian fascia more conspicuous, discal spot minute but usually present; fringes of both wings usually chequered.

E. haworthiata (Fig.1e): base of abdomen with orange patch; forewing costa more arched; forewing generally browner, dark geniculate median line usually present, postmedian fascia and subterminal line more conspicuous; less uniform overall.

E. valerianata (Fig.1c): forewing with dentate or incomplete subterminal line more conspicuous; fore- and hindwings with prominent pale tornal spot; overall more uniform and glossy.

E. pygmaeata (Fig.1g): forewing costa straighter; fore- and hindwings almost uniform dark brown, tornal spot more prominent.

E. indigata (Figs 1f,i): forewing more elongate, apex more pointed; forewing discal spot most prominent feature.

LIFE HISTORY

OVUM Cream; 0.5mm long, ovoid. Laid on the foodplant. May and June.

LARVA (Fig.30d) *First and second instars*. Translucent yellowish cream or pale greenish grey; dorsal and subdorsal lines red but sometimes absent. The gut is visible through the integument, resembling a dorsal line. It feeds inside the corolla of the foodplant, eating the contents. *Third and fourth instars*. Short and stumpy, tapering at each end; length 10mm when full-grown; head and thoracic legs reddish brown; integument smooth and translucent, pale ochreous yellow; dorsal line well defined, reddish, broadening at the centre of each segment; subdorsal and spiracular lines fine, red; segmental divisions pale yellow. Feeds inside a corolla with the anal end protruding. In its final instar it eats the whole flower. It is a very cryptic larva which imitates perfectly a withering flower.

Feeds mainly on common cow-wheat (*Melampyrum pratense*) but stated to feed on field cow-wheat (*M. arvense*) and yellow-rattle (*Rhinanthus minor*) (Allan, 1949). Allan also states that the larvae accept eyebright (*Euphrasia* spp.) and red bartsia (*Odontites vernus*) in captivity. July and August.

PUPA Length 6–7mm; thorax, wings and abdomen golden yellow; segmental divisions and cremaster darker. Usually in a cocoon amongst ground-debris but, in captivity, occasionally inside a dried corolla. August to May. Sometimes overwinters twice.

FLIGHT PERIOD AND HABITAT Univoltine, May to July, with a peak during the third week of June. Found in woodland areas where the foodplant is common and occasionally on sandhills. Although Simson (1980;1981a,b) states that it often occurs at very low density, RIS light-trap catches indicate that adults come readily to light at night, the males sometimes in very large numbers. Further, Tutt (1906c) states that it is sometimes abundant in woodland clearings at dusk. King (1953) records that adults fly naturally during the afternoon. It is easily disturbed from undergrowth during the day.

DISTRIBUTION (Map 4) A local moth, usually associated with mature woodland and occasionally sandhills (Skinner, 1984). Its range is possibly becoming more contracted due to a decline in its foodplant caused by the absence of coppicing. It would be interesting to monitor populations on nature reserves where more traditional coppicing techniques have been employed in an attempt to strengthen colonies of the heath fritillary butterfly (*Mellicta athalia* (Rottemberg): Nymphalidae). Where this has been done, common cow-wheat (also the foodplant of *M. athalia*) has flourished.

In England this species is most often recorded south of a line from Lincolnshire to the Severn estuary with a few colonies elsewhere. In Wales, the distribution is scattered but it appears to be locally common. In Scotland large colonies are known from Stirlingshire and East Ross. In Ireland local but widely distributed; also recorded from the Channel Islands. This species is easily overlooked and it probably remains undiscovered in many localities within its range.

COLLECTING AND REARING A captive female will lay eggs but it is probably easier to search for the larvae as large quantities of wet, black frass are produced which render occupied flowers conspicuous. It is important to note that the foodplant does not do well in water nor, being semi-parasitic, planted in pots. Consequently, a good supply of local wild cow-wheat is required.

Eupithecia abietaria (Goeze, 1781)   Cloaked Pug
= *pini* (Retzius, 1783), nec (Linnaeus, 1758)
= *strobilata* (Borkhausen, 1794)
= *togata* (Hübner, 1814–17)
= *togaria* Boisduval, 1840

Pl.1, figs 7,8; Pl.6, figs 16,17; Pl.A, fig.5

HISTORY First recorded in Britain from Black Park, Buckinghamshire in June 1844 as *Eupithecia togata* (Stevens, 1845) and described by Newman (1845). Larvae were first described under the same name by Crewe (1872d,e).

IMAGO *Characteristic features* (Fig.2a). Wingspan *c*.27mm; the large size and distinctive forewing markings preclude confusion with any other eupitheciine found in the British Isles.

Genitalia. Figs 13f,27e ♂; 20f ♀

Variation. Old Scottish specimens in the national collection (BMNH) show only slight variation. However, specimens caught from 1985–87 in the RIS light-trap at Kielder, Northumberland, vary considerably both in size and form.

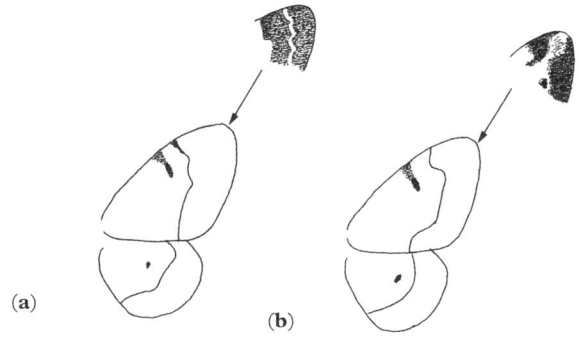

Text figure 2 Wing shapes and patterns (Similar species)
(a) *Eupithecia abietaria* (b) *Alcis jubata* (Ennominae: Geometridae)

Some are very pale with large conspicuous markings whereas others are very dark brown with most of the markings obliterated. In some individuals there are broad claret-brown marginal bands which make the pale subterminal line appear very bright. The same range of variation was noted in adults reared from stock collected in 1987 by Riley in Hamsterley Forest, Co. Durham, but those from Kyloe Forest, Northumberland, vary little. The reasons for such localized variation are unknown. The only named form is f. *constrictata* Prout (1914) which has the median area reduced to 1–2mm in width. The post- and antemedian lines are connected by black nervures. Specimens of this form from Perthshire and Morayshire are in the national collection.

Similar species (See Fig.2, p. 43)
Alcis jubata (Thunberg) (Ennominae: Geometridae) (Fig.2b): male antennae bipectinate; forewing apex with pale patch; hindwing discal spot larger; usually larger and whiter, lacking liver-brown tinge.

LIFE HISTORY

OVUM Pale yellow, later bright red; 0.6mm, ovoid. On the unripe cones of the foodplant in June.

LARVA (Fig.30e) *First and second instars.* Pale greenish brown; head, thoracic and anal plates black. *Third and fourth instars.* Grub-like, length 15mm full-grown; integument waxy, wrinkled, sparsely covered with black tubercles and hairs; pinkish red to deep brownish red; dorsal line white and very faint.

Feeds principally on Norway spruce (*Picea abies*), eating the young shoots and the ripening seeds inside the cones. Hellins (1872) records that, in captivity, larvae will feed by burrowing into the inner bark at the cut ends of spruce stems. It has also been noted on Sitka spruce (*Picea sitchensis*) (Barbour, 1985), noble fir (*Abies procera*) (Rutherford, 1988) and European silver-fir (*A. alba*) (Styles, 1961; Winter, 1983). Old cones of Douglas fir (*Pseudotsuga menziesii*) found by Riley and D. Davies in 1989 near Rhandirmwyn, Dyfed, also appeared to have been infested. M. Townsend (pers.comm.) states that larvae can be reared successfully on the cones of Scots pine (*Pinus sylvestris*), though the resulting moths are diminutive. July to early September.

PUPA Length 8.5–9.5mm; bright reddish brown. In a cocoon on the surface of the ground or in fallen cones. Overwinters in this stage from September to the end May. Anderson (1890) cites a record of a pupa from Shetland passing two winters (1888–90) in this stage.

FLIGHT PERIOD AND HABITAT Univoltine, mid-May to late August. The main flight period appears to be during June and July with the peak emergence in mid-July. Occasional individuals have been recorded in September. Woods or woodland clearings in the vicinity of spruce. Most common in mature spruce plantations. During the day it rests on tree-trunks; both sexes come readily to light.

DISTRIBUTION (Map 5) During the nineteenth century, this species was recorded throughout Britain and it was formerly common in the New Forest and parts of Scotland, particularly Perthshire. Until 1984 it was recorded singly in widely scattered localities. Since that time, strong colonies have been discovered in Montgomeryshire, Durham and Northumberland. There are no known colonies in Scotland but this is likely to be due to a lack of recording. Searches in mature spruce plantations may reveal its presence there. Recently recorded from Co. Down, Ireland (McLoughlin, pers.comm.).

COLLECTING AND REARING A captive female will oviposit if supplied with fresh branches of the foodplant complete with young cones, but the authors do not know of a successful attempt to rear this species from egg to adult. It is easier to rear from collected larvae. The presence of feeding larvae can be detected by using binoculars to search for reddish brown frass extruding from between the scales of the spruce cones while they are still on the tree. Tenanted cones must then be collected from the branches. Alternatively, fallen cones can be inspected after storms in late July and August, though this method is never as successful.

Eupithecia linariata ([Denis & Schiffermüller], 1775)
Toadflax Pug
= *linariaria* (Borkhausen, 1794)
= *linaria* Boisduval, 1840

Pl.1, fig.9; Pl.5, fig.9; Pl.A, fig.6

HISTORY First recorded in Britain by Haworth (1809) as the Beautiful Pug. Larvae were first described in detail by Crewe (1860a;1861k).

IMAGO *Characteristic features* (Fig.3a). Wingspan *c*.20mm; forewing median band uniform, without extensive pale irroration, postmedian edge of median band curved, dark basal patch reduced to bar along costa.

Genitalia (Figs 13g,27f ♂; 20g ♀). The genitalia of this species are very similar to those of *E. pulchellata*, though the aedeagi of the males are relatively easy to distinguish. The male abdominal plates take the form of two weakly sclerotized prongs which, in *E. pulchellata*, are set further apart and are comparatively more robust. In both species, the bursa copulatrix is very fine in

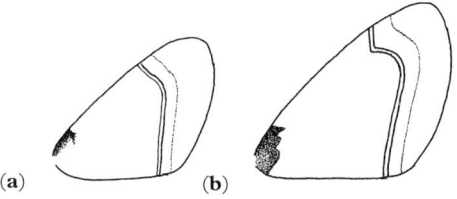

Text figure 3 Wing shapes and patterns (Similar species)
(a) *Eupithecia linariata* (b) *E. pulchellata* (p. 47)

texture and lacks obvious signa; the inner wall is partially covered with minute chitinous spots. However, the area covered by these is greater in *E. linariata* than in *E. pulchellata*. The bursa copulatrix of *E. pulchellata* is more slender than that of *E. linariata*.

Variation. Size varies little. Ab. *pallescens* Dietze (1910;1913) is a pale, washed-out form, possibly of pathological origin. There is considerable variation in the form of the median striae, which can be summarized as follows:

Forewing generally darkened with median fascia almost black: ab. *nigrofasciata* Dietze (1910;1913).

Median fascia broken into a short curved bar and two minute dots: ab. *praeruptata* Richardson (1952).

Median fascia reduced to a central spot and a small mark on the dorsum; basal line reduced to a spot on the costa: ab. *punctata* Cockayne (1953). (Type bred from specimens collected in Brighton, Sussex, in 1919.)

Median area much lighter due to strong reduction in dark markings: ab. *reducta* Lempke (1947;1951).

Median area whitish as far as the small costal spot; median fascia very narrow: ab. *reducta* Foltin (1938).

Median band almost obsolete; hindwing with grey markings only slightly developed: ab. *flavofasciata* Foltin (1938).

Two transverse striae bordering the median fascia close together: ab. *approximata* Lempke (1947;1951).

Similar species (See Fig.3, p. 45)
Eupithecia pulchellata (Fig.3b): forewing with postmedian edge of median band biangulate, median band with more extensive pale irroration, particularly in dorsal half, basal dark patch usually complete, rarely reduced to bar along costa.

LIFE HISTORY

OVUM Greenish white when laid, later turning to deep yellow; 0.5mm, ovoid. June to August.

LARVA (Fig.30f) *First and second instars.* Bright yellow with faint black dorsal spots. *Third and fourth instars.* Short and stumpy, tapering towards the head; length 12mm when full-grown; head and thoracic legs pale brown; integument wrinkled, covered with sparse black tubercles and hairs; bright yellow, grass green or yellowish green; dorsal markings resemble outstretched wings of a bird or bat, brown or dark green extending to the spiracles, occasionally absent.

Feeds on common toadflax (*Linaria vulgaris*), snapdragon (*Antirrhinum majus*) and possibly other cultivated varieties of *Antirrhinum*. At first the larva feeds inside the flower-head, eating the stamens and pistil; later boring into the seed-capsule and eating the ripening seeds. June to the beginning of October.

PUPA Length 7–8mm; abdomen brown, thorax and wings bright green, turning to olive-green. Late September to May.

FLIGHT PERIOD AND HABITAT The moth is on the wing from late May to mid-September. There appear to be peaks of emergence during late June and mid-August, suggesting two protracted and overlapping broods. By day it is difficult to find the adult but at night both sexes come readily to light.

DISTRIBUTION (Map 6) Widespread and locally common throughout England and Wales and likely to occur wherever the larval foodplant is established. In Scotland it is more or less restricted to the lowlands of the south though colonies are also known from West Inverness. There are old, unconfirmed records from Aberdeenshire. Also found in the Isle of Man and the Channel Islands, and there are a few doubtful records from Ireland.

COLLECTING AND REARING This common and widespread moth seems to be found most often in urban areas and even in the centre of cities such as London. Its larval foodplant is often found growing near disused railway-sidings and on waste ground. The larva is easy to collect by searching the *Linaria* corollae. Its presence is often indicated by a round exit-hole in the side of the flower on which it has fed. Rearing in captivity is straightforward, the best results being obtained by keeping the foodplant in water.

Eupithecia pulchellata (Stephens, 1831) Foxglove Pug
= *pulchellaria* Doubleday, 1849

Pl.1, figs 10,11; Pl.5, figs 6–8; Pl.A, figs 7,8

HISTORY First recorded in Britain by Stephens (1831). After several appeals for livestock by Crewe (1862i,l), he published the first descriptions of the larva and pupa (Crewe, 1864a). F. *hebudium* was originally described by Sheldon (1899) from specimens collected on one of the smaller 'almost treeless' Inner Hebridean islands.

IMAGO *Characteristic features* (Fig.3b). Wingspan *c.*22mm; median band with extensive pale irroration, particularly in basal half, postmedian edge of median band biangulate, basal patch usually complete.

Genitalia. Figs 13h,27g ♂; 20h ♀. See *E. linariata.*

Variation. F. *hebudium* (Pl.1, fig.11; Pl.5, fig.8) is much greyer, the reddish brown areas of the type almost totally replaced by greyish white. It is found in parts of western and north-western Wales, Anglesey, the Outer Hebrides and western Ireland. Some from the Inner Hebrides and western Scotland appear intermediate (Agassiz *et al.*, 1981).

Variation in the markings of the typical form are more or less restricted to general lightening of the wings, most notably the median fascia, which reaches its extreme in ab. *defasciata* Metschl & Salzl (1935) in which this band is completely absent. In ab. *reducta* Bastelberger (1907) the band is very light with the darker scaling confined mostly to a small area around the discal spot, and in ab. *guttata* Cockayne (1953) it is slightly paler, as are all the other markings. The type specimen of ab. *guttata* was caught at Lydd, Kent, in 1932 by A. J. Bowes

(Cockayne, 1953). Ab. *iberica* Dietze (1910;1913) is generally pale with whitish ground colour and sparse markings, and in ab. *approximata* Lempke (1947;1951) the two lines bordering the median area are close together.

Similar species (See Fig.3, p. 45)
Eupithecia linariata (Fig.3a): median band more uniform without extensive pale irroration, postmedian edge of median band curved, basal dark patch incomplete, reduced to bar along costa.

LIFE HISTORY

OVUM Cream turning to deep yellow; 0.5mm, ovoid. Laid on the underside of the leaves and in the crevices of the leaf ribs. May to early July.

LARVA (Fig.30h) *First instar.* Head and thoracic legs black; integument greenish brown or green, dorsal markings absent. *Second instar.* Dirty greenish black; segmental divisions paler. *Third and fourth instars.* Grub-like, tapering to head and tail; length 16mm when full-grown; head and thoracic legs brown; integument wrinkled, covered with sparse hairs; dirty greyish green to purplish pink; dorsal markings deep purplish brown, resembling the outstretched wings of a bird, extending to the spiracles and connected by purplish brown dorsal, subdorsal and spiracular lines. The markings are usually so extensive that only small areas of the ground colour remain.

The larva of f. *hebudium* differs in colour in all its instars – *First and second instars.* Head and thoracic legs black; integument smoky grey; dorsal markings absent. *Third instar.* Dark grey or greyish brown; dorsal markings cruciform, black, sometimes developed into a broad band covering the whole of the dorsal area; subdorsal line interrupted, greyish yellow, spiracular line irregular, black. *Fourth instar.* Uniform dark brown, without markings.

The larva feeds inside the flowers of foxglove (*Digitalis purpurea*), at first on the stamens and then on the unripe seeds. The opening of the corolla is often closed with strands of silk and as the larva moves from flower to flower it leaves a small exit-hole at the base of those in which it has fed. From the end of June to August.

PUPA Length 7–8mm; abdomen deep orange with the anal tip red; thorax and wings green. August to May in a slight cocoon just below the surface of the ground.

FLIGHT PERIOD AND HABITAT Univoltine, mid-April to the end of August, with a peak during the third week of June. It is likely to be found wherever the foodplant occurs. During the day it rests on tree-trunks, fences, etc., near the foodplant. At night both sexes come readily to light.

DISTRIBUTION (Map 7) Locally common throughout the British Isles, including the Isle of Man, as far north as the Orkneys; Channel Islands. Its density depends upon that of the larval foodplant and in some areas, particularly parts of Wales and Scotland, it is found in all but the most urban localities.

COLLECTING AND REARING Not a difficult species to rear, though greater success can be achieved by rearing larvae from the wild rather than by trying to obtain ova from a captive female, as the early instars are difficult to maintain. The easiest method of collecting larvae is to find a closed corolla and then search inside this and adjacent corollae. As they do not always spin up the opening of the flower, many larvae will be missed if only closed corollae are examined.

Eupithecia irriguata (Hübner, 1813)   Marbled Pug
= *variegata* (Haworth, 1809), nec (Scopoli, 1763)
= *irriguaria* Boisduval, 1840

Pl.1, fig.12; Pl.2, fig.10; Pl.A, fig.9

HISTORY First recorded in Britain by Haworth (1809) as *Phalaena variegata*. He stated that he had seen only three examples of it and that it was extremely rare in the London area. A larva was described by Dietze (1870) but it was presumably continental; a British larva was first described by Crewe (1871).

IMAGO *Characteristic features* (Fig.4c). Wingspan *c.*20mm; ground colour white; forewing marginal band dark and broken by white patches, costal markings and discal spot prominent and distinctive; hindwing discal spot conspicuous.

Genitalia. Figs 13d,27h ♂; 20d ♀

Variation. This species varies very little in size or markings. The only known named aberration is ab. *franconica* Dietze (1910;1913) which is stated to be poorly marked; the black markings, especially in the female, dissipated into a regular grey irroration.

Similar species
Eupithecia subumbrata (Fig.4a): forewing discal spot minute or absent, dark costal margin less conspicuously broken by pale postmedian line, marginal band more complete, dark subbasal fascia absent; hindwing discal spot minute or absent.

E. tantillaria (Fig.4b): forewing costal markings not forming complete bar, striae present in median area giving more uniform general appearance, subbasal fascia less conspicuous; hindwing with more conspicuous postmedian line, particularly on underside; abdomen with pale basal and dark subbasal band.

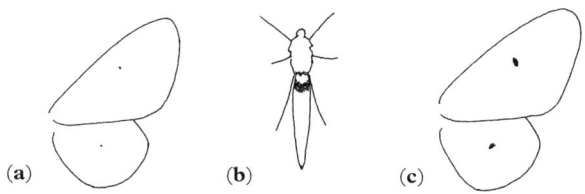

Text figure 4 Wing and abdominal markings (Similar species)

(a) *Eupithecia subumbrata* (p. 98) (b) *E. tantillaria* (p. 129) (c) *E. irriguata*

LIFE HISTORY

OVUM Bright pale yellow when first laid, turning yellow, then red and finally, just prior to hatching, dark brown; 0.4 × 0.5mm (J. Reid, pers.comm.). April and May.

LARVA (Fig.30g) *First instar.* Pale translucent yellow with the gut visible through the integument; fine dorsal line reddish. *Second and third instars.* Yellowish green; dorsal markings triangular, forward pointing, red edged yellowish white; subdorsal line fine, red; on anal segments dorsal markings replaced by a continuous red line; a series of red lateral patches is present between the thoracic segments. *Fourth instar.* Slender, tapering slightly towards the head; 22mm when full-grown; head reddish yellow; thoracic legs red; integument smooth and velvety; bright grass green; dorsal marks triangular, red outlined in yellow, merging into yellow segmental divisions; anal segment and claspers with single red spots.

Feeds on the leaves of pedunculate oak (*Quercus robor*). On the Continent it is stated to feed on evergreen oak (*Quercus ilex*) (Dietze, 1913). At first it feeds on the edges of the leaf, leaving small holes. Later it eats the whole leaf. When not feeding it stands erect, resembling a leaf stalk. June and early July.

PUPA Length 5.5–6.0mm (J. Reid, pers.comm.); thorax and abdomen brown, wings dark green. Larvae usually make a slight cocoon on or just beneath the ground surface, however, they will sometimes pupate under the bark or under moss growing on the tree on which they have fed (Tutt, 1906c). July to April.

FLIGHT PERIOD AND HABITAT Univoltine, mid-April to mid-June with a peak during the second week of May. It flies in the vicinity of oak trees and rests on the trunks of the trees during the day. It is most often found in mature oak woods where it comes freely to light at night.

DISTRIBUTION (Map 8) Very local in England and Wales south of a line from Caernarvon to the Thames estuary, occurring in ancient oak woods. Outside this range it has been recorded only singly or from historical literature. Absent from Scotland and Ireland.

COLLECTING AND REARING Although often common where it occurs, it is a very localized species which is generally difficult to find. This is especially the case in the larval stage, as it usually feeds high up in mature oaks. However, captive females oviposit readily and once eggs have been obtained the larvae are easily reared.

Eupithecia exiguata (Hübner, 1813)   Mottled Pug
= *trimaculata* (Haworth, 1809), nec (Villers, 1789)
= *ochreata* Stephens, 1831
= *exiguaria* Boisduval, 1840
= *lanceolaria* Wood, 1854

Pl.1, figs 13,14; Pl.4, figs 9,10; Pl.A, fig.10

HISTORY British specimens were first described by Stephens (1831) as *ochreata*. He referred to it as scarce and restricted. Further, Tutt (1906d) stated that, although it was widely distributed, it was never abundant. However, this species is now known to be generally distributed and common. The larva was first fully described by Crewe (1859m; 1861k).

Form *muricolor* was first described by Prout (1938) from specimens collected in eastern Aberdeenshire by Cockayne and at the time was given subspecific status.

IMAGO *Characteristic features*. Wingspan c.24mm; forewing slightly elongate, costa arched, apex pointed; ground colour pale grey; forewing discal spot conspicuous, submarginal band broken into three characteristic dark patches, postmedian fascia biangulate, postmedian line with distinctive small dark basally-pointing triangular marks, especially near discal spot; hindwing discal spot conspicuous.

Genitalia. Figs 12c,13i,7i ♂; 21a ♀

Variation. F. *muricolor* Prout (Pl.1, fig.14; Pl.5, fig.10) has a pale grey ground colour with stronger markings than the type. It is stated to be restricted to eastern Aberdeenshire but intermediate forms occur in north-eastern England (Agassiz *et al.*, 1981) and similar forms may be found throughout the species range. Ab. *albofasciata* Lempke (1947;1951) has the median area of the forewing whitish.

Similar species
The characteristic forewing markings should preclude confusion with other species found in the British Isles.

LIFE HISTORY

OVUM Cream when first laid, later turning to pale green; 0.7mm, ovoid. Laid on the leaves and twigs of the foodplant in late May and June.

LARVA (Fig.30i). *First instar*. Pale greyish green; dorsal line faint, dark green. *Second instar*. Head and thoracic legs brown; integument pale grass green; dorsal line faint, dark green; small red dorsal dots. *Third and fourth instars*. Slender, tapering slightly towards head; length 22mm when full-grown; head pale green, outlined red; thoracic legs green; integument smooth, velvety, with sparse fine hairs; bright green, dorsal spots eliptical, deep red with a yellow centre; dorsal line fine, dark green; two spiracular lines, the uppermost yellow, the lower red, continuing as an outline to the anal flap; segmental divisions yellow.

Feeds on the foliage of a wide range of deciduous shrubs and trees. Hawthorn (*Crataegus monogyna*), goat willow (*Salix caprea*), pedunculate oak (*Quercus robur*), blackthorn (*Prunus spinosa*), rowan (*Sorbus aucuparia*) and field maple (*Acer campestre*) are the most commonly recorded. When feeding, the larva often bridges the hole which it has eaten into the edge of the leaf. In this position the spiracular lines closely mimic the missing leaf edge, thus disguising the feeding damage. When at rest it either lies along the midrib on the underside of a leaf or stalk or mimics the latter by standing erect at an angle to the branch. Late May or early June to early October.

PUPA Length 7–8mm; slender; thorax and abdomen olive-green; wings light green; abdominal divisions yellow. Pupation occurs just beneath the ground surface in a slight cocoon. October to May.

FLIGHT PERIOD AND HABITAT Univoltine, May and June, with occasional individuals recorded as early as late March and as late as mid-July. The peak appears to be during the second week of June. Generally distributed. Rests on tree-trunks and fences during the day, and the males especially come readily to light.

DISTRIBUTION (Map 9) Widespread and generally common throughout England, Wales and the Isle of Man. It is likely to occur in all of those counties not represented on the map. In Scotland it is much more localized but is found throughout the country. Widespread and common in Ireland.

COLLECTING AND REARING A female will usually lay freely in captivity. Alternatively, the larvae can be beaten easily from the foodplants, even in urban areas. It is easy to rear and seems to prefer dry leaves. Remarkably, it remains in the larval state for up to four months.

Eupithecia insigniata (Hübner, 1790)　　Pinion-spotted Pug
= *consignata* (Borkhausen, 1794)
= *consignaria* Boisduval, 1840

Pl.1, fig.15; Pl.4, fig.23; Pl.A. fig.11

HISTORY First recorded in Britain by Haworth (1809). Larvae were first described by Crewe (1868). From about 1866 to 1905 this species was continuously bred by Mrs Emma Hutchinson of Leominster, Herefordshire. It was from her that many of the important collectors of the period such as Doubleday and Stainton obtained their specimens, so elusive was the species at that time (Tutt, 1906g).

IMAGO *Characteristic features.* Wingspan *c.*23mm; ground colour pale grey; forewing discal spot large and conspicuous, dark costal spots and basal patch prominent and distinctive.
　　Genitalia. Figs 12b,14a,27j ♂; 21b ♀
　　Variation. A remarkably invariable species.

Similar species
None.

LIFE HISTORY

OVUM Greenish white; 0.6mm long, ovoid. In captivity the ova are usually laid singly or in pairs on the petals of the foodplant during May.

LARVA (Figs 30k,m). *First instar.* Translucent dirty greenish white with very faint dorsal spots. *Second instar.* Pale yellowish green; dorsal line light brown; subdorsal lines white; segmental divisions yellow. *Third instar.* Light green; dorsal line reddish brown; subdorsal lines white. *Fourth instar.* Slender; length 25mm when full-grown; head light brown with tiny black dots; thoracic legs light brown or green; integument smooth and velvety; dorsal area bright grass-green; dorsal markings variable, bright purplish red, outlined white; lateral and ventral surfaces light green; subdorsal line white, sometimes interrupted; a series of lateral red dots; anal claspers with lateral heavy red line outlined in white; segmental divisions white.

Feeds on the foliage of hawthorn (*Crataegus monogyna*) and apple (*Malus domestica*). Tutt (1906b) also lists blackthorn (*Prunus spinosa*), and Dietze (1913) and Skou (1986) confirm that this foodplant is also used in mainland Europe. Late May to July.

PUPA Length 7.5–8.0mm (J. Reid, pers.comm.); abdomen and thorax golden brown; wings olive-green. July to April in a cocoon just beneath the soil surface. Sterling (1985) cites an individual spending two winters (1983–85) in the pupal stage.

FLIGHT PERIOD AND HABITAT Univoltine, May and early June. Comes to light in small numbers.

DISTRIBUTION (Map 10) In England this species is found mainly south of a line from Herefordshire to the Humber estuary but extending to north-east Yorkshire in the east. In the south-west it has been recorded only singly in South Devon and East Cornwall. In Wales there are no confirmed colonies though its presence may be revealed by further research. It is apparently absent from Scotland and Ireland.

COLLECTING AND REARING This is a rather local species though it may be more common than supposed as the young larvae are very similar to many other geometrid species and may be overlooked. All small green larvae falling on to the beating tray or sheet should be retained, as it is only when they reach the final instar that those of *E. insigniata* are unmistakable. Larvae of *Lomographa bimaculata* (Fabricius) (Geometridae) closely resemble *E. insigniata* in their last instar but are much larger and lack the red stripe on the anal claspers. *E. insigniata* is most often found in areas where apple and hawthorn grow together. Larvae are not difficult to collect by beating, though they often tend to feed high up in the trees. A captive female will oviposit readily on the flowers of the foodplant. It is an easy species to rear.

Eupithecia valerianata (Hübner, 1813)   Valerian Pug
= *viminata* Doubleday, 1858
= *viminaria* Morris, 1861

Pl.1, fig.16; Pl.7, fig.8; Pl.A, fig.12

HISTORY First recorded in Britain by Doubleday (1858b) as *E. viminata* from specimens collected by N. Greening, F. Bond and P. H. Newnham. This record is reviewed by Stainton (1859b). The larva was first described by Crewe

(1862e;1863a), but it was not until 1885 that the native foodplant was found. This discovery was so fortuitous that the account is worth quoting verbatim. 'Until the larvae were accidentally discovered by Mr Baker, later of Derby, *E. valerianata* was considered a rare species. It was then known as *E. viminata*; and one day Mr Baker had thrashed some osiers long and valiantly, but to no purpose, thinking, as the name implied, the willow leaves were the food. Tired and disgusted he gathered a bunch of the common valerian, then in the full beauty of its delicate lavender flowers, to ornament his home. On looking them over for a chance larva, nothing was to be seen; but later that night, on suddenly entering the darkened room where they had been placed in water, to his astonishment, the whole bunch of flowers was alive with larvae of this moth.' (Carrington, 1885a).

IMAGO *Characteristic features* (Figs 1c,8i). Wingspan *c*.19mm; forewing apex pointed; forewing generally uniform in appearance, fresh specimens glossy, discal spot obsolete or inconspicuous, geniculate postmedian fascia inconspicuous, pale dentate or incomplete subterminal line evident, tornal spot elongate and prominent, small dark costal markings present though often absent in worn individuals; hindwing uniform, discal spot absent, minute pale tornal spot usually present.

Genitalia. Figs 14b,27k ♂; 20i ♀
Variation. Markings and colour vary little.

Similar species (See Fig.1, p. 35; Plate 7)
Eupithecia tenuiata (Fig.1b): forewing costa more arched; forewing with distinct discal spot, tornal spot obsolete, fine striae numerous, subterminal line obsolete, marginal band usually darker than rest of wing; hindwing with minute discal spot; overall less uniform and greyer.

E. haworthiata (Fig.1e): base of abdomen with distinctive orange patch; forewing costa more arched, termen more rounded; forewing generally less uniform, striae and fasciae more prominent with dark median fascia conspicuous, postmedian fascia curved, pale tornal spot absent or obsolete; hindwing with striae more prominent, pale tornal spot absent.

E. inturbata (Fig.1a): forewing costa more arched; fore- and hindwings less uniform; forewing with dark submarginal patches, discal spot more prominent, tornal spot less so.

E. plumbeolata (Fig.1d): forewing with subterminal line and tornal spot obsolete or inconspicuous. When worn, these two species are very similar and examination of the genitalia is recommended.

E. pygmaeata (Fig.1g): forewing costa straight; fore- and hindwings much darker, almost uniform brown; forewing discal spot absent.

E. vulgata (Fig.1k): pale or worn individuals sometimes similar but forewing postmedian fascia biangulate rather than geniculate.

E. indigata (Fig.1f): forewing more elongate, discal spot large and prominent, subterminal line and tornal spot absent.

LIFE HISTORY

OVUM Pinkish cream; 0.4mm long, ovoid. Laid on the foodplant in June and July.

LARVA (Fig.30j) *First instar.* Pale yellowish green; dorsal line fine, faint green or brown. *Second instar.* Similar to first but dorsal line stronger, dark red or brown. *Third and fourth instars.* Thickset, tapering slightly towards the head; length 16mm when full-grown; head and thoracic legs yellow; integument pale green; dorsal line fairly broad, dark brown or green; subdorsal lines dark brown or green; spiracular lines white.

Feeds on the seed-heads of common valerian (*Valeriana officinalis*) and marsh valerian (*V. dioica*). Hammond (1952) states that in captivity the larvae will accept red valerian (*Centranthus ruber*). July and August.

PUPA Length 6mm; thorax and abdomen golden brown; wings green. September to late May or early June in a cocoon spun amongst ground debris.

FLIGHT PERIOD AND HABITAT Univoltine, June and July, with the peak during the last week of June. It is usually found in damp open localities though it is known to occur in dry limestone quarries at Portland, Dorset. Flies in the late afternoon and also after dark when it comes to light in small numbers.

DISTRIBUTION (Map 11) Widely distributed and locally common, like its foodplant, throughout Britain including the Isle of Man, though seemingly absent from high altitudes and the western and northern isles. Local in Ireland. Easily overlooked and almost certainly under-recorded.

COLLECTING AND REARING Larvae are difficult to find in the field, but can be shaken off the foodplant on to a small tray. They are sometimes present in large numbers. A good supply of the foodplant is essential for this species as the larvae are voracious eaters. The larvae are best reared on flower-stalks kept in water. Captive females mate and oviposit readily.

Eupithecia pygmaeata (Hübner, 1799)   Marsh Pug
= *pygmaearia* Boisduval, 1840
= *palustraria* Doubleday, 1850

Pl.1, figs 17,18; Pl.7, figs 13,14; Pl.A, fig.13

HISTORY First recorded in Britain by Doubleday (1850b) as *E. palustraria*. Larvae were first described by Crewe (1872b,c).

IMAGO *Characteristic features* (Fig.1g). Wingspan c.16mm; forewing costa straight, apex pointed; colour dark leaden brown, most forewing markings usually inconspicuous or absent, discal spot absent, tornal spot conspicuous; hindwing almost uniform, its only marking usually a conspicuous whitish tornal spot.

Genitalia. Figs 14c,271 ♂; 21c ♀

Variation. Restricted to slight variation in the intensity of the ground colour. Ab. *grabei* Cornelsen (Goodson, unpubl.) is a light fawn colour and is particularly pale in the underside basal area. There is slight variation in the boldness of the forewing cross lines. In some specimens the wing fringes are not as noticeably chequered and occasionally they are unicolorous. Form *pseudozibellinata* Dietze (1910; 1913) is copiously sprinkled with light scales. This variation reaches its extreme in f. *zibellinata* Chrétin, which is illustrated by Weigt (1985).

Similar species (See Fig.1, p. 35; Plate 7)
E. *tenuiata* (Fig.1b): forewing costs arched; fore- and hindwings with numerous dark striae on paler ground not uniform dark brown; forewing discal spot distinct, pale tornal spot inconspicuous or absent; hindwing tornal spot absent.

E. *inturbata* (Fig.1a): forewing costa more arched; generally paler, not uniform dark brown; forewing with distinct dark submarginal patches; fore- and hindwing tornal spots much less prominent.

E. *haworthiata* (Fig.1e): forewing costa more arched; abdomen with orange basal patch; fore- and hindwings paler, less uniform, fasciae and striae more prominent, pale tornal spot inconspicuous or absent; forewing with dark median line.

E. *plumbeolata* (Fig.1d): generally much paler; fore- and hindwings with tornal spots obsolete or absent.

E. *valerianata* (Fig.1c): forewing costa slightly more arched; generally much paler, not dark brown.

LIFE HISTORY

OVUM Cream; 0.6mm long, ovoid. Attached to buds, flowers and seed-capsules of the foodplant in June.

LARVA (Fig.30l). *First and second instars.* Long, thin and tapering towards the head; pale green or greenish yellow, dorsal line fine, brown. *Third instar.* Long, thin and tapering towards the head. Colour similar to previous instars but dorsal line wider; subdorsal lines fine; a row of spiracular dots. All markings dark grey, dark green or brown. *Fourth instar.* Short and grub-like; length 14mm when full-grown; head pale yellow; thoracic legs pale brown; integument smooth, translucent, pale greyish green or brown; dorsal blotches faint, oval, bead like, connected by fine dark green or brown subdorsal line; spiracular line irregular, cream; ventrally pale greyish green, central line white. The larva is remarkable for the change in its shape in the last instar.

Feeds on the flowers and seed-heads of field mouse-ear (*Cerastium arvense*). Scorer (1913) also lists snow-in-summer (*C. tomentosum*) and Allan (1949) cites greater stitchwort (*Stellaria holostea*). June and July.

PUPA Length 6mm; abdomen, thorax and wings golden brown. Occasionally the thorax and wings are green and the tip of the abdomen is golden yellow. Late July to May, sometimes delaying emergence for a further year (Farren, 1892; Agassiz, *et al.*, 1981).

FLIGHT PERIOD AND HABITAT Usually univoltine and diurnal, flying near its foodplant in sunshine in May and early June (Skinner, 1984). It does not usually fly in dull weather but occasionally comes to light. Atmore (1891) states that adults were found in Norfolk from 15 June to the end of June and again on 26 August 1891. This suggests that a second brood can occur in favourable years. It inhabits rough fields, meadows, coastal sand-dunes and heathland. It is not, as its vernacular name might suggest, restricted to marshland.

DISTRIBUTION (Map 12) Widely distributed throughout the British Isles except the southernmost English counties, where it is apparently very uncommon or absent, and the western islands of Scotland and the Shetlands. In Ireland, though widespread, it is local and rare. The present preference for recording with light-traps may have led to this mainly diurnal species being overlooked.

COLLECTING AND REARING This very local moth is not easily found. The larvae are also difficult to find by searching the foodplant as they usually feed inside the seed-capsules with only the anal end protruding. In this position they closely resemble an elongated seed-capsule. The larvae are straightforward to rear but it is essential to ensure a supply of fresh food. Daily supplements of fresh seed-capsules are recommended but the larvae should not be disturbed by removing them from tenanted capsules. Alternatively, the foodplant can be kept in water. The larvae should be allowed to move to the fresh food as and when they want; they should never be transferred by hand.

Eupithecia venosata (Fabricius, 1787)   Netted Pug

subsp. *venosata* (Fabricius, 1787)
= *decussata* (Donovan, 1799)
= *venosaria* Boisduval, 1840

subsp. *ochraceae* Gregson, 1886

subsp. *fumosae* Gregson, 1887

subsp. *plumbea* Huggins, 1962

Pl.1, figs 19–23; Pl.4, figs 1–5; Pl.A, figs 14,15

Note: Represented in the British Isles by the nominate and the three other subspecies. There is also one geographical form, f. *hebridensis*, Parkinson Curtis.

HISTORY The nominotypical subspecies (subsp. *venosata*), was first recorded in England and described by Donovan (1799) as *decussata*, The Pretty Widow moth. Specimens were stated by him to have been caught at Faversham, Kent. Larvae were first fully described by Crewe (1861k).

E. venosata subsp. *ochracae* was first described by Gregson (1886) from specimens caught by E. Roper Curzon in Orkney.

E. venosata subsp. *fumosae* was also first described by Gregson (1887). The type specimens of this form were caught by Curzon in Shetland.

E. venosata subsp. *plumbea* was first described by H. C. Huggins from specimens caught at Inishvickilaun, on the Blasket Isles, off the coast of Co. Kerry, Ireland, on 22 May 1962 (Huggins, 1962a).

E. venosata f. *hebridensis* Parkinson Curtis was first recorded as British from specimens in the collection of A. Druitt which were caught on the Isle of Lewis (Curtis, 1944a). This form was originally given subspecific status.

IMAGO *Characteristic features*. Wingspan c.25mm; forewing marked boldly and distinctively.

Genitalia. Figs 14d,27m ♂; 21e ♀

Variation. In subsp. *venosata* (Pl.1, fig.19; Pl.4, fig.1), which is widespread over mainland Britain, variation is slight and is restricted to a general darkening of the ground colour in northern England and Scotland. Ab. *basinigrata* Cockayne (1954) has the area between the basal and antemedian lines smoky. There is a small degree of variation in the intensity and form of the black cross-lines. Ab. *circumfluxa* Kitt (1925) has all the black cross-lines thickened, especially at the costa. In f. *confluens* Dietze (1913) the median line is thickened and partly confluent with the antemedian, though sometimes only by black shading.

Subspecies *ochracae* (Pl.1, fig.21; Pl.4, fig.3) is restricted to the Orkneys and has an ochreous-brown ground colour with less distinct cross-lines. This subspecies is very variable, some populations producing forms close in appearance to subsp. *fumosae*.

Subspecies *fumosae* (Pl.1, fig.22; Pl.4, fig.4) has a dark ochreous-grey ground colour with finer black cross-lines and less conspicuous white bands than subsp. *venosata*. It is restricted to the Shetlands though similar forms occur in Orkney. Ab. *bandanae* Gregson (1887) of subsp. *fumosae* retains the conspicuous white bands of the nominotypical subspecies.

Subspecies *plumbea* (Pl.1, fig.23; Pl.4, fig.2) has a leaden grey ground colour, often with a purplish tinge. It is restricted to the Blasket Islands, Co. Kerry, and part of the adjacent mainland where it is allopatric and constant in appearance. Individuals from other parts of the western mainland of Ireland are stated to approach this form (Agassiz *et al.*, 1981). Ab. *sepiata* Huggins (1968) of this subspecies, taken on the Dingle Peninsula, Co. Kerry, is very dark sepia with brownish black (not black) netted markings.

Form *hebridensis* (Pl.1, fig.20; Pl.4, fig.5) is known to occur on Canna, Islay, Skye and Lewis. Here the species is very variable and appears superficially intermediate between subsp. *ochracae* and subsp. *fumosae*. Some individuals from Skye are stated to appear intermediate between subsp. *ochracae* and subsp. *venosata* (Agassiz *et al.*, 1981). However, nowhere does this form appear to be constant in appearance and allopatric and we can find no evidence to support subspecific status.

Similar species
None.

LIFE HISTORY

OVUM Greenish white when first laid, turning black just prior to emergence when the shell becomes transparent, revealing the larva; 0.7mm long, ovoid. It is laid on the flowers of the foodplant during May and June.

LARVA (Fig.31a) *First instar.* Pale greenish white with a large black head. *Second instar.* Entirely dark brown, almost black in some individuals. *Third and fourth instars.* Stoutly built with deep segmental divisions; length 22mm when full grown; head and thoracic legs black; integument wrinkled, waxy in appearance, lightly sprinkled with hairs; pale greenish white, dorsal band broad, deep grey or black; spiracles forming a row of small black dots.

Feeds on the flowers and ripening seed-capsules of several *Silene* species, mainly bladder campion (*Silene vulgaris*) and sea campion (*S. uniflora*) The stamens and pistil are eaten first, and later the seeds. Late June to early August.

PUPA Length 7–8mm; thorax, abdomen and wings brown. In an earthen cocoon from August to May. Anderson (1890) cites a record of the pupa overwintering twice (1888–90).

FLIGHT PERIOD AND HABITAT Univoltine, mid-May to mid-July, in the late afternoon and evening. The peak emergence appears to be during early and mid-June. Tutt (1906b) states that the moth is reluctant to fly if disturbed during the day, preferring to crawl to safety. At night it comes readily to light.

DISTRIBUTION (Map 13) Locally common, especially in coastal areas, throughout the British Isles, but absent from high altitudes. In Ireland, recorded only from the north-east and south-west.

COLLECTING AND REARING This widespread species is most commonly found in the larval state, but is often subject to attack by insect parasitoids. Riley found approximately 80 per cent parasitization in a colony in Bedfordshire in 1985.

The female will oviposit in captivity, though perhaps the best method of obtaining stock is to search for larvae. They are usually very easy to find and should then be removed and kept on a foodplant which has been checked to see that no other lepidopterous larvae are present. Several carnivorous species of the genus *Hadena* (Noctuidae) also feed on *Silene* and, if not separated, most of the *E. venosata* larvae will be eaten. Once segregated, they can be reared successfully in small boxes or on foodplant kept in water.

Eupithecia egenaria Herrich-Schäffer, 1848
Pauper or Fletcher's Pug
= *undosata* Dietze, 1875

Pl.1, fig.24; Pl.8, fig.8; Pl.B, fig.1

HISTORY Discovered in the Wye Valley on 15/16 June 1962 by R. M. Mere, E. C. Pelham-Clinton and J. D. Bradley, and first recorded by Mere (1962) who stated that the specimens had been caught amongst 'broad-[=large-]leaved lime'. However, Haggett & Mere (1964) found this to be erroneous – small-leaved lime being the correct foodplant in that area – and they published the first description of the larva. The species had, in fact, been caught in 1953 in the Breck district of Norfolk by J. Fenn (Agassiz *et. al.*, 1981; G. M. Haggett, pers.comm.) but had not then been identified.

IMAGO *Characteristic features* (Figs 5a, 8 l). Wingspan *c.*25mm; abdomen with pale basal band; forewing elongate, discal spot elongate and conspicuous, ante- and postmedian lines double and smoothly geniculate, marginal band slightly darker than rest of wing, subterminal line and tornal spot inconspicuous; hindwing with pale curved postmedian fascia, discal spot inconspicuous.

Genitalia. Figs 12d,14f,27n ♂; 21d ♀

Variation. Specimens from Norfolk are stated to be slightly darker and larger than those from the Wye Valley (Haggett, 1981a). Within each of the two regions, variation is limited to slight differences in the intensity of the ground colour.

Similar species (See Figs 5,8, pp. 61,69; Plate 8)
E. intricata subsp. *arceuthata* (Figs 5d,8b): abdomen with dark subbasal band; forewing with dark median line strongly developed between costa and discal spot, antemedian line more strongly geniculate, double ante- and postmedian lines absent.

E. satyrata (Figs 5d,8j): usually smaller; forewing less elongate, double ante- and postmedian lines absent, nervures spotted alternate light and dark, pale tornal spot more prominent; more uniform overall, fresh specimens more glossy.

E. tripunctaria (Figs 5f,8h): metathorax often with white spot; forewing without double ante- and postmedian lines, subterminal row of white spots usually conspicuous, white tornal spot much more conspicuous, hindwing with conspicuous white tornal spot.

E. denotata subsp. *jasioneata* (Figs 5i,8f): forewing postmedian fascia biangulate, double ante- and postmedian lines absent, median striae more prominent.

E. subfuscata (Figs 5 l,8g): forewing less elongate, double ante- and postmedian lines absent, postmedian fascia more strongly geniculate, pale tornal spot more prominent; hindwing often with pale tornal spot.

E. virgaureata (Figs 5 l,8g): forewing less elongate, double ante- and postmedian lines absent, postmedian fascia more strongly geniculate, pale tornal spot more prominent, nervures sometimes spotted alternate light and dark; hindwing often with pale tornal spot; metathorax in pale forms sometimes with white spot.

E. lariciata (Fig.5k): metathorax usually with white spot; forewing postmedian fascia and line more strongly geniculate, double ante- and postmedian lines absent.

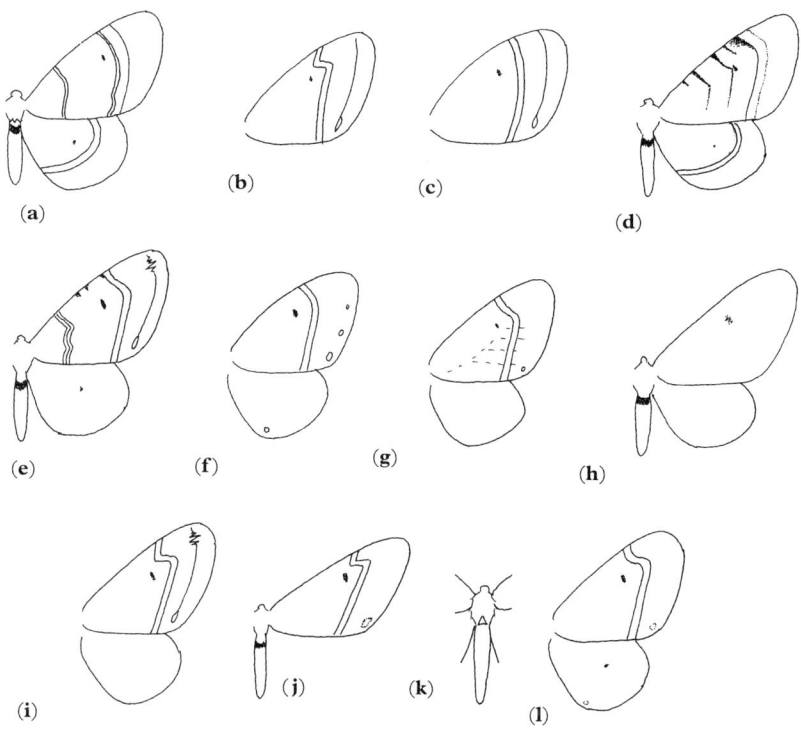

Text figure 5 Wing shapes and patterns (Similar species)

(a) *E. egenaria* (p. 59) (b) *E. vulgata* (p. 83) (c) *E. simpliciata* (p. 101)
(d) *E. intricata* (p. 67) (e) *E. pimpinellata* (p. 106) (f) *E. tripunctaria* (p. 86)
(g) *E. satyrata* (p. 72) (h) *E. millefoliata* (p. 99) (i) *E. denotata* (p. 88)
(j) *E. nanata* (p. 108) (k) *E. lariciata* (p. 127) (l) *E. subfuscata/virgaureata* (pp. 91/114)

LIFE HISTORY

OVUM Bluish white; 0.6mm, ovoid. Laid on the buds and bracts of the food-plant in late May and June.

LARVA (Fig.31b) *First instar*. Head and thoracic legs dark brown or black; integument pale yellowish green or grey, markings absent. Nibbles around leaf edges and eats small holes into the bracts. *Second instar*. Head pale yellow; thoracic legs pale brown; integument pale greyish green or pinkish white; dorsal line faint green. Feeds on the bracts and buds. *Third instar*. Similar to third but dark green dorsal line more pronounced. Eats holes into the flower-buds. *Fourth instar*. Slender; length 25mm when full-grown; head pale yellowish-grey; thoracic legs green; integument smooth and translucent, revealing the gut, pale grass green or yellowish green; dorsal line fine, dark green; spiracular lines fine, yellow;

segmental divisions pale yellow. Feeds by excavating the flower-buds or, if the flowers are open, by standing stretched out (rather than buried in the flower) and eating the stamens and pistil.

Feeds on small-leaved, large-leaved and common lime (*Tilia cordata*, *T. platyphyllos* and *T.* × *vulgaris*). June and July.

PUPA Length 8–9mm; stout; thorax and abdomen light brown; wings tinged green. July to May of the following year in a tough cocoon amongst ground debris.

FLIGHT PERIOD AND HABITAT Univoltine, late May and June (Skinner, 1984), in the vicinity of the foodplant. During the day it can occasionally be found resting on trunks of lime. At night it comes readily to light, sometimes in quite large numbers. Wakeley (1957) states that adults frequent the tops of lime trees and Haggett (1968d) suggests that this species requires a high canopy and that optimum conditions occur where limes reach 60ft. (18.5m) tall and have been left undisturbed for many years.

In the Wye Valley this species may be expected to occur wherever small-leaved lime forms woodland. In the Breckland district of Norfolk, Haggett (1981b) states that it is also associated with large-leaved and common lime.

DISTRIBUTION (Map 14) Until recently, known in the British Isles only from Monmouthshire and Gloucestershire in the west and Norfolk and Lincolnshire in the east where it formed strong colonies in association with the larval foodplant (Waring, 1997). Recent records from Worcestershire, Somerset, Berkshire (VC22) [including parts now in Oxfordshire] and Suffolk indicate an extension of its British distribution (Waring, 2002, 2003). There are also records from Surrey (Collins, 1997), suggesting the presence of a colony. The single record from West Sussex (C. Pratt, pers.comm, confirmed by Riley) presumably represents a vagrant individual.

Whether this species had been overlooked in Britain prior to its discovery in 1962 or whether it is a relatively recent arrival to this country is a matter for debate: Wakely (1957), Mere (1963), Emmet (1981) and Haggett (1981b) discuss the pros and cons. The apparent morphological differences between specimens from the two original, widely-separated localities suggest either that it has been resident at least long enough for subspeciation to begin or that there were two separate colonizations (cf. Haggett, 1981b).

COLLECTING AND REARING In captivity, *E. egenaria* is straightforward to rear. Females are attracted to light and will oviposit freely on sprigs of lime or pieces of netting. Larvae can be beaten easily from the wild foodplant.

Eupithecia centaureata ([Denis & Schiffermüller], 1775)
Lime-speck Pug

= *oblongata* (Thunberg, 1784)
= *centaurearia* Boisduval, 1840

Pl.1, fig.25; Pl.4, fig.22; Pl.B, fig.2

HISTORY Moses Harris (1766) was the first to describe a British example of this species. As well as accurate illustrations of two adults, he figured and described two green forms of the larvae reared on African marigold, the ova and pupa, and briefly outlined the life history. A more detailed description of the early stages was given by Crewe (1859k;1861k).

IMAGO *Characteristic features.* Wingspan c.22mm; ground colour white; forewing with characteristic dark median costal patch uniting with conspicuous black discal spot.
   Genitalia. Figs 12e,14e,270 ♂; 21f ♀
   Variation. Variation in the markings is limited to general darkening of the ground colour: f. *obscura* Dietze (1910;1913), in which there is a smoky suffusion; or a weakening of the normal markings: ab. *albidior* Heinrich (1916), in which the mid costal spot is greatly reduced); ab. *punctata* Hannemann (1916), in which the same spot is reduced to a small dot; and ab. *centralisata* Staudinger (1892), in which all the forewing markings are greatly reduced or obsolete.

Similar species
None.

LIFE HISTORY

OVUM Creamy white and glossy when first laid, later turning orange; 0.57mm long, ovoid. On the flowers and leaves of the foodplant, June to September.

LARVA (Fig.31c) *First instar.* Pale yellow or very pale green; dorsal line faint, dark green or brown. *Second and subsequent instars.* Slender, tapering slightly towards the head; length 18mm full-grown; head and thoracic legs brown or dark green; integument wrinkled, rough, covered with sparse short fine hairs; ground colour variable, usually matching that of the flowers on which the larva has fed, markings also very variable, often restricted to dorsal and spiracular lines of a darker shade than the ground colour; dorsal line occasionally resembles a string of beads. In this form the subdorsal and spiracular lines are usually strong and darker shade of the ground colour. Probably the most common form has a series of five dark trident-shaped dorsal markings linked by a fine dark dorsal line with spiracular lines irregular, strong, dark.
   Polyphagous on the flowers of herbaceous plants from late June to October.

PUPA Length 7mm; thorax and abdomen brownish yellow; wings dark green. In a cocoon spun amongst debris on the surface of the ground. July to May of the following year, sometimes producing a second generation of adults in September.

FLIGHT PERIOD AND HABITAT Adults may be found any time between early May and early October. It is unclear whether this species is univoltine, with a protracted emergence, or partially bivoltine. During the day it rests rather conspicuously on fences, walls and tree-trunks. At night it comes readily to light. It is found in many habitats, often occurring commonly in gardens and on waste ground.

DISTRIBUTION (Map 15) Widespread and generally common throughout

England and in the Isle of Man. In Wales it is apparently not found in upland areas. In Scotland it also appears to be restricted to the lowlands and is absent from the Orkneys and Shetlands. In Ireland it is common and generally distributed.

COLLECTING AND REARING One of the easiest of the pugs to rear. Both sexes come to light and the female oviposits freely in captivity. The larvae are very easy to rear and readily transfer from one foodplant to another. In the wild they can be found by searching or by sweeping low herbage but they are often heavily attacked by parasitoid insects.

Eupithecia trisignaria Herrich-Schäffer, 1848
Triple-spotted Pug
= *trisignata* Herrich-Schaffer, 1861
Pl.1, figs 26–28; Pl.6, fig.1; Pl.7, fig.17; Pl.8, fig.12; Pl.B, fig.3

HISTORY First recorded in Britain by Doubleday (1861) from larvae collected by Henry Harpur Crewe and Joseph Greene in Derbyshire in 1860 and identified by Herrich-Schäffer. Larvae were first fully described by Crewe (1861d (as *trisignata*); 1862a). The first British records were reviewed by Stainton (1862).

IMAGO *Characteristic features* (Figs 6a,7j,8k). Wingspan *c.*22mm; forewing almost uniform, most fasciae and striae obsolete or inconspicuous, costa with two characteristic dark spots forming corners of equilateral triangle with prominent oval discal spot (fig.8k), postmedian fascia geniculate when present, subterminal line dentate towards costa; hindwing with fasciae and striae obsolete, discal spot elongate and prominent. Melanic form with prominent forewing discal spot.

Genitalia. Figs 12f,14g,27p ♂; 21g ♀
Variation. The melanic f. *angelicata* Prout (1938) (Pl.6, fig.1) is found mainly in the eastern counties; notably at Wicken Fen, Cambridge (Agassiz *et. al.*, 1981).

Similar species (See Figs 6,7,8, pp. 65,67,69; Plate 6)
E. vulgata f. *atropicta* (Figs 6b,7e,8e): forewing narrower, discal spot minute or absent, costal spots absent, tornal spot often prominent, postmedian fascia biangulate.

E. tripunctaria (Figs 6c,7i,8h): general appearance less uniform; forewing with row of subterminal white spots and prominent tornal spot; hindwing with prominent white tornal spot; metathorax often with white spot. Examination of the male abdominal plate or the genitalia is recommended for separation of the melanic forms.

E. denotata (Figs 7h,8f): overall appearance less uniform, fore- and hindwings with more conspicuous fasciae and striae and pale tornal spot; forewing with prominent costal spots absent, postmedian fascia biangulate.

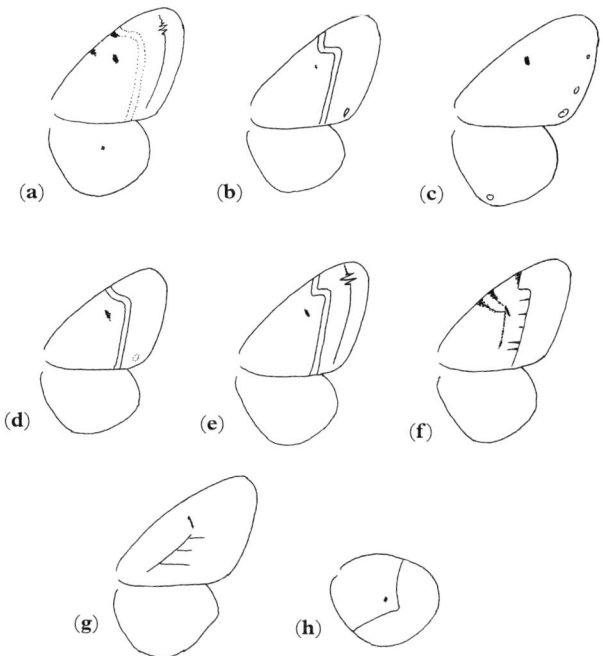

Text figure 6 Wing shapes and patterns (Similar species)
(**a**) *E. trisignaria* (p. 64) (**b**) *E. vulgata* (p. 83) (**c**) *E. tripunctaria* (p. 86)
(**d**) *E. subfuscata/virgaureata* (pp. 91/114) (**e**) *E. denotata* (p. 88) (**f**) *E. pusillata* (p. 122)
(**g**) *E. abbreviata* (p. 117) (**h**) *P. rectangulata*, hindwing underside (p. 135)

E. subfuscata (Figs 6d,8g): overall appearance less uniform, fore- and hindwings with more conspicuous fasciae and striae; forewing with pale tornal spot, prominent costal spots absent. Examination of the male abdominal plate or the genitalia is necessary to separate the melanic forms.

E. virgaureata (Figs 6d,8g): overall appearance less uniform, fore- and hindwings with more conspicuous fasciae and striae; forewing with pale tornal spot, prominent costal spots absent, nervures sometimes spotted alternate light and dark; pale forms sometimes with white metathoracic spot. Examination of the male abdominal plate or the genitalia is required to separate the melanic forms.

E. pusillata (Figs 6f,8p): some dark forms similar but forewing narrower, vestiges of fasciae usually remain.

E. lariciata f. *nigra* (Pl.6, fig.8): similar to f. *angelicata* (see above). Usually distinguished by slightly more pointed forewing but examination of the male abdominal plate or the genitalia is recommended.

E. abbreviata f. *nigra* (Pl.6, fig.19): similar to f. *angelicata* but forewing much more elongate, discal spot linear, nervures intensely black (Figs 6g,8r); hindwing termen concave, apex pointed.

Pasiphila rectangulata f. *anthrax* (Pl.6, fig.10): similar to f. *angelicata* but forewing broader, costa relatively shorter and more arched; black generally more intense; forewing sometimes with vestiges of green coloration, discal spot less prominent; underside of hindwing usually with prominent angulate postmedian line (Figs 6h,8q).

LIFE HISTORY

OVUM Cream; 0.7mm long, ovoid. Laid on the foodplant in June and July.

LARVA (Fig.31d) *First and second instars.* Pale yellowish or greyish green; dorsal line and thicker subdorsal lines dark green; spiracular line cream or pale yellow. *Third and fourth instars.* Short and stumpy, tapering slightly towards the head; length 15mm when full-grown; head dark brown divided dorsally by a green line; thoracic legs yellowish green; integument smooth, slightly wrinkled; pale grass green; dorsal and subdorsal lines fine, dark green; spiracular line undulate, cream; segmental divisions yellow; anal tip with a rounded triangular red dot; ventrally dark green with a yellow central line. Usually feeds at night, hiding by day beneath the umbels of the foodplant.

Feeds on the flowers and seeds of angelica (*Angelica sylvestris*). It is also stated to feed on hogweed (*Heracleum sphondylium*) (G. M. Haggett, pers.comm.), burnet-saxifrage (*Pimpinella saxifraga*) (Allan, 1949) and greater burnet-saxifrage (*Pimpinella major*) (Firmin *et al.*, 1975). Late August to early October.

PUPA Length 6–6.5mm; thorax and wings dark green; abdomen reddish yellow; cremaster deep red. October to June.

FLIGHT PERIOD AND HABITAT Univoltine, late June to mid-August with a peak during mid-July. It comes sparingly to light but is rarely found flying by day. Fens, marshes, damp woodlands and ditches.

DISTRIBUTION (Map 16) Widely distributed over England and Wales including the Isle of Man but seemingly rather localized. In Scotland it is even more sporadic and is apparently absent from most of the Western Isles and the Orkneys and Shetlands. Local in Ireland (Cos Fermanagh, Tyrone, Armagh and Dublin). This species is difficult to identify in the adult stage and is likely to have been overlooked over much of its range. It is probably more common than the map suggests.

COLLECTING AND REARING Larvae seem to prefer the seeds of the foodplants to the flowers and can easily be found, often along with those of *E. tripunctaria*, by gently tapping the seed-heads over a sheet. *E. tripunctaria* has a larval form that resembles that of *E. trisignaria* but they can usually be distinguished by the colour of their heads (black or dark brown in *E. trisignaria* and yellowish green in *E. tripunctaria*). Further, *E. trisignaria* lacks the brown dorsal markings of *E. tripunctaria*. The larvae are straightforward to rear in captivity.

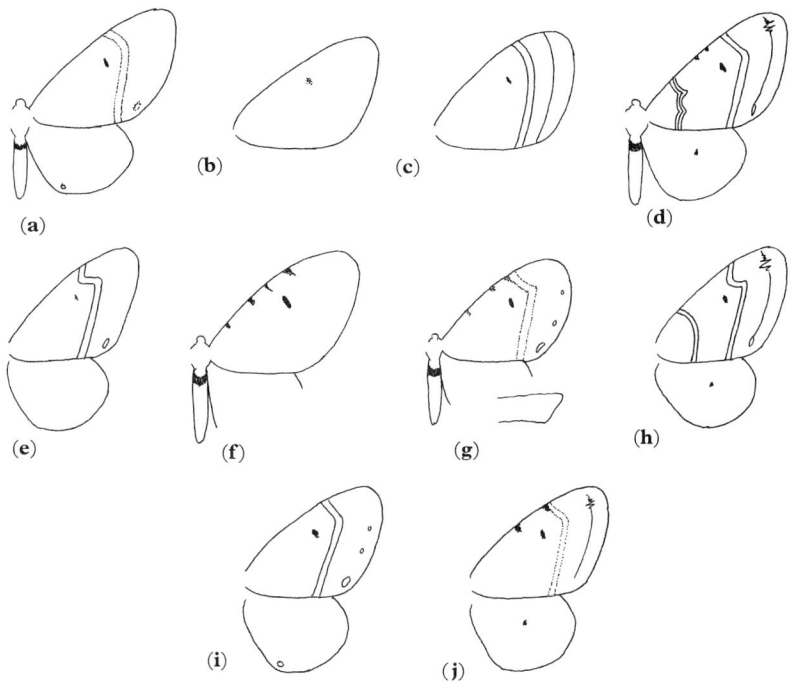

Text figure 7 Wing shapes and patterns (Similar species)

(**a**) *E. absinthiata* (p. 76) (**b**) *E. millefoliata* (p. 99) (**c**) *E. simpliciata* (p. 101)
(**d**) *E. pimpinellata* (p. 106) (**e**) *E. vulgata* (p. 83) (**f**) *E. expallidata* (p. 79)
(**g**) *E. assimilata* with profile of ♂ abdomen (p. 81) (**h**) *E. denotata* (p. 88)
(**i**) *E. tripunctaria* (p. 86) (**j**) *E. trisignaria* (p. 64)

Eupithecia intricata (Zetterstedt, 1839)

E. intricata subsp. *arceuthata* (Freyer, 1842)	Freyer's Pug
E. intricata subsp. *millieraria* Wnukowsky, 1929	Edinburgh Pug
= *anglicata* Millière, 1869, nec Herrich-Schäffer, 1863	
= *helveticaria* sensu auctt.	
E. intricata subsp. *hibernica* Mere, 1964	Mere's Pug

Pl.1, figs 29–31; Pl.6, fig.18; Pl.8, figs 6,16; Pl.B, fig.4

Note: The nominate subspecies does not occur in the British Isles where it is represented by subspecies *arceuthata* Freyer, *millieraria* Wnukowsky and *hibernica* Mere.

HISTORY Subspecies *arceuthata* Freyer (1842) (Freyer's Pug) was first recorded in Britain by Crewe (1860n) from larvae found in Buckinghamshire. However, these were identified as *E. helveticata* (= *E. helveticaria* Boisduval), and it was not until 1862 that Crewe (1862d) questioned the status of English and Scottish specimens. He later (1862p;1863a) raised subsp. *arceuthata* to specific level (*E. arceuthata*), describing the larvae of both this and subsp. *millieraria*. Further larval descriptions of subsp. *arceuthata* were published by Crewe (1863b,c).

Subspecies *millieraria* Wnukowsky (1929) (Edinburgh Pug) was first recorded in Britain by Doubleday (1857) from specimens found by R. F. Logan on the Pentland Hills near Edinburgh. They were identified by Guenée as *E. helveticaria*. Larvae were described by Crewe (under the name *helveticata*) (1860 l,1861k,1862p,1863d). The subspecies was subsequently described under the name *anglicata* by Millière (1869). However, this name was preoccupied and *millieraria* was therefore proposed by Wnukowsky (1929).

Subspecies *hibernica* (Mere's Pug) was first recorded and described by Mere (1964) from specimens caught at Doughbrannan in Co. Clare, Ireland. The larva was described by Haggett (1968c).

IMAGO *Characteristic features* (Figs 5d,8b). Subsp. *arceuthata* (Pl.1, fig.29; Pl.8, fig.6; Pl.B, fig.4) and *hibernica* (Pl.1, 31; Pl.6, fig.18). Wingspans *c*.25mm and 22mm respectively; abdomen with dark subbasal band; forewing costa almost straight, ground colour grey, discal spot conspicuous, ante- and postmedian lines conspicuous at costa, often extending across whole wing, antemedian line smoothly geniculate or curved, postmedian line usually curved, median line strongly developed at costa, usually extending to discal spot, often forming a complete sharply geniculate line across whole wing, postmedian fascia curved, subterminal line and tornal spot inconspicuous; hindwing discal spot faint, postmedian fascia curved.

Subsp. *millieraria* (Pl.1, fig.30; Pl.8 fig.16). Wingspan *c*.22mm; ground colour brown; abdomen with dark subbasal band; forewing discal spot conspicuous, median, ante- and postmedian lines usually only strongly developed at costa, median line often extends to discal area, postmedian fascia usually curved, subterminal line inconspicuous or obsolete; hindwing uniform, fasciae and striae usually obsolete, discal spot prominent.

Genitalia Figs 15a,27q ♂; 21i ♀

Variation. Subsp. *arceuthata* differs from the nominate subspecies (which is not found in Britain) in having a less brown ground-colour, slightly more ample wings and more conspicuously spotted nervures. Subsp. *millieraria* is smaller and browner with dark brown, almost black, markings. Subsp. *hibernica* is also smaller than *arceuthata* but with a cold, steely-grey ground-colour. Occasionally, very dark specimens of subsp. *millieraria* occur and these are referrable to f. *suffusa* Dietze (1910;1913). Ab. *mediofasciata* Dietze (1910;1913) of subsp. *arceuthata* has the median area of the forewing partly or completely darkened and the distal area relatively weakly marked.

Similar species (See Figs 5,8, pp. 61,69; Plates 6,8)

E. satyrata (Figs 5g,8j): abdomen without dark subbasal band; forewing without

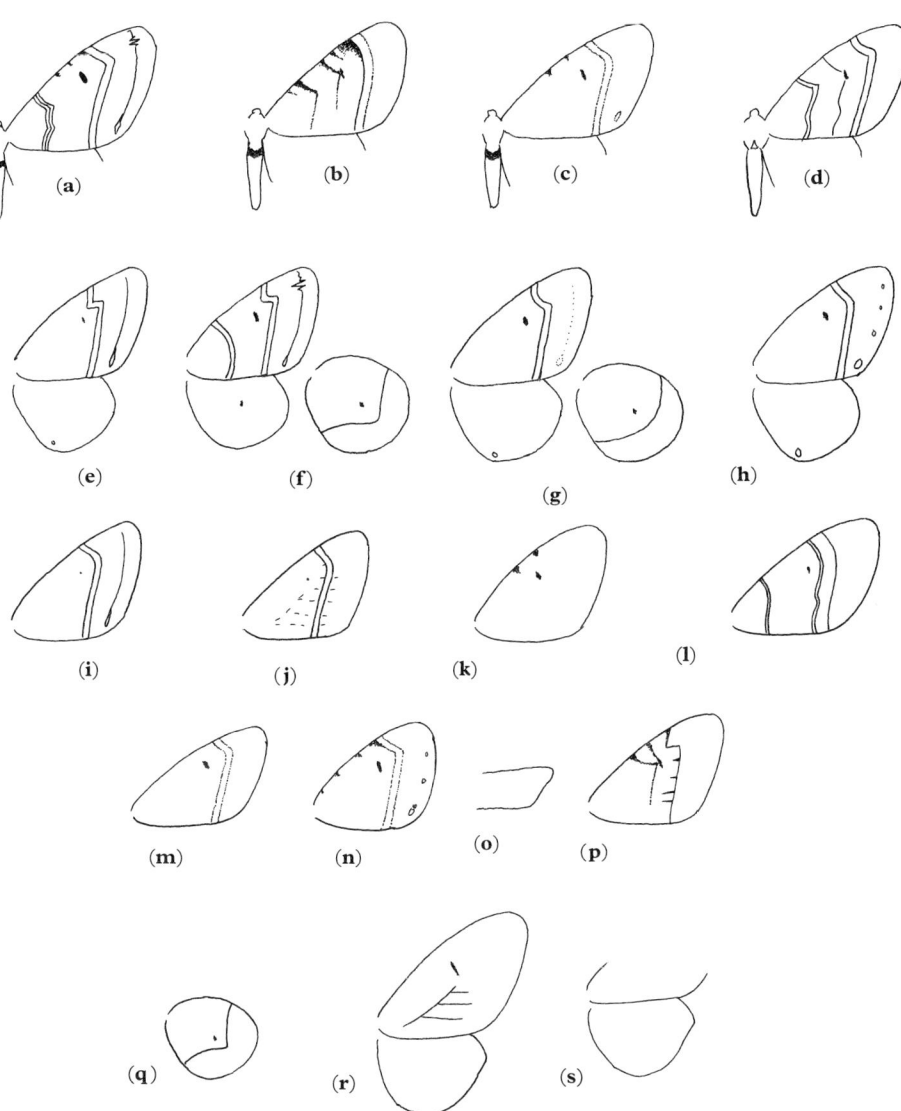

Text figure 8 Wing shapes and patterns (Similar species)
(a) *E. pimpinellata* (p. 106) (b) *E. intricata* (p. 67) (c) *E. absinthiata* (p. 76)
(d) *E. lariciata* (p. 127) (e) *E. vulgata* (p. 83) (f) *E. denotata* (p. 88)
(g) *E. subfuscata/virgaureata* (pp. 91/114) (h) *E. tripunctaria* (p. 86)
(i) *E. valerianata* (p. 53) (j) *E. satyrata* (p. 72) (k) *E. trisignaria* (p. 64)
(l) *E. egenaria* (p. 59) (m) *E. indigata* (p. 105) (n) *E. assimilata* (p. 81)
(o) *E. assimilata*, profile of ♂ abdomen (p. 81) (p) *E. pusillata* (p. 122)
(q) *Pasiphila rectangulata*, hindwing underside (p. 135) (r) *E. abbreviata* (p. 117)
(s) *E. dodoneata*, hindwing (p. 119)

dark median, ante- and postmedian lines, discal spot much smaller and less elongate, nervures usually spotted alternate light and dark; hindwing with dorsal origins of striae less conspicuous. F. *callunaria* paler than subsp. *millieraria*; forewing greyer rather than brown. F. *curzoni* similar to subsp. *hibernica* but forewing discal spot minute or absent, postmedian line geniculate, median line less prominent; hindwing paler, less uniform, striae prominent, discal spot minute or absent. *E. satyrata* f. *curzoni* does not occur in the same geographical range as *E. intricata* subsp. *hibernica*.

E. egenaria (Figs 5a,8 l): abdomen with pale basal band, dark subbasal band less prominent; forewing with distinctive double ante- and postmedian lines; hindwing paler, striae and fasciae more prominent.

E. vulgata (Figs 5b,8e): sometimes similar to subsp. *millieraria* but abdomen without dark subbasal band; forewing postmedian fascia biangulate, discal spot minute, pale subterminal line more conspicuous, tornal spot prominent.

E. tripunctaria (Figs 5f,8h): abdomen without dark subbasal band; metathorax sometimes with white spot; forewing with subterminal row of white spots; fore- and hindwings with prominent white tornal spot.

E. denotata (Figs 5i,8f): abdomen without dark subbasal band; forewing postmedian fascia biangulate, dark median, ante- and postmedian lines less prominent.

E. subfuscata (Figs 5 l,8g): abdomen without dark subbasal band, abdomen ventrally conspicuously paler; forewing costa more arched, apex less pointed, dark median, ante- and postmedian lines not prominent, postmedia fascia geniculate; hindwing paler basally.

E. millefoliata (Fig.5h): forewing tinged ochreous, dark median, ante- and postmedian lines absent, discal spot less well defined; hindwing paler basally.

E. virgaureata (Figs 5 l,8j): abdomen without dark subbasal band, abdomen ventrally conspicuously paler; metathorax sometimes with white spot; forewing costa more arched, apex less pointed, dark median, ante- and postmedian lines not prominent, postmedian fascia geniculate, nervures sometimes spotted alternate light and dark; hindwing paler basally.

E. lariciata (Fig.5k): metathorax usually with white spot; abdomen without dark subbasal band, abdomen ventrally conspicuously paler; forewing costa more arched, dark median, ante- and postmedian lines not prominent.

LIFE HISTORY

OVUM Cream when first laid, later turning greenish yellow; 0.7mm long, ovoid. Late May to the end of June.

LARVA (Fig.31g). *First instar*. Dark green, dorsal markings darker, obscure. *Second instar*. Head and thoracic legs yellowish green; integument bluish green; dorsal line fine, dark green or black; subdorsal line uneven, pale yellow, running the length of the body and continuing round the back of the head; ventrally dark green. *Third and fourth instars.* Large, robust; length 20mm when fullgrown; head and thoracic legs yellowish green; integument glossy, dorsally grass-green, laterally darker green; dorsal line fine, dark green; subdorsal lines cream; spiracular lines thicker than the subdorsal lines, cream or pale yellow; segmental divisions yellow; ventrally grass-green with a pale yellow central line.

Subsp. *arceuthata* has been found on juniper (*Juniperus communis*), Lawson's cypress (*Chamaecyparis lawsoniana*), Monterey cypress (*Cupressus macrocarpa*), common savin (*Juniperus sabina*), Chinese thuja (*Thuja orientalis*) and Leyland cypress (×*Cupressocyparis leylandii*). Hammond (1952) also cites Douglas fir (*Pseudotsuga menziesii*). Subsp. *millieraria* is usually stated to be solely associated with juniper but the larvae have also been found on Lawson's cypress (Prior, pers.obs.). Subsp. *hibernica* is possibly associated only with the prostrate *Juniperus communis nana*. Late July to late September.

PUPA Length 8–9mm; thorax and abdomen usually bright green; wings darker green, occasionally with reddish edges; cremaster red. September to May.

FLIGHT PERIOD AND HABITAT Univoltine, May to mid-July with a peak during the second week of June. It is found in the vicinity of its food plant and comes freely to light at night. Often found in gardens. Subsp. *hibernica* is stated by Haggett (1968c) to occur only at high altitude on exposed hills in open limestone fissures.

DISTRIBUTION (Map 17) In England, this species is represented by subsp. *arceuthata* and is locally very common northwards to Cheshire and North Yorkshire. It is apparently absent from much of Wales, though where it occurs it is represented by the same subspecies. In Scotland there are no records for the south-west but otherwise it is widely distributed and is represented by subsp. *millieraria*. However, in the North Ebudes, K. Bland (pers.comm.) found specimens which resembled the Irish subsp. *hibernica*. It is apparently absent from the Outer Hebrides, the Orkneys and the Shetlands. Subsp. *millieraria* and subsp. *arceuthata* are separated by a gap in north-western and northern England. There are single records for Cumberland and Westmorland but the subspecific identification is unknown. Indeed, as there are no voucher specimens, their identification as this species cannot be guaranteed. In Ireland, *E. intricata* is restricted to Co. Clare where it is represented by subsp. *hibernica*, although there was one 19th-century record, 'referred to the var. *arceuthata*', from Co. Kerry.

COLLECTING AND REARING This species is easy to find and rear. Females will lay freely in captivity. In the south of England, subsp. *arceuthata* is found most frequently in urban gardens. Larvae are easy to rear in the standard way, though leaves of juniper and cypress dry out very quickly. They seem to prefer the fresh growth at the tips of the leaves.

Eupithecia satyrata (Hübner, 1813)   Satyr Pug
= *satyraria* Boisduval, 1840
= *curzoni* Gregson, 1884

Pl.2, figs 1-4; Pl.6, figs 14,17; Pl.8, figs 9,18; Pl.B, fig.5

HISTORY The nominotypical form of *Eupithecia satyrata* (Hübner) was first recorded in Britain from Aylesbury, Buckinghamshire, in 1853 as *E. fagicolaria* (Greene, 1854). However, Stainton (1855) stated that it had been caught prior to this by J. W. Douglas and J. J. Weir at Mickleham, Surrey, in 1849, but had not been recorded in the literature. *E. satyrata* f. *callunaria* was also first recorded as a distinct species, *E. callunaria*, by Doubleday (1850b). *E. satyrata* f. *curzoni* Gregson was first described as British from specimens caught at Balta Sound, Shetland, in June, 1884, by E. Roper Curzon and it too was described as a distinct species (*E. curzoni* Gregson, 1884b,1885). However, the first British specimens had actually been caught in 1880 in the Shetlands by H. McArthur (1884), but were misidentified as *E. nanata*. Hodgkinson (1885) was the first to suggest that *curzoni* was a form of *E. satyrata*. The larva of *E. satyrata* was first fully described by Crewe (1860k;1861k).

IMAGO *Characteristic features* (Figs 5g,8j). Wingspan *c.*22mm; f. *callunaria* (Pl.2, fig.3; Pl.8, fig.18) and f. *curzoni* (Pl.2, fig.4; Pl.6, fig.17) *c.*20mm; forewing costa almost straight, only slightly arched, apex pointed, discal spot minute but usually conspicuous, postmedian fascia geniculate, marginal band usually darker than rest of wing, subterminal line incomplete or absent, small white tornal spot conspicuous, nervures usually spotted alternate light and dark; hindwing discal spot inconspicuous or absent. F. *curzoni* conspicuously striped; forewing discal spot conspicuous, postmedian fascia curved or geniculate, all other fasciae and lines geniculate, subterminal line prominent, complete or a series of conspicuous white spots, tornal spot conspicuous; hindwing with markings strongly developed, postmedian fascia curved.

Genitalia. (Figs 15c,27s ♂; 21h ♀). The female bursa copulatrix is very similar in general form to that of *E. cauchiata* but in that species the signa are larger and more robust in appearance.

Variation. Nominotypical form of *E. satyrata* (Pl.6, fig.14; Pl.8, fig.9): variation in the shade of the whole or part of the ground colour is considerable and is summarized as follows:

Considerably lighter with strikingly black discal spot: ab. *medionotata* Dietze (1910;1913);

Median area of forewing broadly darkened: ab. *nigrofasciata* Dietze (1910;1913);

Median area of forewing partly darkened leaving pale transverse bands: ab. *transversa* Dietze (1910;1913);

Boundary lines (ante- and postmedian) of median area darkened: ab. *bistrigata* Dietze (1910;1913);

Whole of wings brownish, approaching f. *callunaria*: f. *fagicolaria* Robson & Gardner (1886);

Ground colour grey, sometimes with brownish tint, forewing with many undulate striae, nervures with small black dots, discal spots distinct: ab. *griseata* Lempke (1947;1951);
Ground colour dark grey, row of prominent pale subterminal spots: ab. *limbopunctata* Dietze (1910;1913);
Unicolorous violet grey, discal spot obsolete: ab. *concolor* Dietze (1910;1913);
Unicolorous dark grey, central line visible: ab. *subatrata* Staudinger (1871);
Unicolorous black, markings just visible: ab. *nigra* Cockayne (1953).

Ab. *caeca* Dietze (1913) lacks the forewing discal spot. Ab. *constrastata* Dannehl (1925) and ab. *trilineata* Cockayne (1953) somewhat resemble f. *curzoni*, the latter being very similar to *E. satyrata* f. *curzoni* ab. *trifasciata*. The former has been recorded from Abbot's Wood, Sussex (Cockayne, 1953).

F. *callunaria* is smaller and browner than the nominotypical form. In ab. *strandi* Fuchs (1901) the forewings are dusted with white scales and the markings are stronger than typical *callunaria*. Very light individuals of this form occur occasionally.

F. *curzoni* has a paler, whitish ground colour with very dark cross-bands. Ab. *trifasciata* Wolff (1929) is whitish ochreous with a brown basal patch, three distinct transverse brown bands and a weakly marked termen.

Similar species (See Figs 5,8, pp. 61,69; Plate 6,8)

E. egenaria (Figs 5a,8 l): abdomen usually with pale abdominal band; forewing more elongate with distinctive dark, double ante- and postmedian lines, nervures not spotted alternate light and dark, discal spot more prominent, tornal spot less so.

E. intricata (Figs 5d,8b): abdomen with dark subbasal band. Subsp. *arceuthata* similar to nominotypical form of *E. satyrata* but forewing with median, ante- and postmedian lines prominent, especially at costa, median line usually touching much larger discal spot; hindwing with dorsal origins of striae more conspicuous. Subsp. *millieraria* similar to f. *callunaria* but darker and much browner; forewing median line often prominent at costa. Subsp. *hibernica* similar to f. *curzoni* but fore- and hindwings with more prominent discal spot; forewing postmedian line curved or less strongly geniculate, dark median line more conspicuous, usually touching discal spot; hindwing more uniform, fasciae and striae less conspicuous. These two taxa do not occur in the same geographical range.

E. cauchiata: Very similar to nominotypical form: forewing discal and tornal spots less conspicuous or obsolete, nervures less obviously spotted alternate light and dark, all other markings less well developed giving more uniform overall appearance. Examination of the genitalia is essential in identifying *E. cauchiata*.

E. vulgata (Figs 5b,8e): forewing usually tinged ochreous, postmedian fascia biangulate and more prominent, subterminal line and tornal spot more conspicuous.

E. tripunctaria (Figs 5f,8h): metathorax sometimes with white spot; fore- and hindwings with prominent white tornal spot; forewing with subterminal row of white spots, nervures not spotted alternate light and dark.

E. denotata subsp. *jasioneata* (Figs 5i,8f): forewing discal spot larger and more prominent, postmedian fascia biangulate; hindwing discal spot more conspicuous.

E. subfuscata (Figs 5l,8g): forewing broader, costa more arched, discal spot larger and more prominent, nervures not marked alternate light and dark; hindwing with prominent discal spot.

E. millefoliata (Fig.5h): usually larger; abdomen with dark subbasal band; forewing discal spot less well defined, diffuse, usually larger, tornal spot inconspicuous or absent, nervures not marked alternate light and dark, overall appearance more ochreous; hindwing discal spot usually present.

E. nanata (Fig.5j): similar to f. *curzoni* but forewing termen straighter, postmedian fascia biangulate, discal and tornal spots larger and more conspicuous.

E. virgaureata (Figs 5l,8g): forewing broader, fore- and hindwing discal spot more prominent; pale forms sometimes with white metathoracic spot.

E. lariciata (Figs 5k,8d): metathorax usually with white spot; forewing with prominent fasciae and discal spot, overall appearance more striped than nominotypical form and f. *callunaria*; hindwing discal spot usually present. Generally larger than f. *curzoni*.

LIFE HISTORY

OVUM Pale greenish white; 0.7mm long, ovoid. Laid on the foodplants during May and June.

LARVA (Figs 31e,f). *First and second instars.* Slender; varies in colour according to that of the foodplant; pale green, yellow, buff or pink; dorsal line fine, connecting five narrow, oval, dark green, brown or purple dorsal markings; subdorsal line fine, similarly coloured and touching dorsal markings; spiracular line faint, undulate, matching other markings in colour. *Third and fourth instars.* Plump, length 20mm when full grown; head, thoracic legs and segmental divisions yellow; integument rough, wrinkled and covered sparsely with white tubercles and hairs; colour variable according to that of the foodplant; dorsal line reddish brown, dark brown or dark green; five median dorsal markings bold, chevron-shaped, reddish brown, dark brown or dark green with paler centres; on thoracic and anal segments, dorsal marking ovoid; spiracular lines undulate, cream, brown or dark green; ventral ground colour paler with white central line.

Polyphagous on the flowers of herbaceous plants. Winter (1983) also records it feeding on the soft, young shoots of Sitka spruce (*Picea sitchensis*) and lodgepole pine (*Pinus contorta*). July, August and early September.

PUPA Length 7–8mm; abdomen, thorax and wings golden yellow suffused with red; abdomen brown; segmental divisions and cremaster red. September to May.

FLIGHT PERIOD AND HABITAT Univoltine, May to mid-July with the peak during the second week of June. It is easily disturbed from vegetation during the day and at night comes readily to light. F. *satyrata* prefers open woodland, downland, scrub, rough ground and fenland, whereas f. *callunaria* and f. *curzoni* are associated with moorland. The latter is restricted to the Shetlands and parts of Orkney (R. I. Lorimer, pers.comm.). A few specimens of f. *callunaria* from Aberdeenshire and Inverness-shire in the National collection also resemble f. *curzoni*. In Ireland, the species is predominantly represented by f. *callunaria*. However, specimens from the Burren, Co. Clare, appear closer to the nominotypical form.

DISTRIBUTION (Map 18) This species is locally widespread and common throughout England and Wales though in south-eastern England, where it is usually represented by the nominate form, it is less frequently recorded. In Scotland, *E. satyrata* is found throughout and is often abundant on upland moorland in the north west. Its range extends northwards to, and includes, the Shetlands. In Ireland moderately common in moorland and mountain districts.

COLLECTING AND REARING The larvae of this species can easily be mistaken for those of *E. absinthiata* and *E. tripunctaria* as they all vary considerably and have rather similar dorsal markings. However, *E. satyrata* feeds much earlier and is usually well grown by the end of July, pupating by the end of August, whereas *E. absinthiata* and *E. tripunctaria* feed from September to the end of October. Ova can easily be obtained from a captive female. Alternatively larvae can be collected by sweeping or by gathering the foodplants into bunches and gently shaking or beating them over a sheet. Straightforward to rear. They feed on the flowers of a variety of plants and will easily change from one host to another.

Eupithecia cauchiata (Duponchel, 1831) Doubleday's Pug
= *pernotata* sensu Doubleday, 1858
= *pernotaria* sensu Morris, 1861

Pl.2, fig.5; Pl.6, fig.15

HISTORY William Machin, a nineteenth-century entomologist who lived in the east end of London, sent Doubleday several larvae which he had collected from goldenrod (*Solidago virgaurea*) at an undisclosed locality. Doubleday (1858a) recorded that all were *E. castigata* (= *E. subfuscata*) except one which produced an unusual specimen that he could not identify. He sent this (a female) to Guenée who identified it as the newly described *E. pernotata* (Guenée, 1857). Barrett (1904) considered that Doubleday's specimen was merely a form of *E. satyrata* and, until 1980 when D. S. Fletcher and G. Prior re-examined the specimen, it was known as *E. satyrata* ab. *pernotata* Guenée (=*subatrata* Staudinger). However, their re-examination revealed that it was neither but in fact *E. cauchiata* (Duponchel). It remains the only supposed British specimen.

Some doubt must remain over the origin of this specimen as no locality was cited by Doubleday. It seems from Machin's obituary (Tutt, 1894) that he collected mainly in Essex and Kent, but we cannot be certain that he found

E. cauchiata there. Further doubt is raised by Doubleday's (1858a) statement 'I believe mine [meaning his specimen of *cauchiata*] was reared from larvae kindly sent to me by Mr Machin ...'. We may deduce from this that Doubleday was not absolutely certain where the specimen came from. As he was obviously in close contact with continental authorities at that time, one could speculate that it came from a foreign source and had been accidentally labelled as British. The specimen is currently in the national collection at the Natural History Museum.

IMAGO *Characteristic features.* Wingspan *c.*23mm; apart from inconspicuous geniculate postmedian fascia, most forewing markings obsolete, discal spot often absent, marginal band darker than rest of wing, subterminal line inconspicuous, tornal spot faint; hindwing with numerous inconspicuous striae, discal spot usually absent. Examination of the genitalia is essential to identify this species.

Genitalia. (Figs 15b,27r ♂; 22e ♀). The female genitalia are very similar to those of *E. satyrata*. The differences are discussed under that species.

Similar species
E. satyrata. Forewing discal spot more prominent, nervures spotted alternate light and dark, other markings well developed giving a less uniform overall appearance.

LIFE HISTORY

OVUM Not known.

LARVA Green with darker dorsal line, lateral lines white; slender, length 19mm when full grown.

Foodplant. Feeds on goldenrod (*Solidago virgaurea*), standing outstretched on the edges and undersides of leaves. July to September.

PUPA September to May in a cocoon just below the ground-surface.

A more detailed account of the biology is given by Weigt (1990).

FLIGHT PERIOD AND HABITAT Late May to mid-July. Rough ground, meadows and open woodland.

Eupithecia absinthiata (Clerck, 1759) Wormwood Pug; Ling Pug
= *minutata* ([Denis & Schiffermüller], 1775)
= *elongata* (Haworth, 1809)
= *notata* Stephens, 1831
= *innotata* sensu Wood, 1835
= *minutaria* Boisduval, 1840
= *elongaria* Doubleday, 1849
= *goossensiata* Mabille, 1869
= *knautiata* Gregson, 1874

Pl.1, figs 32,33; Pl.7, figs 15,16; Pl.B, figs 6,7

HISTORY First recorded in Britain by Haworth (1802) who later (1809) gave a full description of *E. absinthiata* and in the same work also described *E. elongata* (Long-winged Pug) which was subsequently found to be conspecific. The larva was first described by Crewe (1859f;1861k). F. *goossensiata* (Mabille, 1869a), the Ling Pug, which is still considered by some authorities to be a distinct species (*E. goossensiata* Mabille), was first recorded in Britain as *E. notata* Stephens, 1831. The larva of this form was first described by Crewe (1860f;1861k).

Much has been published in the entomological literature on the status of f. *goossensiata* (e.g. Fibiger, 1979; Weigt, 1980; Kaaber, 1980;1981; Skou, 1986). This debate was summarized by Riley (1986a) who suggested that *goossensiata* is merely a heathland form of *E. absinthiata*. (See also brief review on p.142.)

IMAGO *Characteristic features* (Figs 7a,8c). Wingspan c.24mm; abdomen with dark subbasal band; forewing elongate, apex pointed, uniform brown, black discal spot prominent, costa often with dark spots, fasciae, lines and striae often obsolete, when present postmedian fascia geniculate, subterminal line incomplete or absent, tornal spot present but usually not prominent; hindwing uniform, striae and fasciae usually obsolete, pale tornal spot sometimes present.

Genitalia (Figs 15f(ii),27t ♂; 22b ♀). The female corpus bursae of this species and of *E. expallidata* appear to be identical and we can find no constant differences. Moreover, the male abdominal plates are unsatisfactory as characters for identification: although their apex is generally more notched in *E. expallidata* (Fig.28a), variation makes this character unreliable. The male genitalia differ in that the second cornutus in the aedeagus is smoother in *E. absinthiata* (Fig.27t). Otherwise they appear to be identical.

Variation. Subtle differences from the nominotypical form are restricted to the shade of the ground colour. The extreme ab. *obscura* Dietze (1910;1913) has dark, sepia-coloured wings, while f. *goossensiata* (Pl.1, fig.33; Pl.7, fig.15; Pl.B, fig.7) is generally smaller (18–21mm) and greyer. Offspring from a nominotypical female will be of this form if the larvae are fed on *Calluna*. Variation in f. *goossensiata* is again restricted to a general darkening of the ground colour, as in ab. *obscura* Cockayne (1951a). Ab. *mediofasciata* Lempke (1947;1951) has only the median area of the forewing darkened.

Similar species (See Figs 7,8, pp. 67,69; Plate 7)

E. trisignaria (Figs 7j,8k): forewing broader, ground colour greyer, pale tornal spot absent, discal and two costal spots only prominent markings forming corners of equilateral triangle; hindwing with distinct discal spot.

E. expallidata (Fig.7f): forewing broader, costa more arched, ground colour much paler, discal spot larger and more prominent, costal spots more prominent, tornal spot less conspicuous; hindwing discal spot more often present.

E. assimilata (Figs 7g,8o): usually smaller; forewing costa more arched, forewing notably broader, discal spot larger and more prominent, white subterminal spots and tornal spot more prominent; male abdomen truncate in profile.

E vulgata (Figs 7e,8e): forewing postmedian fascia biangulate and prominent, discal spot smaller, pale subterminal line more conspicuous, white tornal spot more prominent; hindwing less uniform, striae more conspicuous, especially at their dorsal origins.

E. *denotata* subsp. *denotata* (Figs 7h,8f): forewing postmedian fascia biangulate, subterminal line dentate near costa; hindwing less uniform, discal spot conspicuous, underside with discal spot prominent, usually absent in E. *absinthiata*.

E. *pimpinellata* (Figs 7d,8a): generally paler; forewing with more numerous and conspicuous striae, antemedian fascia more prominent, pale subterminal line dentate near costa; hindwing much paler, particularly basally, discal spot prominent, striae more conspicuous, generally less uniform.

LIFE HISTORY

OVUM Pale greenish or yellowish white, turning to deep yellow after about ten days; 0.6mm long, ovoid. Laid on the flowers of the foodplant in June and July.

LARVA (Figs 31h,i) *First instar*. Buff, pale yellow, pale green or pale pink; dorsal and subdorsal lines a darker shade than ground colour. *Second instar*. Coloration similar to that of the first instar, matching the flowers on which it is feeding. Freer (1892) states that the brighter forms are found only on plants which are still in bloom. Those feeding on senescent inflorescences are duller. Central five segments with blunt chevron dorsal marking, darker shade of ground colour. *Third and fourth instars*. Short, stout, tapering towards the head; length 13–18mm when full-grown; integument wrinkled and rough with sparse tubercles and short hairs.

Probably the most variable of all the pug larvae. Ground colour usually similar to that of the flowers on which the larva feeds (e.g. yellowish green on ragwort, pink on heather); dorsal markings very variable, sometimes absent.

Polyphagous on the flowers of herbaceous plants, most commonly Compositae. F. *goossensiata* usually results from larvae that have fed on heather (*Calluna vulgaris*). From late August to October, occasionally into early November.

PUPA Length 7–8mm; thorax yellowish green or golden yellow; abdomen reddish yellow, occasionally with a dark green dorsal line; wings bright green or golden yellow. October to June, in a tough cocoon amongst plant-debris.

FLIGHT PERIOD AND HABITAT Univoltine, from late May to early September with the main emergence during June and July. The peak period appears to be during the last week of July. Generally distributed and found in many habitats including waste ground, waysides, gardens and woodlands, perhaps most commonly in the two former. The *Calluna*-feeding form is common on heathlands and has occasionally been recorded in gardens where it presumably feeds on cultivated heathers. Flies at dusk and throughout the night, when both sexes come freely to light and are attracted to flowers of Compositae and heather.

DISTRIBUTION (Map 19) Widespread and generally common throughout the

British Isles, though absent from the Shetlands. F. *goossensiata* is also widespread, being locally common in suitable habitats.

COLLECTING AND REARING A species that is easy to collect and rear, though Hellins (1867) records cannibalism by larvae of f. *goossensiata* which, in captivity, have also been recorded feeding on the pupae (Riley, pers.obs.). The larvae can be swept or beaten in quantity from the foodplants, although those collected in the wild are often heavily parasitized. The females will oviposit readily in captivity.

Eupithecia expallidata Doubleday (1856) Bleached Pug
= *expallidaria* Morris, 1861

Pl.2, fig.6; Pl.7, fig.19; Pl.B, fig.8

HISTORY First described by Doubleday (1856). British larvae were first described by Crewe (1860i;1861k).

IMAGO *Characteristic features* (Fig.7f). Wingspan c.26mm; abdomen with dark subbasal band; forewing broad, costa arched, termen rounded, uniform pale brown, discal spot large, oval and prominent, costa with clearly defined and prominent black spots, tornal spot inconspicuous or obsolete, most other markings obsolete; hindwing almost uniform, discal spot usually present, discal spot on underside conspicuous.

Genitalia Figs 15f(i),28a ♂; 22b ♀. See *E. absinthiata* (p. 77).

Variation. Restricted to a general darkening or lightening of the ground colour. The extremes are f. *pseudoabsinthiata*, which is as dark as *E. absinthiata*, and ab. *pallida*, which is whitish grey with markings almost absent (Schwingeschuss, 1953).

Similar species (See Fig.7, p. 67; Plate 7)

E. absinthiata (Fig.7a): forewing narrower, darker brown, discal spot much smaller, costal spots less clearly defined, less prominent, sometimes absent, pale tornal spot sometimes conspicuous; hindwing discal spot inconspicuous, often absent; underside of forewing lacking costal spots; underside of hindwing lacking conspicuous discal spot.

E. assimilata (Fig.7g): forewing darker brown, costal spots much less conspicuous, white tornal spot prominent; underside of hindwing with much smaller discal spot, often absent; male abdomen truncate in profile.

E. pimpinellata (Fig.7d): forewing costa straighter; fore- and hindwings less uniform, striae and fasciae well developed, often conspicuous; forewing without prominent costal spots, discal spot smaller.

LIFE HISTORY

OVUM Shiny greenish ivory when first laid, later turning bright yellow; 0.6mm long, ovoid. Laid on the flowers of the foodplant in July and August.

LARVA (Fig.31k) *First instar.* Pale yellow; dorsal line fine, pale brown, broadening at the centre of each segment; subdorsal line undulate, pale brown; spiracular line pale brown. *Second instar.* Pale yellow or light green; dorsal, subdorsal and spiracular lines similar to first instar but darker and more conspicuous. *Third instar.* Pale green or deep yellow; dorsal line dark brown, expanding into an elongate oval marking on each of the five central segments; subdorsal lines dark brown, flanked by oblique brownish orange stripes; spiracular lines broad, dark brown, turning to red towards the anal segments where they terminate as a reddish outline to the anal claspers. *Fourth instar.* Stout, tapering slightly towards the head and tail; length 18mm when full grown; head and thoracic legs reddish brown; integument covered densely with white tubercles, giving a frosted appearance; pale green or deep yellow; dorsal markings dark reddish brown, on four central segments forming pairs of fine posteriorly-pointing chevrons, on four anterior segments represented by single bead-like marks. These are joined by a pale brown dorsal line which runs the entire length of the body; subdorsal lines interrupted, dark reddish brown; spiracular lines pale reddish brown; laterally a series of oblique orange stripes; ventrally pale grey. Crewe (1860f) also describes forms which are green with few markings and canary yellow with most of the body suffused with chocolate brown.

Feeds on the flowers and unripe seeds of goldenrod (*Solidago virgaurea*) and, in captivity, common ragwort (*Senecio jacobaea*), Canadian goldenrod (*S. canadensis*) and hairy Michaelmas-daisy (*Aster novi-belgii*). September and October.

PUPA Length 8–9mm; thorax and abdomen yellow, suffused with red; wings greenish yellow. In a tough cocoon amongst ground debris or slightly below the surface. October to June.

FLIGHT PERIOD AND HABITAT Univoltine, from late June to early August (Skinner, 1984) on scrubland near the coast and in open woodlands and woodland rides. Flies naturally at dusk and through the night when it comes to light, usually in small numbers. Adults are also attracted to flowers of goldenrod and ragwort.

DISTRIBUTION (Map 20) Locally common in England, Wales and the Isle of Man, following approximately the distribution of *Solidago virgaurea* (Perring & Walters, 1962). In Scotland the only known colonies are in Argyll and on the Isle of Arran. Very local and uncommon in Ireland.

COLLECTING AND REARING Larvae can be found by beating the flower-heads of the foodplant over a sheet or tray, or by careful searching. Females oviposit readily in captivity. Easy to rear in the normal way.

Eupithecia assimilata Doubleday (1856) Currant Pug
= *assimilaria* Morris, 1861

Pl.2, figs 7,8; Pl.7, figs 18,20; Pl.B, fig.9

HISTORY First described by Doubleday (1856); larvae first described by Crewe (1859a). This larval description was followed by much discussion and amendment by Crewe (1859h,n,o;1860j;1861k), Gregson (1859a,b) and Greene (1859).

IMAGO *Characteristic features* (Figs 7g,8n,o). Wingspan *c*.22mm; abdomen with dark subbasal band, male abdomen truncate in profile; forewing broad, costa arched, uniform brown, discal spot large and prominent, postmedian fascia inconspicuous, strongly geniculate, costa with dark spots, subterminal line a series of white spots, white tornal spot prominent; hindwing almost uniform, striae and fasciae weak or obsolete, small tornal spot usually present, discal spot usually absent.

Genitalia. Figs 15d,28b ♂; 22a ♀

Variation. The ground colour of the first generation is often a richer brown than that of the second, the latter being duller and greyer. Specimens of the occasional third emergence tend to be greyer still, lacking the warm brown of the first two. Dietze (1910;1913) ascribes these grey forms to f. *grisescens*.

Specimens from Scotland tend to be paler with brighter markings than those from southern England.

The variation in ground colour was illustrated well by Prout (1894) who exhibited adults of four broods from different females. Although fed on the same foodplants, each brood was consistently distinct, producing adults which were variously (i) small, dull and unicolorous, (ii) dull and unicolorous but larger, (iii) large, well marked and reddish, and (iv) delicate greyish.

Similar species (See Figs 7,8, pp. 67,69; Plate 7)

E. trisignaria (Figs 7j,8k): forewing greyer, discal spot much smaller, white tornal spot absent; hindwing tornal spot absent.

E. absinthiata (Figs 7a,8c): usually larger; forewing narrower, discal spot smaller, less elongate, white tornal spot much less conspicuous; hindwing discal spot often absent.

E. expallidata (Fig.7f): usually larger; forewing much paler accentuating prominence of costal and discal spots, tornal spot inconspicuous; hindwing tornal spot absent or obsolete.

E. vulgata (Figs 7e,8e): forewing narrower, discal spot minute or absent, striae well defined, postmedian fascia conspicuous and biangulate, white tornal spot smaller; hindwing usually paler basally.

E. tripunctaria (Figs 7i,8h): forewing grey, discal spot less prominent, striae more conspicuous; hindwing tornal spot more conscpicuous, metathorax sometimes with white spot.

E. denotata subsp. *denotata* (Figs 7h,7f): forewing discal and tornal spots much less prominent, postmedian fascia biangulate, dark costal spots absent.

E. pimpinellata (Figs 7d,8a): usually larger; forewing costa straighter, forewing less uniform, tornal spot not prominent, antemedian fascia more conspicuous; hindwing not uniform, basal two-thirds paler, discal spot more prominent.

LIFE HISTORY

OVUM White at first, later turning yellow; 0.6mm long, ovoid. Usually laid on the underside edges of the foodplant's leaves in early June and September.

LARVA (Fig.31j) *First instar.* White or cream with no discernible markings. *Second instar.* Pale green, without markings. *Third instar.* Deep yellowish green; dorsal markings rhomboid or oval, brown or greenish brown, darkest on anterior half of each segment; dorsal line strong, dark green or greenish brown. *Fourth instar.* Slender, tapering sightly at each end; length 17mm full grown; head and thoracic legs brown or dark green; integument variable in colour: some similar to third instar with more pronounced markings, others pinkish brown with dark brown markings, or pale yellowish green with only a fine green dorsal line, or pale green with a dark green dorsal line flanked by single rows of green subdorsal dots. Prior to pupation the ground colour turns pink with faded markings.

Feeds on the leaves of hop (*Humulus lupulus*), red currant (*Ribes rubrum*) and black currant (*R. nigrum*). Larvae from hop are usually more brightly marked than those from currant. June and July, and again in September and October.

PUPA Length 6.5–7.5mm; thorax, abdomen and wings yellowish green. Amongst ground litter in a tough cocoon. July and August, and October to May.

FLIGHT PERIOD AND HABITAT The adult has been recorded between early May and mid-September. This species is usually bivoltine, with the first brood peaking during the first week of June and the second during the last week of July. An occasional partial third emergence has been noted in October in RIS light traps in southern England and a single individual was caught in the RIS trap at Sherwood Forest on 30 December 1985. Often found in gardens where currant is grown. Also common on wasteland and rough ground in the vicinity of hop. During the day adults can be found resting on fences, etc. near the foodplant and are easily disturbed. At night they are attracted to flowers and frequently come to light.

DISTRIBUTION (Map 21) Widespread throughout the British Isles though apparently absent from upland areas and from the Outer Hebrides. In Ireland frequent where currants are grown.

COLLECTING AND REARING Captive females oviposit readily, either on the foodplant, on muslin or on the sides of the container in which they are kept. Alternatively, wild larvae can be collected from the foodplant. In the early stages they eat tiny holes into the leaves, creating very distinctive feeding damage. Searching the undersides of affected leaves often reveals several larvae though careful examination is required as they closely resemble the leaf veins. Larger larvae can easily be beaten from the plants.

This is perhaps one of the easiest species to rear in captivity. The adults pair very quickly and produce large numbers of eggs. Red or black currant is preferable as a cut foodplant (Riley & Prior, pers.obs.), as hop tends to desiccate more quickly.

Eupithecia vulgata (Haworth, 1809) Common Pug
= *clusterata* (Hübner, 1813)
= *austerata* (Hübner, 1825)
= *austeraria* Boisduval, 1840
= *vulgaria* Morris, 1861

Pl.2, figs 9–15; Pl.5, fig.12; Pl.6, fig.5; Pl.7, figs 9,22,23; Pl.8, figs 10,11,17; Pl.B, figs 10,11

HISTORY First recorded in Britain and described by Haworth (1803); larva first described by Crewe (1859e;1861k). F. *scotica* was first recorded from Aviemore, Inverness-shire, as a distinct subspecies by Cockayne (1951b), and f. *clarensis*, which was first found on the Burren, Co. Clare, Ireland in 1956 and described by Huggins (1962b), was also originally given subspecific status.

IMAGO *Characteristic features* (Figs 1k,5b,6b,7e,8e). Wingspan c.22mm; forewing narrow, apex pointed, usually tinged ochreous, discal spot minute or absent, postmedian line biangulate, subterminal line a series of white dots, white tornal spot prominent; hindwing with pale tornal spot, discal spot absent or inconspicuous. Melanic form with forewing discal spot absent or inconspicuous.

Genitalia. Figs 15e,28c ♂; 22d ♀

Variation. Varies little in size. The form found in southern England usually has a greyish brown ground colour, more or less tinged with orange, sometimes superficially resembling a small striated form of *E. absinthiata*. In Scotland the ground colour becomes more ochreous with the striae more pronounced. This form has been referred to as subsp. *scotica* Cockayne. However there appears to be a continuous cline from Scotland to southern England without the allopatric divisions normally required for subspecific status (Riley, 1988). It is therefore the opinion of the present authors that subspecific status should be withdrawn and that the northern taxon should be referred to merely as f. *scotica* Cockayne (Pl.2, fig.14; Pl.7, fig.23). Individuals which correspond most closely to the pale, chalky f. *clarensis* (Pl.2, fig.15; Pl.5, fig.12) are restricted to the exposed limestone pavement of the Burren, Co Clare. Elsewhere in Ireland the pale, whitish grey ground colour is more or less replaced by ochreous grey, thereby superficially resembling f. *scotica*. No named aberrations of the Scottish form have been found by the present authors. Variation seems to be restricted to subtle changes in ground colour. The nominotypical form is prone to a general darkening of the ground colour, reaching its extreme in f. *atropicta* Dietze (1910;1913) (Pl.2, fig.13; Pl.6 fig.5). This is a melanic form frequently recorded from the London area and northern England. Agassiz *et al.* (1981) and others incorrectly state that f. *unicolor* is also a melanic form. However, Lempke (1951) describes this as having the normal ground colour (i.e. not black) with, apart from the discal spot, obsolete markings. F. *nigrofasciata* Dietze (1910;1913) is also weakly marked but has the median area darkened, and f. *impuncta* Lempke (1947;1951) lacks the forewing discal spots.

Similar species (See Figs 1,5,6,7,8, pp. 35,61,65,67,69; Plates 5,6,7,8)

E. *valerianata* (Figs 1c,8i): generally paler; forewing postmedian fascia geniculate; hindwing more uniform.

E. *trisignaria* (Figs 6a,7j,8k): forewing costa more arched, forewing broader, generally more uniform, fasciae and striae much less conspicuous, costa with two characteristic dark spots, discal spot more prominent, pale subterminal line and tornal spot absent or obsolete; hindwing more uniform. Melanic f. *angelicata* with conspicuous forewing discal spot.

E. *intricata* subsp. *millieraria* (Figs 5d;8b): forewing postmedian fascia not biangulate, discal spot larger, white tornal spot absent; abdomen with dark subbasal band.

E. *satyrata* f. *callunaria* (Figs 5g,8j): forewing postmedian fascia geniculate, white tornal spot less prominent; generally not tinged ochreous.

E. *absinthiata* (Fig.8c): abdomen with dark subbasal band; fore- and hindwing more uniform, fasciae and striae weak, forewing postmedian fascia geniculate, white tornal spot much less prominent, discal spot larger and more prominent, dark spots sometimes present on costa, subterminal line often obsolete.

E. *assimilata* (Figs 7g,8n,o): forewing broader, costa more arched; fore- and hindwing more uniform; forewing discal and tornal spots larger and more prominent, costa usually with conspicuous dark spots, postmedian fascia not biangulate; male abdomen truncate in profile.

E. *tripunctaria* (Figs 5f,6c,7i,8h): forewing not tinged ochreous, subterminal line a series of white spots, tornal spot much larger and more prominent, discal spot larger; hindwing with prominent white tornal spot; metathorax sometimes with white spot. Melanic f. *angelicata* with prominent forewing discal spot.

E. *denotata* (Figs 5i,6e,7h,8f): forewing discal spot larger and more prominent, subterminal line dentate near costa, white tornal spot much less conspicuous; hindwing with more conspicuous discal spot. Subspecies *jasioneata* not tinged ochreous.

E. *subfuscata* (Figs 5l,6d,8g): forewing not tinged ochreous, discal spot larger and more conspicuous, postmedian fascia geniculate, pale tornal spot less conspicuous; hindwing discal spot more prominent. Melanic f. *obscurissima* with conspicuous forewing discal spot.

E. *indigata* (Figs 1f,8m): forewing more elongate, costa straighter, apex more pointed, generally paler, more uniform, fasciae and striae less conspicuous or obsolete, postmedian fascia geniculate, discal spot much larger and more prominent; hindwing paler, sometimes with geniculate postmedian line.

E. *virgaureata* (Figs 5l,6d,8g): forewing not ochreous-tinged, discal spot larger and more conspicuous, postmedian fascia geniculate, pale tornal spot less conspicuous, nervures often spotted alternate light and dark; hindwing discal spot more prominent; pale forms sometimes with white metathoracic spot. Melanic f. *nigra* with conspicuous forewing discal spot.

E. pusillata (Figs 6f,8p): dark forms sometimes similar to f. *atropicta* but vestiges of forewing fasciae usually present.

E. lariciata f. *nigra* (Figs 5k,8d): similar to f. *atropicta* but forewing with conspicuous discal spot.

E. abbreviata f. *nigra* (Figs 6g,8r): similar to f. *atropicta* but forewing more elongate, linear discal spot and nervures intensely black; hindwing termen concave, apex pointed.

Pasiphila rectangulata f. *anthrax* (Figs 6h,8q): similar to f. *atropicta* but forewing broader, costa more arched, apex less pointed, black more intense in shade, vestiges of green coloration sometimes present, white tornal spot always absent.

LIFE HISTORY

OVUM Yellowish cream; 0.6mm long, ovoid. On the foodplant in May and June, and possibly also in August.

LARVA (Fig.311). *First instar.* Pale brown, tinged with orange; dorsal and subdorsal lines fine, dark brown. *Second instar.* Similar to first instar but with dorsal line heavier and broadening to form an oval dorsal marking on each of the four or five central segments; spiracular line white. *Third and fourth instars.* Robust, tapering towards the head; length 18mm when full grown; head and thoracic legs pale brown; integument covered with white tubercles and short hairs giving a rough appearance, dorsally pale brown to greyish or greenish brown; ventrally pale grey, central line a series of brown longitudinal dashes; dorsal line dark greenish brown, broadening on median segments into oval or chevron dorsal markings; spiracular lines double, pronounced, sinuate, occasionally branched dorsally, upper line creamy white, lower brown; segmental divisions pinkish yellow or pinkish orange.

Polyphagous on deciduous trees including hawthorn (*Crataegus monogyna*), goat willow (*Salix caprea*) and oak (*Quercus* spp.), and on many shrubs and herbaceous plants such as bramble (*Rubus fruticosus*), raspberry (*Rubus idaeus*), yarrow (*Achillea millefolium*), goldenrod (*Solidago virgaurea*), common ragwort (*Senecio jacobaea*) and dandelion (*Taraxacum officinale*). June and July, and possibly again in September and October.

PUPA Length 7–8mm; abdomen pale reddish brown; thorax and wings olive green. In a tough cocoon on the ground. July to April.

FLIGHT PERIOD AND HABITAT Adults have been recorded between early April and early September, though the main flight period is from May to mid-July. The peak of emergence is around mid-June with northern populations approximately one week later than those in the south. There is sometimes a partial second emergence during August. They fly at dusk and through the night when they come readily to light, the males sometimes in large numbers. The female is less well represented in light-trap catches. This species is very generally distributed and can be found in most habitat types.

DISTRIBUTION (Map 22) Widespread and generally common throughout the British Isles.

COLLECTING AND REARING Larvae are rarely found in the wild. However, adult females oviposit freely in captivity and the resulting larvae are easily reared in small plastic boxes. Care must be taken not to overcrowd as they are sometimes cannibalistic. Withered leaves are preferred to fresh foodplant.

Eupithecia tripunctaria Herrich-Schäffer, 1852
White-spotted Pug

= *albipunctata* (Haworth, 1809), nec (Hufnagel, 1767)

Pl.2, figs 16,17; Pl.6, fig.7; Pl.8, fig.5; Pl.B, fig.12

HISTORY First recorded in Britain by Crewe (1861c) from larvae collected by him in Derbyshire four or five years previously and identified by Herrich-Schäffer. The larva was described in this paper and a further description followed in 1862 (Crewe, 1862a). Stainton (1862) reviewed the first British records.

IMAGO *Characteristic features* (Figs 5f,6c,7i,8h). Wingspan *c.*23mm; forewing costa arched, white tornal spot large and prominent, subtermen usually with a row of small white spots, discal spot conspicuous, postmedian fascia geniculate, this and other markings obscure; hindwing with conspicuous tornal spot, discal spot minute; metathorax usually with white spot. Melanic form with conspicuous discal spot.

Genitalia. Figs 15g,28d ♂; 22g ♀

Variation. In colour, restricted to a general darkening which reaches its extreme in the melanic f. *angelicata* Barrett (1877) (Pl.2, fig.17; Pl.6, fig.7). In this form the usual white tornal spots and metathoracic crest are absent, the only remaining marking being an intensely black discal spot. Intermediate between this and the type is f. *intermedia* Lempke (1951) which has black forewings with an intense black discal spot and a more or less complete row of subterminal white spots including the characteristic white tornal spot. Ab. *privata* Dietze (1913) is rather briefly described as 'poorly marked'. Ab. *sepiata* Parkinson Curtis (1944b) is sepia and irrorate with whitish lavender scales. The type-specimen of this aberration was caught at Farran, Co. Cork.

Similar species (See Figs 5,6,7,8, pp. 61,65,67,69; Plates 6,8)

E. egenaria (Figs 5a,8l): fore- and hindwings without prominent tornal spot; forewing with distinctive double ante- and postmedian lines; metathorax never with white spot.

E. trisignaria (Figs 6a,7j,8k): fore- and hindwings without prominent tornal spot; metathorax never with white spot. Examination of the male abdominal plate or the genitalia is required to separate the melanic forms.

E. intricata (Figs 5d,8b): fore- and hindwings without prominent tornal spot; abdomen with dark subbasal band; metathorax never with white spot.

E. satyrata (Figs 5g,8j): fore- and hindwing tornal spot much smaller and less prominent; forewing nervures often spotted light and dark; metathorax never with white spot.

E. assimilata (Figs 7g,8n,o): ground colour brown rather than grey; forewing discal spot larger and more prominent; metathorax never with white spot; male abdomen truncate in profile.

E. vulgata (Figs 5b,6b,7e,8e): forewing usually ochreous-tinged, white tornal spot much smaller, discal spot much smaller or absent, postmedian fascia prominent and biangulate; hindwing tornal spot less prominent; metathorax never with white spot. Melanic form *atropicta* with forewing discal spot minute or absent.

E. denotata (Figs 5i,6e,7h,8f): forewing postmedian fascia biangulate; fore- and hindwings without prominent tornal spot; metathorax never with white spot. Subsp. *denotata* brown rather than grey.

E. subfuscata (Figs 5l,6d,8g): fore- and hindwing tornal spot much less prominent; metathorax never with white spot. Examination of the male abdominal plate or the genitalia is required to separate the melanic forms.

E. virgaureata (Figs 5l,6d,8g): fore- and hindwing tornal spot much less prominent; forewing nervures sometimes spotted alternate light and dark. Examination of the male abdominal plate or the genitalia is required to separate the melanic forms.

E. pusillata (Figs 6f,8p): some dark forms similar to f. *angelicata* but forewing narrower with darker fasciae; metathorax never with white spot.

E. lariciata (Figs 5k,8d): fore- and hindwings without conspicuous tornal spot; forewing with contrasting light and dark geniculate fasciae giving more striped appearance. Examination of the male abdominal plate or the genitalia is required to separate the melanic forms.

E. abbreviata f. *nigra* (Figs 6g,8r): similar to f. *angelicata* but forewing more elongate, nervures intensely black; hindwing termen concave, apex pointed.

Pasiphila rectangulata f. *anthrax* (Figs 6h,8q). Similar to f. *angelicata* but forewing broader, sometimes with vestiges of green coloration; hindwing underside often with prominent angulate postmedian line.

LIFE HISTORY

OVUM Cream; 0.7mm long, ovoid. On the flowers of foodplants in May and June, and again in August and September.

LARVA (Figs 31m,n) *First instar*. Very pale greenish white; dorsal and subdorsal lines fine, darker green. *Second and third instars*. Pale green, pale yellow or buff; dorsal and subdorsal lines fine, pale brown; dorsal markings on five central segments similar in shape to an heraldic fleur-de-lis, pale brown; laterally a series of oblique pale yellow streaks separated by brownish blotches, ventrally greenish yellow. *Fourth instar*. Stout, tapering slightly towards the head; length 15mm when full grown; integument folded, wrinkled and covered with white tubercles and short fine hairs; colour and markings similar to those of the second and third instars but usually brighter and more pronounced, though occasionally they are reduced; dorsal line broadens and darkens on the anal segment. Immediately prior to pupation the dark brown dorsal markings fade, sometimes disappearing entirely.

Feeds on the flowers of elder (*Sambucus nigra*) and possibly cow parsley (*Anthriscus sylvestris*) in June and July, and again in September and October on the flowers of angelica (*Angelica sylvestris*), hogweed (*Heracleum sphondylium*), common ragwort (*Senecio jacobaea*), goldenrod (*Solidago virgaurea*) and wild parsnip (*Pastinaca sativa*). In captivity it has been reared on bramble (*Rubus fruticosus*) and also fennel (*Foeniculum vulgare*) (Hammond, 1952) and wild cherry (*Prunus avium*) (Allan, 1949).

PUPA Length 7-8mm; thorax yellowish green; abdomen pale brown to dull red; wings dark green. In a cocoon on or just below the ground surface. July and August, and October to May.

FLIGHT PERIOD AND HABITAT Partially bivoltine, May and June, and August and September. Although only some pupae from the first generation hatch in August and September, this second emergence is often the larger. The phenology of *E. tripunctaria* is discussed in detail by Riley and Prior (1990); Riley (1990d) records an adult caught in a RIS light-trap in Devon in December 1989. This species is often common in damp localities, roadside verges and on waste ground. It is also regularly recorded from gardens. The larvae are particularly common but are usually heavily parasitized. The adults fly at night and come to light readily in suitable localities, females often outnumbering the males.

DISTRIBUTION (Map 23) Widespread and locally common throughout England and Wales. In Scotland its range extends northwards to Ross-shire though it appears to be absent from upland areas. Scarce in Ireland, with most records from the west. The Channel Islands

COLLECTING AND REARING Larvae can be collected easily by shaking or beating the foodplant over a sheet or tray but these are often heavily parasitized. It is perhaps more efficient to obtain ova from a captive female. This, and the rest of the rearing process is straightforward to achieve in the standard way.

Eupithecia denotata (Hübner, 1813)

subsp. *denotata* Hübner, 1813   Campanula Pug
= *campanulata* Herrich-Schäffer, 1861

subsp. *jasioneata* Crewe, 1881   Jasione Pug

Pl.2, figs 18,19; Pl.7, fig.21; Pl.8, fig.1; Pl.B, figs 13,14

Note: Represented in the British Isles by the nominate subspecies and by subspecies *jasioneata* Crewe.

HISTORY *E. denotata denotata* Hb. was first recorded as British by Crewe

(1864c;1865a) from specimens found near his home at Drayton Beauchamp, Buckinghamshire. This was reviewed by Knaggs (1865) who figured the adult for the first time. The larva and pupa were described by Crewe (1864c;1865a).

The subspecies, *E. denotata jasioneata* Crewe (Jasione Pug) which feeds on sheep's-bit, was at first described as a new species by Crewe (1881) from larvae found by R. Ficklin in north Devon in 1878. The larva was described in the same article.

IMAGO *Characteristic features* (Figs 5i,6e,7j,8f). Wingspan c.23mm, forewing discal spot prominent, postmedian fascia biangulate, antemedian fascia curved in *jasioneata* (Fig.8f) but usually absent or faint in nominal subspecies (Figs 5i,6e), subterminal line dentate near costa, tornal spot elongate; hindwing discal and tornal spots not prominent, underside with geniculate postmedian line.

Genitalia. Figs 12g,16a,28e ♂; 22c ♀

Variation. Subsp. *jasioneata* (Pl.2, fig.19; Pl.8, fig.1; Pl.B, fig.14) greyer and sometimes smaller. There is little morphological variation, though two aberrations of the nominotypical subspecies are described. Ab. *solidaginis* Fuchs (1900) is clear dark slate-grey without the usual brown tinge and ab. *ochraceata* Fuchs (1904) has both fore- and hindwings ochre yellow and almost devoid of markings.

Similar species (See Figs 5,6,7,8, pp. 61,65,67,69; Plates 7,8)

E. egenaria (Figs 5a,8 l): similar to subsp. *jasioneata* but forewing postmedian fascia smoothly geniculate or curved, with characteristic double ante- and postmedian lines.

E. trisignaria (Figs 6a,8k): general appearance more uniform, without conspicuous fasciae and striae; fore- and hindwing tornal spot absent or obsolete.

E. intricata (Figs 5d,8b): forewing postmedian fascia not biangulate; abdomen with dark subbasal band.

E. satyrata (Figs 5g,8j): forewing discal spot much smaller, postmedian fascia geniculate; hindwing discal spot absent.

E. absinthiata (Figs 7a,8c): similar colour to subsp. *denotata* but forewing more uniform, most fasciae and striae inconspicuous, postmedian fascia geniculate, subterminal line not dentate near costa, discal spot more prominent; hindwing more uniform, discal spot usually absent.

E. assimilata (Figs 7g,8n): similar colour to subsp. *denotata* but forewing with more prominent discal and tornal spots, dark spots present on costa.

E. vulgata (Figs 5b,6b,7e,8e): forewing discal spot minute or absent, subterminal line not dentate near costa, white tornal spot more conspicuous; hindwing discal spot often absent.

E. tripunctaria (Figs 5f,7i,8h): fore- and hindwing tornal spot much larger and more prominent, subtermen usually with row of white spots, postmedian fascia geniculate; metathorax sometimes with white spot.

E. subfuscata (Figs 5l,6d,8g): forewing postmedian fascia geniculate; underside of hindwing with postmedian line curved.

E. pimpinellata (Figs 5e,8a): abdomen with dark subbasal band; forewing postmedian fascia geniculate, discal spot larger, more elongate and more prominent, antemedian striae more acutely geniculate, costal spots more prominent.

E. virgaureata (Figs 5l,6d,8g): forewing postmedian fascia geniculate; metathorax of pale individuals sometimes with white spot.

E. lariciata (Figs 5k,8d): worn individuals sometimes similar to subsp. *jasioneata* but metathorax usually with white spot; forewing ante- and postmedian fascia geniculate.

E. dodoneata (Fig.8s): uncommon melanochroic forms similar to small *E. denotata* but hindwing termen concave, apex pointed.

LIFE HISTORY

OVUM Cream when first laid, later turning orange; 0.6mm long, ovoid. On the flower-heads of the foodplant in July.

LARVA (Fig.32a) *First instar.* Pale greenish white; dorsal line faint, dark brown. *Second instar.* Dirty yellowish brown or deep buff; dorsal markings a single, dark brown, roughly circular ring on each of the central segments joined by a fine, dark brown, dorsal line; subdorsal lines fine, dark brown. *Third and fourth instars.* Stout, tapering slightly at head and tail, length 18mm when full grown; head and thoracic legs very dark brown; integument wrinkled, covered with tubercles and short hairs, giving a rough texture; light brown; markings similar to those of the previous instar but darker and heavier; segmental divisions pale buff.

Subsp. *denotata* usually feeds on nettle-leaved bellflower (*Campanula trachelium*) and giant bellflower (*C. latifolia*) but has been recorded from clustered bellflower (*C. glomerata*) (Prior, 1980). Crewe (1870) lists several other *Campanula* species. During the first two instars the larva eats the stamens and pistils. It later eats a ragged hole in the top of the seed-capsule and devours the unripe seeds. Subsp. *jasioneata* feeds on the withering flower-heads of sheep's-bit (*Jasione montana*). The larva forms a chamber by eating the contents of the flower-head. Occasionally it stands with its head buried in the flower-head and the rest of the body exposed. The larvae of the two subspecies are indistinguishable and both normally feed from August to October. It is noteworthy that Prout (1907b) records larvae of both subspecies feeding well into November and cites an instance of a larva still feeding on 15 December 1906 (J. Peed, in litt.).

PUPA Length 7.5–8.5mm; thorax and wing cases deep yellow; abdomen reddish brown with red cremaster. P. M. Waring (pers.comm.) states that subsp. *denotata* pupates in dead seed-heads of bellflower but, in captivity, the larva leaves the dead flower-heads to pupate amongst ground debris. It overwinters from September to late June.

FLIGHT PERIOD AND HABITAT Univoltine, late June and July (Skinner, 1984). Subsp. *jasioneata* is found only near the coast and is stated to emerge slightly earlier than the nominotypical form (Prout, 1907b). Subsp. *denotata* may be found wherever the foodplant is well established. Both are infrequently seen in the adult state and rarely come to light. However, Tutt (1906b) states that they can be found flying over the foodplant at dusk.

DISTRIBUTION (Map 24) In south-western England, Wales, the Isle of Man and Ireland, *E. denotata* is locally common in coastal localities and is represented by subsp. *jasioneata*. In the rest of England, inland in southern Wales and in the Channel Islands, it is represented by subsp. *denotata*. It is locally common south of a line from Herefordshire to the Thames estuary. Elsewhere there are sporadic records northwards to Nottinghamshire with known colonies in North Lincolnshire. It is apparently absent from large areas, including most of East Anglia, but this may be due to under-recording.

COLLECTING AND REARING Young larvae of the *Campanula*-feeding subspecies can be found feeding on the flowers or by collecting withered flower-heads. Alternatively seed-capsules, inside which older larvae feed, can be gathered in the autumn. If these are kept free from mould, adults often emerge in due course. However, care must also be taken to ensure the pupae do not dessicate. Larvae of subsp. *jasioneata* can be found by searching withered (not green) flower-heads of sheep's-bit or tapping the heads over a tray. They can then be reared on in the usual way.

Eupithecia subfuscata (Haworth, 1809) Grey Pug
= *singulariata* sensu Haworth, 1809
= *castigata* (Hübner, 1813)
= *castigaria* Boisduval, 1840
= *blancheata* Cooke, 1881

Pl.2, figs 20,21; Pl.6, fig.2; Pl.8, fig.3; Pl.B, fig.15

HISTORY First recorded in Britain by Haworth (1809) as two species: *subfuscata*, the Brown Grey Pug, and *singulariata*, the Grey Pug. British larvae were first described by Crewe (1860e;1861k).

IMAGO *Characteristic features* (Figs 5l,6d,8g). Wingspan c.23mm, forewing broad, discal spot conspicuous, postmedian fascia geniculate, subterminal line and elongate tornal spot well developed; hindwing discal spot prominent, pale tornal spot often present. Melanic form with conspicuous forewing discal spot.
  Genitalia. Figs 16b,28f ♂; 22f ♀
  Variation. There is little variation in size. Specimens from the north of Scotland tend to be slightly paler than the nominotypical form. Otherwise, variation is usually limited to a general darkening of the ground colour which reaches its extreme in f. *obscurissima* Prout (1914) (Pl.2, fig.21; Pl.6, fig.2) – a melanic form which occurs throughout the species range, but most frequently in

the south east, the Midlands and northern England. This form was first recorded by Tugwell (1892) under the name of the Paisley Pug. F. *obscura* Dietze (1910;1913) is strongly darkened with only the subterminal line remaining. Richardson & Mere (1958) refer to a sandy-coloured form caught on the Isles of Scilly, during the period 22–31 April 1958. However, no formal description of this form has been published and the specimens cannot be found.

Similar species (See Figs 5,6,7,8, pp. 61,65,67,69; Plates 6,8)

E. egenaria (Fig.5a): forewing more elongate, ante- and postmedian lines double, postmedian fascia curved or less accutely geniculate; fore- and hindwing tornal spot faint or absent.

E. trisignaria (Fig.6a): fore- and hindwings more uniform, lacking conspicuous fasciae and striae, tornal spot absent or less conspicuous; forewing with two characteristic dark costal spots. Examination of the male abdominal plate or the genitalia is necessary to separate the melanic forms.

E. intricata (Figs 5d,8b): abdomen with dark subbasal band, abdomen unicolorous, not paler ventrally; forewing costa straighter, apex more pointed, median, ante- and postmedian lines often prominent, especially near costa, postmedian fascia curved; hindwing not usually paler basally.

E. satyrata (Figs 5g,8j): forewing narrower, costa straighter, discal spot much smaller, nervures usually spotted alternate light and dark; hindwing discal spot often absent.

E. vulgata (Figs 5b,6b,7e,8e): forewing narrower, usually ochreous-tinged, postmedian fascia biangulate, tornal spot more prominent; hindwing discal spot minute or absent. Melanic f. *atropicta* without conspicuous discal spot.

E. tripunctaria (Figs 5f,6c,7i,8h): fore- and hindwings with large prominent white tornal spot; metathorax sometimes with white spot. Examination of the genitalia is required to separate the melanic forms.

E. denotata subsp. *jasioneata* (Figs 5i,6e,7h): forewing postmedian fascia biangulate; underside of hindwing with postmedian line geniculate rather than curved.

E. virgaureata (Figs 5l,6d,8f): separation by superficial characters alone is not advisable. Antennae of male biciliate, those of *E. subfuscata* simple (see Text Figures, p. 115); genitalia and male abdominal plate diagnostic. Forewing nervures spotted alternate light and dark and presence of small basally-pointing dark triangles on postmedian line suggest *E. virgaureata*.

E. abbreviata f. *nigra* (Figs 6g,8r): similar to f. *obscurissima* but forewing more elongate, nervures intensely black; hindwing termen concave, apex pointed.

E. dodoneata (Fig.8s): uncommon melanochroic forms similar to small *E. subfuscata* but hindwing termen concave, apex pointed.

E. pusillata (Figs 6f,8p): dark forms similar to f. *obscurissima* but forewing narrower, vestiges of dark fasciae usually present.

E. lariciata (Figs 5k,8d): metathorax usually with white spot; forewing apex more pointed, tornal spot inconspicuous or absent, alternate light and dark fasciae giving more striped appearance. Examination of the male abdominal plate or the genitalia is required to separate the melanic forms.

Pasiphila rectangulata f. *anthrax* (Figs 6h,8q): similar to f. *obscurissima* but forewing costa relatively shorter and more arched, apex more rounded, vestigates of green coloration sometimes present; hindwing underside often with prominent angulate postmedian line.

LIFE HISTORY

OVUM Smooth, pale green; 0.6mm long, ovoid. On the leaves of the foodplant in May and June.

LARVA (Fig.32b) The ground colour varies from yellowish brown to deep buff or greenish buff, though all larvae reared on a particular foodplant are usually similar. Changing the foodplant can change the colour. *First instar.* Long, cylindrical; dorsal line faint, brown. *Second and third instar.* Similar to first instar; dorsal line strong, broadening to form an oval marking on each of the central segments; subdorsal streaks oblique, white, joining laterally to form the spiracular line. *Fourth instar.* Slender, tapering slightly towards the head, length 18mm when full grown; head dark brown or green; thoracic legs light brown; integument rough, covered with white tubercles from each of which arises a short hair; colour similar to previous instars; dorsal markings on median five segments dark brown with paler oval centre, resembling a letter V or Y pointing anteriorly; segmental divisions pink; ventrally a fine, central red line. Sometimes similar to larva of *E. vulgata* but in that species red ventral line finer and interrupted at segmental divisions. When feeding, the larva adopts an omega-shaped position; when resting, it usually stands stretched out.

Polyphagous on the leaves and flowers of trees, shrubs and herbaceous plants. The following are commonly recorded: hawthorn (*Crataegus monogyna*), blackthorn (*Prunus spinosa*), bramble (*Rubus fruticosus*), heather (*Calluna vulgaris*), common ragwort (*Senecio jacobaea*), goldenrod (*Solidago virgaurea*), angelica (*Angelica sylvestris*), yarrow (*Achillea millefolium*) and mugwort (*Artemisia vulgaris*). Late July to October.

PUPA Length 7.5–8.5mm; abdomen reddish or yellowish green; wings greenish yellow. In a fairly tough cocoon amongst ground debris. October to May.

FLIGHT PERIOD AND HABITAT Usually univoltine, May to late July with a peak during the last week of June. There is occasionally a partial second emergence in August. During the day it can sometimes be found resting on walls, fences or tree-trunks. Barrett (in Tutt, 1906c) states that scores may be found together under large boughs and are reluctant to fly if disturbed. It flies freely at dusk and at night both sexes come readily to light. Very generally distributed and found in many habitat types including woodland, waste ground, heathland and gardens.

DISTRIBUTION (Map 25) Widespread and generally common throughout the British Isles, though seemingly absent from the Shetlands.

COLLECTING AND REARING The larvae can be beaten easily from trees and shrubs and swept from low vegetation. However, as they are often parasitized, the best way to rear this species is from ova obtained from a captive female. The adults lay freely on sprigs of foodplant, netting or on the sides of the container in which they are placed. It is a very easy species to rear, the larvae transferring readily from one foodplant to another.

Eupithecia icterata (Villers, 1789)   Tawny-speckled Pug
Pl.2, figs 23–25; Pl.4, figs 6–8; Pl.C, fig.1

HISTORY First described in Britain by Haworth (1809). The larva and pupa were first described by Crewe (1860b;1861i;1861k). Comparative descriptions with *E. succenturiata* (with which the larva had previously been confused) were given by Hellins (1861) and Crewe (1862a).

IMAGO *Characteristic features.* Wingspan *c.*26mm; median area of forewing orange. Extreme examples of f. *cognata* (Pl.2, fig.24; Pl.4, fig.7) usually retain vestiges of orange.

Genitalia (Figs 16c(ii),28g ♂; 23a(ii) ♀). The male and female genitalia of this species are very similar to those of *E. succenturiata*. In *E. icterata*, the third cornutus of the male aedeagus, which is shaped like the head of a shepherd's crook, possesses a definite 'heel'. This is absent in *E. succenturiata*. The apex of the male abdominal plate is more deeply notched in *E. succenturiata* than in *E. icterata*. In the female bursa copulatrix, the signa near the base of the ductus seminalis are short and stout in *E. icterata*;. in *E. succenturiata* they are longer and finer.

Variation. Wingspan 20–24mm. F. *subfulvata* (Haworth, 1809) (Pl.2, fig.23; Pl.4, fig.6) is the type found commonly over most of the British mainland. A large, pale, whitish grey form, in which the typical fulvous colour is absent, often predominates in parts of Scotland and western Ireland. This was described by Stephens (1831) as subsp. *cognata* but, as it is sympatric with the typical form over part of its range, it should not be regarded as a subspecies but as f. *cognata* Steph. Ab. *excelsa* Dietze (1910;1913) appears synonymous with this form.

Close to f. *cognata* are f. *oxydata* Treitschke (1828) (Pl.2, fig.24; Pl.4, fig.7) and f. *grisescens* Lempke (1951) (Pl.2, fig.25; Pl.4 fig.8). The former retains much of the fulvous colour of f. *subfulvata* but it is less well defined. In the latter, the fulvous colour is reduced to a small area beneath the discal spot. A further modification of f. *oxydata* is ab. *dietzii* Prout (1914) in which the thorax and basal part of the costa are whitish. Intermediate between f. *oxydata* and f. *subfulvata* is f. *intermedia* Dietze (1910;1913) which is more richly marked with fulvous colour than the former and less so than the latter forms. F. *goodsoni* Cockayne (1953) was described from a specimen from Tring, Herts. It was bred by Goodson in 1945 and is a slight modification of the nominotypical f. *subfulvata*, in which the forewing costa is entirely dark, lacking the usual pale striae. Below the median nervure and on each side of nervure 2 is a longitudinal band of fuscous crossed by transverse pale lines. The submarginal line is indistinct and the pale marginal lunules are absent. The hindwing is darker with only indistinct markings. The ground colour reaches extremes of light and darkness in ab. *flavescens* and ab. *melaena*, the former being loam yellow instead of fulvous and the latter much suffused with black (Dietze, 1910;1913). Ab. *impuncta* Lempke (1947;1951) lacks the forewing discal spots.

Similar species
None.

LIFE HISTORY

OVUM Cream when first laid, later turning to deep golden yellow; 0.6mm long, ovoid. On the flowers and leaves of the foodplant in July and August.

LARVA (Fig.32c) *First and second instars.* Pale buff, pale greenish grey, reddish brown or pale brown; dorsal line dark brown or green, broadening on each of the central segments to form an oval blotch, subdorsal line fine, indistinct dark brown or green, spiracular line whitish. *Third and fourth instars.* Elongate, slender and tapering towards the head; length 20mm when full grown; head and thoracic legs light brown; integument covered sparsely with white tubercles and short hairs giving a rough texture; ground colour similar to first two instars; ventrally white with a purplish central line; dorsal markings a series of distinct rhomboid spots, dark brown, olive or green with oval centres of the ground colour, or slightly darker interspaced with indistinct patches of green or brown; dorsal line often absent; subdorsal line fine, dark brown or green, spiracular line white. There is a great superficial resemblance between the larva of this species and that of *E. succenturiata* which led some early entomologists to suspect that they were merely forms of one species (cf. Hellins, 1861; Freer, 1896; Prout, 1896; Sheldon, 1896; Tutt, 1906f; Dadd, 1906b; Gardner, 1907; Prout, 1907a). However, *E. icterata* often lacks the fine dorsal line and is not usually as dark in its later stages. Further, *E. icterata* usually rests stretched out, resembling a stalk of its foodplant, whereas *E. succenturiata* usually adopts an omega-like (Ω) position.

Feeds on the flowers of yarrow (*Achillea millefolium*). West (1994) cites feverfew (*Tanacetum parthenium*), and C. W. Plant (pers.comm.) states that in the London area it also feeds on southernwood (*Artemisia abrotanum*). September and October.

PUPA Length 8.5–9.5mm; abdomen and thorax reddish brown; wings reddish brown and deeply furrowed. In a tough cocoon amongst ground debris. From the end of October to early July.

FLIGHT PERIOD AND HABITAT Univoltine, recorded between late May and late September, the main flight period is from June to August with a peak during the second week of August. Often common in urban areas, hedgerows and waste ground in the vicinity of the foodplant. Generally distributed. It can be found during the day resting on fences, walls, tree-trunks, etc. Flies at dusk and comes readily to light at night. It is also attracted to flowers such as ragwort and yarrow.

DISTRIBUTION (Map 26) Widespread and generally common in lowland Britain, becoming more localized in upland areas. Apparently absent from the Outer Hebrides and the Northern Isles. Moderately common in Ireland, wherever the foodplant occurs, usually as f. *cognata*. Common in the Channel Islands,

COLLECTING AND REARING The larvae are easily dislodged from the foodplant by shaking it over a sheet. Alternatively, they can be searched for at night when they feed openly on the flower-heads. Females oviposit readily in captivity. Captive larvae will eat cultivated tansy (*Tanacetum* spp.). Prout (1896) notes that captive larvae can be reared on chrysanthemum (*Dendranthema* spp.). Easy to rear as the foodplant does not wither quickly.

Eupithecia succenturiata (Linnaeus, 1758) Bordered Pug
= *disparata* (Hübner, 1799)
= *succenturiaria* Boisduval, 1840

Pl.2, figs 26,27; Pl.4, figs 11,12; Pl.C, fig.2

HISTORY First recorded in Britain by Haworth (1809) from specimens collected at Coombe Wood, near Maidstone, Kent, possibly by Haworth himself. The larva was first described by Crewe (1861j,k). Comparative notes with *E. icterata* were published by Hellins (1861) and Crewe (1862a).

IMAGO *Characteristic features*. Wingspan *c.*26mm; forewing median area white.
 Genitalia. Figs 16c(i),28h ♂; 23a(i) ♀
 Variation. Apart from f. *disparata* Hübner (1796), which has a rust-coloured suffusion along the dorsum, variation is more or less restricted to general differences in intensity of colouring. Exceptionally light forms, in which even the terminal dark patches are weakened, are referrable to f. *extrema* Dietze (1910;1913) and f. *exalbidata* Staudinger (1901). Ab. *malaisei* Djakanov (1929), has the forewing wholly white with a strong dark costa, discal spot and median fascia and strongly chequered fringes. The hindwing of this aberration is also white with strongly chequered fringes. The commonest form is perhaps f. *obscurata* Lempke (1951) (Pl.2, fig.27; Pl.4, fig.12) in which the white of the forewing is heavily obliterated by dark markings, remaining only as a few transverse lines and a small basal patch. The hindwing of this form is also darkened. Illustrations of f. *obscurata* have been erroneously labelled by Skinner (1984), Agassiz *et al.* (1981) and others as ab. *disparata* Hübner. It occurs not infrequently in populations of *E. succenturiata* throughout Britain. Another dark form has both sides of the central area of the forewing bordered by sharp, dark, transverse lines and is described as ab. *bistrigata* by Lempke (1947;1951).

Similar species
None.

LIFE HISTORY

OVUM Cream when first laid, later turning orange; 0.6mm long, ovoid. On the leaves of the foodplants during August and early September.

LARVA (Fig.32d) *First and second instar*. Pale pinkish brown or greyish green; dorsal and subdorsal lines faint, dark brown or dark green; spiracular line white; dorsal line interrupted at segmental divisions and broader towards the centre of each segment. In the first instars, the larvae eat small holes into the upper surface of the leaves, leaving tiny transparent sections of the lower epidermis. *Third and fourth instars*. Elongate and slender, tapering towards the head; length 20mm when full grown; head and thoracic legs light brown or yellowish brown; integument wrinkled and covered sparsely with white tubercles from each of which arises a short white hair; ground colour pinkish or purplish brown or grey-green; ventrally white with purple central line; dorsal markings a series of rhomboid spots reducing in size towards the head and tail, dark brown or dark green with oval centres of the ground colour, usually connected by a fine similarly-coloured dorsal line; subdorsal line fine, dark brown or dark green; spiracular line broad, white, interrupted, reaching from the thoracic segments to the tip of the anal flap, edged ventrally with a heavily undulate black line. Some larvae feed at the base of the foodplant on dead leaves which have turned black. These individuals are usually very dark in colour, sometimes almost black.

Feeds on the leaves of mugwort (*Artemisia vulgaris*) and yarrow (*Achillea millefolium*). C. W. Plant (pers.comm.) also cites wormwood (*Artemisia absinthium*) and southernwood (*A. abrotanum*) in the London area. Prout (1896) and Allan (1949) state that larvae in captivity accept chrysanthemum (*Dendranthema* spp.). September and October.

PUPA Length 8.5–9.5mm; thorax and abdomen yellowish green; wings dark green. In a tough cocoon amongst ground debris. October to early July.

FLIGHT PERIOD AND HABITAT Univoltine, mainly in July and August, though it has been recorded from mid-June to mid-September. The peak of emergence appears to be during the third week of July. Found wherever the foodplants occur. Infrequently seen during the daytime but both sexes come readily to light at night.

DISTRIBUTION (Map 27) In England this species is widespread and generally common in the south east, becoming more localized in the south, south west and north. In Wales it is locally common in lowland areas. In Scotland it is found at lower altitudes as far north as Morayshire. Recorded from the Channel Islands but it is absent from the Western and Northern Isles. In Ireland it is scarce and very local.

COLLECTING AND REARING Larvae can be beaten easily from the foodplants but are often heavily parasitized. Captive females oviposit readily and the larvae are easy to rear in the usual way.

Eupithecia subumbrata ([Denis & Schiffermüller], 1775)
Shaded Pug

= *scabiosata* (Borkhausen, 1794)
= *piperitata* Stephens, 1829
= *piperata* Stephens, 1831
= *subumbraria* Boisduval, 1840
= *piperaria* Doubleday, 1849

Pl.2, figs 28,29; Pl.5, fig.13; Pl.8, fig.13; Pl.C, fig.3

HISTORY First described in Britain from Riddlesdown, Surrey, by Stephens (1831) as *piperata* though it was included in Stephens' list of 1829 as *piperitata*. Crewe (1854) recorded rearing this species as *piperaria* but the larva was not fully described until 1860 (Crewe, 1860m;1861k).

IMAGO *Characteristic features* (Fig.4a). Wingspan *c*.21mm; forewing apex pointed, ground colour white with numerous fine striae, discal spot minute or absent, postmedian fascia geniculate, marginal band dark and very prominent, broken at angle of postmedian fascia by white streak reaching subterminal line towards apex; hindwing white, discal spot minute, marginal band dark and prominent.

Genitalia. Figs 16d,28i ♂; 22h ♀

Variation. F. *obscurata* Lempke (1951) (Pl.8, fig.13) is a very dark form which has only vestigial white colouring. This form occurred frequently in adults bred from a Bedfordshire population by Riley in 1987. Similar to this, but not as dark, is f. *aequistrigata* Staudinger (1871). It is stated to be 'more equally marked throughout' without the usual sharp distinction between the distal area and the rest of the wing. Lighter forms are represented by f. *obrutaria* Herrich-Schäffer (1848) in which the basal and central areas are less copiously marked than in the type, leaving a sharply contrasting distal band, and ab. *bistrigata* Dietze (1910;1913) which is weakly marked throughout apart from conspicuous ante- and postmedian lines. F. *impuncta* Lempke (1947;1951) has the discal spot of the forewing absent. This form occurs occasionally throughout the species' range.

Similar species (See Fig. 4, p. 49; Plate 5)
E. irriguata (Fig.4c): ground colour white but forewing discal spot prominent, marginal band broken by white patches; hindwing with conspicuous discal spot.

LIFE HISTORY
OVUM Deep yellow, glossy; 0.6mm long, ovoid. Laid on the foodplant in July.

LARVA (Fig.32e) *First and second instars.* Either pinkish brown with fine, dark brown dorsal, subdorsal and spiracular lines, or pale yellowish green with dark green dorsal, subdorsal and spiracular lines. *Third and fourth instars.* Slender, tapering towards the head, length 22mm when full-grown; head and thoracic legs dull ochreous yellow; integument smooth, covered sparsely with fine hairs; pinkish buff with a wide, dark reddish brown dorsal stripe and fine subdorsal and spiracular lines of the same colour or pale yellowish green with dark green dorsal and subdorsal lines and white spiracular lines, both forms with a triangular red spot on the anal segment and a central red line on first two segments.

Feeds on the flowers of a wide variety of plants. The following are most commonly recorded: common ragwort (*Senecio jacobaea*), goldenrod (*Solidago virgaurea*), field scabious (*Knautia arvensis*) and burnet-saxifrage (*Pimpinella saxifraga*). July to the beginning of September.

PUPA Length 6–7mm; abdomen reddish brown; thorax and wings yellow or green; cremaster dusky red. In a cocoon amongst ground debris. September to June.

FLIGHT PERIOD AND HABITAT Univoltine, late May to late July with a peak during the last week of June. It inhabits rough grassland. Can be disturbed from its foodplant during the day and flies naturally at night, coming to light in small numbers.

DISTRIBUTION (Map 28) This is a locally common species in south-eastern England south of a line from the Severn estuary to The Wash, extending northwards in the east to South-east Yorkshire. There are records of outlying colonies in north Lancashire. In Wales it appears to be restricted to Glamorgan, Pembrokeshire and Cardiganshire. It is locally common in parts of western Scotland. Locally common in Ireland; the Channel Islands.

COLLECTING AND REARING Larvae of this localized and often elusive species can be obtained by dedicated searching of the foodplant or by sweeping. However, it is probably easier to obtain ova from a captive female. It is straightforward to rear, will feed on a variety of easily-available foodplants, and will readily change from one to another.

Eupithecia millefoliata Rössler, 1866   Yarrow Pug
= *achilleata* Mabille, 1869

Pl.2, fig.22; Pl.8, fig.19; Pl.C, fig.4

HISTORY Specimens were discovered in Britain by Baron de Worms at Hamstreet, Kent, on 7 August 1933 (de Worms, 1952) and by A. Richardson on the Kent coast at Sandwich in July 1939. However, these were not identified correctly until 1947 when S. Wakely found further specimens in south-east Kent. These records are reviewed by de Worms (1951;1952). The early stages were first recorded and described by Haggett (1955).

IMAGO *Characteristic features* (Figs 5h,7b). Wingspan *c.*26mm; abdomen with dark subbasal band; forewing costa straight; forewing tinged ochreous, discal spot with perimeter characteristically diffuse, postmedian fascia geniculate with pale streak from angle to subterminal line towards apex, marginal band darker than rest of wing, subterminal line and tornal spot present but not prominent, terminal half of marginal band near apex paler, in fresh individuals grey; hindwing discal spot inconspicuous, marginal band darker than rest of wing.

Genitalia. Figs 16e,28k ♂; 23b ♀

Variation. Uncommon. Only one major aberration has been found in the literature: ab. *uniformis* Dietze (1910;1913) has the ground colour and markings unicolorous sepia grey with only the discal spot remaining conspicuous.

Similar species (See Figs 5,7, pp. 61,67; Plate 8)
E. intricata (Fig.5d): forewing postmedian fascia curved or more smoothly geniculate, discal spot more clearly defined, median, ante- and postmedian lines more prominent, especially at costa.

E. satyrata (Fig.5g): usually smaller; abdomen without dark subbasal band; forewing discal spot smaller and clearly defined, tornal spot usually prominent, nervures usually spotted alternate light and dark; hindwing discal spot usually absent.

E. pimpinellata (Figs 5e,7d): forewing discal spot larger and clearly defined, costal spots more conspicuous; hindwing discal spot conspicuous. Generally glossier and smoother appearance in fresh specimens.

LIFE HISTORY
OVUM Pale greenish or yellowish white; 0.55 × 0.45mm (J. Reid, pers.comm.), ovoid. Laid on the foodplant in June and July.

LARVA (Fig.32f). *First and second instars.* Pale yellowish brown; dorsal markings V-shaped, faint, light chocolate brown. *Third instar.* Light to medium chocolate brown; on each of the seven or eight central segments a pale pink or buff dorsal patch containing an anterior pointing, dark purplish brown chevron, dorsal line fine, pale, connecting dorsal chevrons. *Fourth instar.* Robust, 16mm full grown; thoracic legs pale brown; head dark grey, speckled with black; integument wrinkled, covered sparsely with tubercles and hairs giving a rough texture; chocolate brown; dorsal markings on ten central segments similar to previous instar but patches usually pinkish white, chevrons darker purplish brown, almost black, becoming smaller towards the head, on posterior segments reduced to a short line; spiracular line fine, white, sometimes edged ventrally with dark green or a series of alternating red and dark green streaks beneath which is a row of dark brown spots; ventrally pale brown with a dark brown central line.

Feeds on Yarrow (*Achillea millefolium*), at first on the flowers and later on the seed-heads of the foodplant. In the late afternoon the larva is stated to come to the top of the seed-heads to feed. During the rest of the day it remains hidden (de Worms, 1953). August to early November.

PUPA Length 8–9mm; thorax and abdomen yellowish brown; wings greenish brown (J. Reid, pers.comm.). In a cocoon amongst ground debris. October to early June.

FLIGHT PERIOD AND HABITAT Univoltine, late May to late July with a peak during the first week of July. Most regularly found in coastal habitats, it has been recorded from dry heathland, downland, waste ground, commons and farmland but the precise inland ecological requirements are not presently known. The adults are rarely recorded except at light.

DISTRIBUTION (Map 29) Locally common east of a line from Sussex to Norfolk, extending along the south coast westwards to Dorset and along the east to Lincolnshire. Absent from Wales, Scotland, Ireland and the Channel Islands.

COLLECTING AND REARING A very cryptically coloured larva, *E. millefoliata* is difficult to find merely by searching yarrow seed-heads. The most efficient way of collecting larvae is to beat or sweep the brown and withered-looking seed-heads. They are straightforward to rear but are prone to a fungal disease which kills the larva but leaves it attached to the foodplant as if still alive.

Eupithecia simpliciata (Haworth, 1809) Plain Pug
= *subnotata* (Hübner, 1813)
= *subnotaria* Boisduval, 1840

Pl.2, fig.30; Pl.8, fig.21; Pl.C, fig.5

HISTORY First described and recorded as British by Haworth (1809). The larva was first described by Crewe (1859i;1861k) under the name *E. subnotata*.

IMAGO *Characteristic features* (Figs 5c,7c). Wingspan *c.*25mm; forewing broad, costa arched, termen rounded, ground colour characteristic pale ochreous, discal spot prominent, postmedian fascia smoothly and distinctly curved, subterminal line conspicuous; hindwing discal spot absent or minute.

Genitalia Figs 16f,281 ♂; 23c ♀

Variation. In Britain, usually restricted to a general lightening or darkening of the ground colour. Extreme examples of the latter, which are dark brown, are referable to ab. *brunnea* Lempke (1951) though the present authors are not aware of this aberration occurring in Britain.

Similar species (See Figs 5,7, pp. 61,67; Plate 8)
E. pimpinellata (Figs 5e,7d): abdomen with dark subbasal band; forewing narrower, costa less arched, postmedia fascia geniculate; hindwing with prominent discal spot.

LIFE HISTORY

OVUM Glossy cream when laid, later turning brownish yellow; 0.5mm long (J. Reid, pers.comm.), ovoid. On the foodplant in late June and July.

LARVA (Fig.32g) *First and second instars.* Pale greenish yellow; dorsal line interrupted, faint darker green. *Third instar.* Pale green or dusky yellow; dorsal line faint, dark green; dorsal markings a series of faint oval dark green rings. *Fourth instar.* Stout, tapering towards the head, 18mm when full grown; head and thoracic legs yellowish green; integument waxy, translucent, covered sparsely with white tubercles and hairs; light grass-green, occasionally reddish grey (Wilson, 1880), ventrally pale green; dorsal markings a series of oval or pointed rings, dark green, brown or grey, often resembling a chain, sometimes coalescing on posterior and anterior segments; connected by a very fine dorsal line of the same colour; subdorsal line fine, of the same colour; spiracular line pale yellow; segmental divisions pale yellowish or reddish green.

Feeds on the flowers and ripening seeds of fat-hen (*Chenopodium album*), stinking goosefoot (*C. vulvaria*), good-King-Henry (*C. bonus-henricus*), frosted orache (*Atriplex laciniata*), common orache (*A. patula*) and possibly other *Chenopodium* and *Atriplex* spp. August and September.

PUPA Length 7–8mm; thorax and abdomen yellowish brown; wings dark green. In a tough cocoon amongst ground debris. September to late June.

FLIGHT PERIOD AND HABITAT Univoltine, mid-June to mid-August with a peak during the last week of July. It is found in a variety of habitats including waste ground, overgrown allotments, waysides, river-banks and salt-marshes. Perhaps most common along rivers and in coastal localities. Flies at dusk and through the night when it comes to light, though usually in small numbers. Very attracted to the flowers of ragwort and marram grass. It can be boxed easily at night as it runs over flowers of the larval foodplant prior to oviposition (Tutt, 1906d).

DISTRIBUTION (Map 30) In England this species is locally common south of Cheshire and south Yorkshire. Elsewhere it is only sporadically recorded. In Wales the only known colonies are in Cardiganshire. There are no confirmed records from Scotland. Rare and local in Ireland. Recorded from the Channel Islands.

COLLECTING AND REARING Larvae can be found by searching the foodplant carefully or by beating over a sheet or tray. However, they drop from the foodplant very easily and great care must be taken when placing the beating receptacle under the plant. The larvae are often heavily parasitized. Simson (1981b) cites 69 per cent in one sample. Captive females oviposit freely. This is a straightforward species to rear though a plentiful supply of food is necessary due to the long larval stage (four months).

Eupithecia sinuosaria Eversman, 1848 — Goosefoot Pug

Pl.2, fig.31; Pl.4, fig.18; Pl.C, fig.6

HISTORY A single female of this species was caught in a M.V. light trap at Berrow, Somerset, on 13/14 June 1992 by B. E. Slade and was identified by D. J. L. Agassiz (Slade & Agassiz, 1991). A second specimen was caught in a R.I.S. light trap on the Rothamsted Estate, Harpenden, Hertfordshire, during the period 19–21 June 1992 (Townsend & Riley, 1991). An intensive search for larvae was made at Rothamsted on 26 August 1992 but none was found.

At the present time this species should be regarded as an immigrant, though further investigation may reveal it to be a newly-established scarce resident.

IMAGO *Characteristic features.* Wingspan c.24mm; forewing elongate, apex pointed, discal spot conspicuous and elongate, subbasal line black, prominent, antemedian line black, geniculate and prominent, postmedian line black,

sinuate and prominent, subterminal line conspicuous; hindwing discal spot conspicuous; abdomen with pale basal band.
Genitalia. Figs 16g,28j ♂; 23d ♀

Similar species
None.

LIFE HISTORY

OVUM Not described.

LARVA 19mm, slender, tapering towards the head; green with white tubercles giving a frosted appearance, anal segment reddish brown.
 Feeds on the flowers and fruits of orache (*Atriplex* spp.) and goosefoot (*Chenopodium* spp.), especially on plants growing in warm microclimates (Skou, 1986). August and September.

PUPA Not described. Overwinters in this stage; September to May.

FLIGHT PERIOD AND HABITAT May and June. Waste ground, gardens, fields and coastal localities.

Eupithecia distinctaria (Herrich-Schäffer, 1848) Thyme Pug
subsp. *constrictata* Guenée, 1856
= *constrictaria* Morris, 1861

Pl.2, fig.32; Pl.7, fig.1; Pl.C, fig.7

Note: The nominate subspecies does not occur in the British Isles.

HISTORY First recorded in Britain as *E. constrictata* Guenée by Doubleday (1856) following correspondence with the French entomologist. The early stages were first described by Crewe (1861f;1862a) from material collected in Derbyshire and Buckinghamshire.

IMAGO *Characteristic features* (Fig.1h,j). Wingspan c.20mm; forewing grey, apex pointed, discal spot elongate and prominent, costa with distinctive dark spots, postmedian fascia geniculate, other fasciae and striae usually inconspicuous, hindwing discal spot small but conspicuous, underside with postmedian line curved.
 Subsp. *constrictata* is stated by Guenée (1857) to differ from the Continental subsp. *distinctaria* in having a less smooth, grey ground colour with more distinct markings. It is also said to be larger than *distinctaria* which has a wingspan of 17–20mm (Skou, 1986).
Genitalia. Figs 12h,17a,28m ♂; 23e ♀
Variation. F. *famelica* Dietze is somewhat smaller and rather poorly marked (Prout, 1938), but the form is thought to result merely from malnutrition. Under normal circumstances this species seems to vary little.

Similar species (See Fig. 1, p. 35; Plate 7)

E. indigata (Figs 1f,i): forewing ochreous or pale brown rather than grey, costa straighter, apex more pointed, generally more uniform, striae and fasciae weak or obsolete, only prominent marking large discal spot; hindwing paler, more uniform, when present underside postmedian line geniculate.

E. dodoneata (Fig.1m): forewing more ochreous, not grey, postmedian fascia biangulate, postmedian line with a series of small basally-pointing dark triangles, discal spot much smaller.

E. ultimaria (Fig.1l): forewing narrower, postmedian fascia biangulate and prominent, discal spot linear; hindwing discal spot linear.

LIFE HISTORY

OVUM Pale yellow; 0.5mm, ovoid. Laid on the flowers of the foodplant in June and July.

LARVA (Fig.32h) *First and second instars.* Pale green, dorsal line faint, purplish red. *Third and fourth instars.* Long and slender, tapering towards the head, 15mm full grown; head and thoracic legs pale brown; integument wrinkled, covered sparsely with tubercles and short hairs; dark grass green, dorsal line strong, purplish red; spiracular line indistinct, pale greenish yellow.

Feeds on the flowers of thyme (*Thymus* spp.). Allan (1949) states that captive larvae can be reared on marjoram (*Origanum vulgare*). August and September.

PUPA Length 6–7mm; thorax and wings green; abdomen brown. In a cocoon amongst ground debris. September to early June.

FLIGHT PERIOD AND HABITAT Univoltine, June and July with the peak during the last week of June. Inhabits rocky places and dunes near the coast and also limestone outcrops, quarries and chalk downland inland, where the foodplant is well established. It is easily disturbed during the day in bright sunshine. Flies naturally at dusk and through the night when it comes to light in small numbers.

DISTRIBUTION (Map 31) Locally common throughout Wales and Scotland. In England it is less frequently recorded though common in parts of the south and the north. In Ireland, common along the coast, rarer inland. Its distribution in the British Isles follows closely that of the larval foodplant (see Perring & Walters, 1962).

COLLECTING AND REARING Being a very local species in the British Isles, often inhabiting inaccessible localities such as precipitous limestone cliffs, collecting larvae can be difficult and often dangerous. Moreover, larvae collected from the wild are often heavily parasitized. The larvae are difficult to find by searching during the day, but Tutt (1906c) states that at night they feed exposed on the thyme flowers. Where the foodplant is accessible the larvae can be found by a careful examination of the flowers, or by tapping the plants over a small sheet or tray. Perhaps the best method of rearing is by obtaining eggs from a captive female; adults can be netted near the foodplant at dusk or caught at light. In captivity the larvae require fresh flowers and it is advisable to use potted rather than cut foodplants. Alternatively, sprigs of thyme in pots of water will suffice.

Eupithecia indigata (Hübner, 1813) Ochreous Pug
= *indigaria* Boisduval, 1840

Pl.3, fig.1; Pl.7, fig.3; Pl.C, fig.8

HISTORY First recorded in Britain by Douglas (1851) from specimens collected in Birch Wood (part of Darenth Wood), Kent, in 1850. The larva was known to Douglas but the first description was published by Crewe (1862f;1863a).

IMAGO *Characteristic features* (Figs 1f,i,8m). Wingspan *c*.20mm; forewing elongate, apex pointed, costa and termen straight, discal spot large and prominent, all other markings obsolete or inconspicuous, when present, postmedian fascia geniculate; hindwing discal spot faint, underside with geniculate postmedian line, sometimes absent.

Genitalia. Figs 17c,280 ♂; 24b ♀

Variation. More or less limited to the shade of the ground colour. Dietze (1913) describes ab. *limbofasciaria* as having the distal three-quarters of the forewing uniformly darkened. In ab. *tristrigata* Fuchs (1904), both fore- and hindwings have three conspicuous cross-lines.

Similar species (See Figs 1,8, pp. 38,69; Plate 7)

E. dodoneata (Figs 1m,8s): forewing costa arched, fore- and hindwing less uniform; forewing postmedian fascia biangulate, postmedian line with a series of small, basally-pointing, dark triangles, discal spot much smaller and not prominent; hindwing termen concave, apex pointed, underside with postmedian line curved.

E. distinctaria (Fig.1h): forewing costa more arched, apex less pointed, ground colour grey, generally less uniform, fore- and hindwing striae and fasciae better developed; forewing costal spots prominent; underside of hindwing with postmedian line curved.

E. ultimaria (Fig.1l): forewing costa more arched, postmedian fascia prominent and biangulate, discal spot linear; hindwing with prominent linear discal spot.

LIFE HISTORY

OVUM White when laid, turning pale brown shortly after; 0.6mm long, ovoid. On the young shoots of the foodplant in May.

LARVA (Fig.32i) *First instar.* Buff or light greenish brown; dorsal line indistinct, dark brown or green; subdorsal and spiracular lines cream. *Second and third instars.* Similar to the first instar but dorsal line more conspicuous. *Fourth instar.* Elongate; 16mm when full-grown; head reddish brown with two small black spots; thoracic legs reddish brown; integument smooth; dorsally pale brown;

dorsal line fine, dark brown or black; subdorsal line bold, cream, widening anteriorly and posteriorly; area between subdorsal and spiracular lines pale brown; ventrally pale grey, central line white.

Feeds on Scots pine (*Pinus sylvestris*), larch (*Larix decidua*) and Norway spruce (*Picea abies*). Winter (1990) also records lodgepole pine (*Pinus contorta*). In early instars, male inflorescences are preferred though Haggett (1992) states that later feeding is restricted mainly to lateral buds on old growth. They will also eat young needles. Leverton (1998) observed larvae in captivity eating aphids. Late June to early September.

PUPA Length 7mm; thorax and abdomen light reddish brown; wings slightly darker. In a cocoon on the ground surface or among leaf-scales *in situ* on the pine branches. From September to April.

FLIGHT PERIOD AND HABITAT Univoltine, late April to early June, with a peak during the third week of May. Can be found resting on tree-trunks during the day. Flies freely at dusk, usually high around the trees, and throughout the night when it comes regularly to light.

DISTRIBUTION (Map 32) Widespread and locally common throughout the British Isles with the exception of the Outer Hebrides and the Shetlands and Orkneys.

COLLECTING AND REARING Larvae can be beaten from the foodplant and are straightforward to rear in plastic boxes or on cut food in water. However, the long larval stage means that a good supply of fresh young shoots should be readily available. Captive females oviposit freely in captivity.

Eupithecia pimpinellata (Hübner, 1813)   Pimpinel Pug
= *pimpinellaria* Boisduval, 1840
= *denotata* sensu Doubleday, 1856

Pl.3, figs 2–4; Pl.7, fig.24; Pl.8, figs 7,20; Pl.C, fig.9

HISTORY First recorded in Britain by Henslow (1852), under the name *E. denotata* Doubleday, from specimens collected from burnet-saxifrage at Hitcham, Suffolk, on 23 August 1850. Although the species was reared by Henslow in 1851, the first larval description, under the name *E. denotata*, was not published until some years later by Crewe (1859g;1861k).

IMAGO *Characteristic features* (Figs 5e,7d,8a). Wingspan *c*.24mm; abdomen with dark subbasal band; forewing pale ochreous, fresh individuals with pale grey wash, discal spot large, intensely black and very prominent, ante- and postmedian fascia geniculate, costa with a series of black spots, subterminal line inconspicuous, dentate near costa, tornal spot elongate; hindwing discal spot usually conspicuous, postmedian fascia curved.

Genitalia. Figs 17d,28p ♂; 24c ♀.

Variation. There is very little variation in this species, the only major aberration recorded being ab. *limbosignata* Dietze (1910;1913) which is brown with conspicuous pale subterminal lines on both fore- and hindwings. G. M. Haggett (pers.comm.) notes an aberration from Lincolnshire in which the median lines are coalesced.

Similar species (See Figs 5,7,8, pp. 61,67,69; Plates 7,8)

E. absinthiata (Figs 7a,8c): forewing more uniform brown, striae and fasciae less conspicuous, subterminal line not dentate near costa; hindwing more uniform, not usually paler basally, discal spot absent or inconspicuous.

E. expallidata (Fig.7f): small individuals similar but forewing costa more arched; fore- and hindwings more uniform, striae and fasciae obsolete; forewing discal and costal spots larger and more prominent.

E. assimilata (Figs 7g,8o): usually smaller; forewing costa more arched; forewing more uniform brown, striae and fasciae less conspicuous, tornal spot prominent, discal spot larger; hindwing darker, discal spot less prominent, male abdomen truncate in profile.

E. denotata (Figs 5i,7h,8f): abdomen without dark subbasal band; forewing postmedian fascia biangulate, discal spot smaller and less elongate, antemedian fascia curved, costal spots inconspicuous.

E. millefoliata (Figs 5h,7b): forewing costa straighter, discal spot with perimeter diffuse, costal spots much less conspicuous, sometimes absent; hindwing discal spot minute or absent. Overall appearance more irrorate.

E. simpliciata (Figs 5c,7c): abdomen without dark subbasal band; forewing broader, costa more arched, postmedian fascia smoothly curved, costal spots less conspicuous or absent; hindwing discal spot minute or absent.

LIFE HISTORY

OVUM Pale yellow; 0.6mm, ovoid. Laid on the flowers of the foodplant in June and July.

LARVA (Fig.32j) *First instar*. Creamy green with no visible markings. *Second instar*. Pale green; dorsal, subdorsal and spiracular lines fine, dark green. *Third and fourth instars*. Long and slender; 18mm when full-grown, tapering towards the head; head and thoracic legs purplish red; integument smooth. There are two colour forms: (i) green, dorsal line purplish brown, broadening posteriorly; if present, spiracular line a series of cream or pale yellow patches; anterior and posterior segments each with a bright purplish red patch; (ii) pale purplish red; dorsal and subdorsal lines fine, darker red or brown, spiracular line bold, cream; each form a paler shade ventrally.

Feeds on ripening seed-capsules of burnet-saxifrage (*Pimpinella saxifraga*), and possibly greater burnet-saxifrage (*P. major*) (Allan, 1949). September and October.

PUPA Length 7–8mm; reddish brown or with reddish brown thorax and abdomen and greenish yellow wings. In a cocoon amongst ground debris. October to June.

FLIGHT PERIOD AND HABITAT Univoltine, June and July (Skinner, 1984). On chalk downland, calcareous grassland and adjacent hedgerows. It can be found flying during the day in the vicinity of the foodplant and comes to light at night.

DISTRIBUTION (Map 33) In England this species is locally common east of a line from Dorset and Herefordshire to north Yorkshire. Elsewhere it is only sporadically recorded, usually as unconfirmed individuals. In Wales it is known to occur only in Caernarvonshire. It is absent from Scotland and, apparently, the Channel Islands. There are a few records from south-west and north-east Ireland and an isolated record from Dublin.

COLLECTING AND REARING A very difficult larva to find as it is so well camouflaged. When flowering, the stalks of the foodplant are green, at which stage the larva rests stretched out along the main stalk; when in seed, the umbel stalks are purplish red, at which stage the larva rests stretched out among the uppermost stalks. When the larva is feeding, it sits in a curled posture with the anal segments brought close to the head, in which position the red anterior and anal spots closely resemble the seed-capsules. It is often found feeding in company with *E. centaureata*, but that species feeds openly on top of the seed-heads. Larvae can be collected by careful searching or by sweeping the foodplant, but they are usually very heavily parasitized; Simson (1980) reported 80 per cent parasitism.. Therefore the most efficient method of rearing *E. pimpinellata* is from ova obtained from a captive female. It can be reared easily in the usual way.

Eupithecia nanata (Hübner, 1813)   Narrow-winged Pug
subsp. *angusta* Prout, 1914
= *angustata* (Haworth, 1809), nec (Gmelin, 1790)

Pl.3, figs 5,6; Pl.4, fig.19; Pl.6, fig.16; Pl.C, figs 10,11

Note: The nominate subspecies does not occur in the British Isles.

HISTORY First recorded in Britain as *Phalaena angustata* by Haworth (1809), probably from specimens collected at Westerham, Kent, as he mentions this area as a locality for the species. The larva was first described by Crewe (1860d;1861k).

IMAGO *Characteristic features* (Fig.5j). Wingspan *c.*19mm; abdomen with dark subbasal band; forewing characteristically elongate, costa and termen straight, apex pointed; alternate light and dark fasciae forming distinctive striped appearance, discal spot conspicuous, postmedian fascia biangulate, tornal spot conspicuous.

Subspecies *angusta* Prout (1938) differs from the Continental subspecies in having more lanceolate forewing and lacking its admixture of reddish or yellowish scales.

Genitalia. Figs 17e,28q ♂; 24d ♀

Variation. There is a great variation in the density of the black cross-lines of the forewing and in f. *mediofaciata* Dietze (= ab. *nigrofasciata* Dietze) (1910;1913) those of the median area are greatly extended to give the appearance of one dark central band. This darkening of the forewing reaches its extreme in f. *oliveri* Prout (1915) where the markings are almost completely obliterated by black scales. A slight tinge of brown usually remains in the areas of the wings where this normally occurs and there are often faint remnants of the white postmedian and subterminal lines. The hindwings are very dark grey. This form occurs mainly in the Midlands but has occasionally been recorded elsewhere.

Generally speaking, summer brood individuals are smaller and darker than those of the first generation and these were named ab. [= f.] *pauxillaria* by Boisduval (1840). Light forms reach their extreme with ab. *bistrigata* Lempke (1947;1951) in which the usual markings are absent apart from the discal spots, fine antemedian and postmedian lines and a dark subterminal band containing a pale submarginal line. Ab. *bicolor* Lempke (1951) has the forewing split into halves, the basal being pale grey with only faint traces of the transverse lines and the distal being dark grey with a pale submarginal line. The hindwing of this form is pale grey with dark discal spots and traces of the transverse lines at the dorsum.

Similar species (See Fig. 5, p. 61; Plate 6)
E. satyrata f. *curzoni* (Fig.5g): forewing termen more rounded, postmedian fascia not biangulate, discal and tornal spots smaller, less conspicuous; abdomen without dark subbasal band.

LIFE HISTORY

OVUM Deep yellow; 0.7mm long, ovoid. Laid on the foodplant in late April and May and probably again in August.

LARVA (Fig.32k) *First and second instars.* White, pale pink or pale green with pointed red dorsal and lateral spots. *Third and fourth instars.* Elongate and slender, tapering slightly towards the head; 18mm when full-grown; head and thoracic legs reddish brown; integument smooth with short, fine hairs; white, pink or bright green; on the four central segments, dorsal markings comprise three angular carmine or purplish red anteriorly-pointing pear-shaped spots; on anterior and posterior segments, these markings smaller and more rounded; subdorsal line inetrrupted, red; ventrally pale pink with a fine red central line.

Feeds almost exclusively on heather (*Calluna vulgaris*), though Chalmers-Hunt (1968–71) also cites cross-leaved heath (*Erica tetralix*), mainly on the internal parts of the flowers. July to October.

PUPA Length 6.0–6.5mm; abdomen reddish yellow; thorax and wings yellow suffused with green. In a cocoon amongst ground debris. August to late April.

FLIGHT PERIOD AND HABITAT Bivoltine, late April to June with a partial second emergence in August. Adults have been recorded as early as the first week of

April and as late as the last week of October. There are two broods throughout Britain peaking in early June and early August. These are protracted and sometimes overlapping. There appears to be no significant delay in the emergence peak of northern populations.

Found most commonly on heathland and heather moorland but is occasionally associated with cultivated heathers. It rests during the day on its foodplant and is easily disturbed. At night it often comes to light in large numbers.

DISTRIBUTION (Map 34) Locally common throughout the British Isles.

COLLECTING AND REARING The larvae are usually easy to obtain by beating the heather over a sheet or tray or by sweeping. They are often found with larvae of *E. absinthiata* f. *goossensiata* to which, in the early stages, they are superficially similar. Captive females oviposit readily on sprigs of foodplant or on muslin. The larvae are straightforward to rear in the usual way.

Eupithecia extensaria (Freyer, 1844)   Scarce Pug
subsp. *occidua* Prout, 1914
= *prolongata* sensu Dietze, 1910

Pl.3, fig.7; Pl.4, fig.14; Pl.C, fig.12

Note: The nominate subspecies does not occur in the British Isles where it is represented, in England only, by subspecies *occidua*.

HISTORY Doubleday (1875) recorded the first capture of this species by W. Prest near Hull, Yorkshire, in 1874. Dobrée (1875) and Sawyer (1875) speculated that it had been accidentally introduced into this country by Baltic steamers using the port at Hull, though the latter believed that the excellent condition of Doubleday's specimen suggested it was a native-bred individual. By the time of publication, the original locality of capture, about a mile from the docks, had been destroyed. Carrington (1882) recorded the existence of a second specimen that had been captured in the Holderness district. Prest reported to Carrington that he had seen a specimen in the collection of J. Buck, an old Hull collector, who told him he had taken it at rest on wormwood at Spurn Head some ten years previously (i.e., c.1872). This therefore became the first British record. Barrett (1887) reported the capture by his eldest son and E. A. Atmore of nine specimens on the Norfolk coast in July 1887. Although no locality was cited for these or for Barrett's later (1889) record of larvae from *Artemisia maritima*, Porritt (1889) gave Hunstanton as the place of capture of several larvae from *A. maritima* on 27 August 1889. This record gave the species its early vernacular name of Hunstanton Pug. The larva was first described by Barrett (1889).

IMAGO *Characteristic features*. Wingspan *c.*25mm; forewing boldly and distinctively marked. The English subspecies *occidua* Prout (1914) is more boldly marked than the nominate subspecies, the ground colour being more silvery grey and the brown bands darker.

Genitalia (Figs 17f,8r ♂; 24f ♀)

Variation. There appear to be no significant morphological differences between Norfolk and Lincolnshire colonies. Cockayne (1953) described the albinistic f. *albescens* as having very pale, weakly marked or obsolescent transverse bands. The type specimen of this form was bred from larvae collected at Thornham, Norfolk. G. M. Haggett (pers.comm.) notes a specimen from Gibraltar Point, Lincolnshire, in which the median bands are coalesced into a solid brown bar. Generally there is little variation in British *E. extensaria*.

Similar species
None.

LIFE HISTORY

OVUM Pale yellow, turning darker with development (J. Reid, pers.comm.); 0.8mm long, ovoid. Laid on the foodplant during June and July.

LARVA (Fig.32n) *First and second instars.* Light silvery green; subdorsal line faint, dark green, dorsal lines faint, dark green, widening slightly at the centre of each segment; spiracular line silvery white. *Third and fourth instars.* Long and stout, tapering slightly towards the head; 25mm full grown; head green with pink mouthparts; thoracic legs pinkish brown or green; integument smooth; colour similar to earlier instars but dorsal markings more conspicuous; spiracular line edged ventrally with an interrupted reddish pink line.

Feeds on the flowers and leaves of sea wormwood (*Seriphidium maritimum*). In captivity it can be reared on southernwood (*Artemisia abrotanum*). Late July to September.

PUPA Length 9.5mm; thorax and abdomen chestnut brown; wings bright green. In a tough cocoon amongst ground debris. September to June.

FLIGHT PERIOD AND HABITAT Univoltine, June and July, with the peak of emergence during the last week of June. Occasional individuals are recorded during the last week of May and the first week of August. Inhabits the edges of saltmarshes where the foodplant is well established. It rests amongst the foodplant during the day and is fairly easily disturbed (see Barrett (1887) who caught some of the first British specimens in this way). It flies at night and comes freely to light.

DISTRIBUTION (Map 35) The only known colonies of this species in the British Isles are along the coasts of south Yorkshire, Lincolnshire, north Norfolk and, until 1983, north Essex. Individuals recorded in Surrey and West Sussex are presumably vagrants.

COLLECTING AND REARING The larvae are so well camouflaged that they are extremely difficult to detect, even in captivity (see also Barrett, 1889), though Tutt (1906b) states that they usually feed openly at night. When resting on the stalks of its foodplant, which have grooves filled with silvery white hairs, the larva blends perfectly with their structure and colour. When feeding, it adopts a curled posture, head and legs meeting the anal claspers, in which position it perfectly mimics a flower-bud. Larvae are, therefore, best obtained by beating bunches of the foodplant over a sheet or tray, or by sweeping. Females oviposit readily in captivity and the larvae are easy to rear in the usual way. However, they are voracious feeders so a good supply of foodplant is essential.

Eupithecia fraxinata Crewe, 1863   Ash Pug
= *innotata* sensu Stephens (1831) nec Hufnagel (1767)
= *tamarisciata* (Freyer, 1836)

Pl.3, figs 8–10; Pl.4, figs 20,21; Pl.6, fig.6; Pl.C, figs 13,14

HISTORY First recorded in Britain by Stephens (1831) as *E. innotata* (Hufnagel, 1767), and the larva was described by Crewe (1859c,j) under the same name. Crewe later recorded the first British *E. fraxinata* (1862o), and this was followed by discussions on the status of these two supposed species, with comparative descriptions of adults and larvae (Crewe, 1863a,e). Two species continued to be recorded in Britain and their specific status was again discussed by Tutt (1906a,e), who also introduced a third species into the literature: *E. tamarisciata* Freyer (1836) (Tutt, 1906a; 1908). Haggett (1963) published a detailed review of this complex and concluded that only *E. fraxinata* occurred in Britain and that all previous records of *E. innotata* were erroneous. See *E. innotata* (p. 143) for further discussion.

IMAGO *Characteristic features.* Wingspan c.24mm; forewing noticeably elongate, costa straight, apex pointed; ground colour dark grey.

Genitalia. Figs 17g,28s ♂; 24g ♀

Variation. Ab. *grisescens* Petersen (1909) and ab. *paupera* Dietze (1910;1913) are dark forms of *E. innotata* in which the usual markings are barely discernible. There appears to be little difference between these (although the former is stated to be bluish grey) and f. *unicolor* Prout (1915) (Pl.3, fig.10; Pl.6, fig.6) which is a dark, slate-grey form found regularly in some British populations. Ab. *suspectata* Dietze (1871b) is a small, weakly marked form, known only from the second generation of Continental *E. innotata*, and ab. *rotundata* Bastelberger (1908) has less elongate forewing with rounded apices. Neither of these taxa have, to our knowledge, been recorded among British *E. fraxinata*.

Similar species
The elongate forewing and dark grey colour should preclude confusion with other species found in the British Isles.

LIFE HISTORY

OVUM Shiny cream when first laid, later turning yellow; 0.7mm long, ovoid. On the leaves of the foodplant in June and August.

LARVA (Fig.33a) *First instar.* Cream or greenish white; markings absent. *Second and third instars.* Pale grass green; dorsal and subdorsal lines darker, the latter interrupted at segmental divisions; spiracular line white or greenish white, edged ventrally with a dark green line; anal segment with a dorsal reddish purple spot.

Fourth instar. Elongate, slightly dorsally compressed, 20mm when full grown; head and thoracic legs green or brown; integument smooth and velvety; two basic colour-forms dark green or pale brown; markings variable; dorsal lines or patches usually a darker shade than ground colour, either a bold dorsal stripe interrupted by the white segmental divisions, or better-defined marks resembling those of *E. nanata*; all degrees of variation between the two forms can occur; subdorsal line fine, a darker shade than ground colour; spiracular line undulate or broken into oblique stripes, conspicuous, cream or pale yellow; below spiracular line a series of reddish brown dots which are largest on the central segments; anal segment with a purplish red, black-centred, dorsal spot.

Feeds on the leaves of ash (*Fraxinus excelsior*), sea-buckthorn (*Hippophae rhamnoides*) and tamarisk (*Tamarix gallica*). In captivity it has been fed on mugwort (*Artemisia vulgaris*), laurustinus (*Viburnum tinus*) (Crewe, 1860q) and lilac (*Syringa vulgaris*) (G. M. Haggett, pers.comm.). Allan (1949) also lists wormwood (*Artemisia absinthium*) and field wormwood (*A. campestris*) but does not state whether these are known natural foodplants. Scorer (1913) and South (1961) cite scabious (*Scabiosa* sp.) but these records are probably attributable to Robson (1902) for which there is some doubt (cf. Prout, 1908). Gregson (1874a) and Prout (1904) record finding larvae on mugwort (*A. vulgaris*) and the latter also cites smooth hawk's-beard (*Crepis capillaris*). Haggett (1963) doubts the authenticity of the mugwort records. June and early July and again in late August and September.

PUPA Length 7.5–8.0mm; abdomen brown; wings and thorax green. In a tough cocoon. It is stated (Mackworth-Praed, 1962) that pupae are sometimes found beneath moss at the base of ash trees. The present authors have no experience of this despite thorough searching. July and early August, and late September to May.

FLIGHT PERIOD AND HABITAT Bivoltine, May and June and again in July and August. The peaks of the two broods appear to be the last week of May and the first week of August. Occasional specimens are recorded in September, and Crewe (1865a) records an adult at ivy blossom in October. Ash-feeding populations are widespread but, apart from those on the limestone pavement of Yorkshire, are generally uncommon. Colonies supported by sea-buckthorn are restricted to coastal sand-hills in east and south-east England and are usually large. Tamarisk-feeding *E. fraxinata* (see *E. tamarisciata*, p. 143) are known only from the coasts of Cornwall and Guernsey (Tutt, 1906a;1908; Riley, 1990c). However, thorough investigation of well-established stands of tamarisk in other localities may reveal more widespread utilization of this foodplant. *E. fraxinata* comes readily to light at night.

DISTRIBUTION (Map 36) Locally widespread throughout most of Britain. It is most common in coastal localities where it is associated with sea-buckthorn. Apart from on the limestone pavements of mid-west Yorkshire, this species is recorded only in small numbers inland or where sea-buckthorn is absent. Not known from the Outer Hebrides and the Shetlands. Very local in western and eastern Ireland where ash is present. Recorded from the Channel Islands.

COLLECTING AND REARING Larvae can be beaten from sea-buckthorn and tamarisk; however, ash-feeding larvae are very difficult to find in this way. Captive females caught at light oviposit freely. An easy species to rear in the usual way.

Eupithecia virgaureata Doubleday, 1861   Golden-rod Pug
= *pimpinellata* sensu Doubleday, 1861
= *virgaurearia* Morris, 1861

Pl.3, figs 11,12; Pl.6, fig.3; Pl.8, fig.2; Pl.C, fig.15

HISTORY First recorded as British by Doubleday (1861), who originally identified and described the specimens as *E. pimpinellata*. They were later correctly determined as *E. virgaureata* by Herrich-Schäffer. Doubleday's description stands as the type-description for this species. The larva was first described by Crewe (1859d;1861k), also under the name *E. pimpinellata*.

IMAGO *Characteristic features* (Figs 5l,6d,8g). Wingspan *c.*23mm; forewing broad, discal spot conspicuous, postmedian fascia geniculate, subterminal line and elongate tornal spot usually well developed, nervures sometimes spotted alternate light and dark, postmedian line sometimes with small basally-pointed dark triangles; hindwing discal spot usually conspicuous, pale tornal spot usually present; pale individuals sometimes with white metathoracic spot. Melanic form with distinct forewing discal spot.

Genitalia. Figs 17h,28t ♂; 24e ♀

Variation. Specimens which are weakly marked, appearing almost unicolorous, are referable to f. *altenaria* Staudinger (1861), whilst those of a unicolorous black with only the forewing discal spot visible are known as f. *nigra* Lempke (1951) (Pl.3, fig.12; Pl.6, fig.3). The latter form occurs sporadically and sometimes commonly in populations throughout the species' range, dominating in some areas. Ab. *nigrofasciata* Dietze (1910;1913) has the median area of the forewing darkened. Dietze (1910;1913) describes two other aberrations: ab. *nigronotata* has unusually large discal spots on both fore- and hindwings, whilst ab. *notata* has unusually prominent discal spots on the forewing only. Specimens caught in the RIS light traps in northern Scotland appear to have a much paler ground colour, somewhat similar in shade to *E. egenaria*. In some of these specimens a pale metathoracic crest is evident.

Similar species (See Figs 5,6,8, pp. 61,65,69; Plates 6,8)

E. egenaria (Figs 5a,8 l): forewing more elongate, generally paler, with characteristic double ante- and postmedian lines; white metathoracic spot always absent.

E. trisignaria (Figs 6a,8k): forewing more uniform, striae and fasciae inconspicuous or absent; fore- and hindwing tornal spot absent; forewing nervures not spotted alternate light and dark; white metathoracic spot always absent. Examination of the male abdominal plate or the genitalia is required to separate the melanic forms.

E. satyrata (Figs 5g,8j): forewing narrower, discal spot minute; hindwing discal spot absent or minute; white metathoracic spot always absent.

E. intricata (Figs 5d,8b): abdomen unicolorous, not conspicuously paler ventrally, with dark subbasal band; forewing costa straighter, apex more pointed, postmedian fascia curved, dark median, ante- and postmedian lines prominent at costa, nervures not spotted alternate light and dark; white metathoracic spot always absent.

E. vulgata (Figs 5b,6b,8e): forewing usually tinged ochreous, postmedian fascia biangulate, discal spot minute, white tornal spot more prominent, nervures not spotted alternate light and dark; white metathoracic spot always absent. Melanic form with forewing discal spot absent or minute.

E. tripunctaria (Figs 5f,6c,8h): fore- and hindwings with conspicuous and prominent white tornal spot. Examination of the male abdominal plate or the genitalia is necessary to separate the melanic forms.

E. denotata (Figs 5i,6e,8f): forewing postmedian fascia biangulate, nervures not spotted alternate light and dark; white metathoracic spot always absent.

E. subfuscata Figs 5l,6d,8g): indistinguishable by superficial characters alone. Male antennae simple rather than biciliate (see Figure below); white metathoracic spot always absent; genitalia and male abdominal plate diagnostic. Forewing nervures spotted alternate light and dark and presence of small basally-pointing dark triangles on postmedian line suggest *E. virgaureata*.

♂ Antennal segments

(a) *subfuscata*, simple
(b) *virgaureata*, biciliate

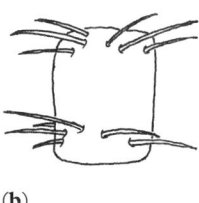

(a)   (b)

E. abbreviata f. *nigra* (Figs 6g,8r): similar to f. *nigra* but forewing more elongate, nervures intensely black; hindwing termen concave, apex pointed.

E. dodoneata (Fig.8s): uncommon melanochroic forms similar but hindwing termen concave, apex pointed.

E. pusillata (Figs 6f,8p): dark forms similar to f. *nigra* but dark forewing fasciae usually present.

E. lariciata (Fig.8d): contrasting light and dark forewing fasciae give more striped appearance, tornal spot inconspicuous or absent. Examination of the male abdominal plate or the genitalia is recommended for separation of the melanic forms.

Pasiphila rectangulata f. *anthrax* (Figs 6h,8q): similar to f. *nigra* but forewing broader, costa relatively shorter and more arched, apex more rounded, vestiges of green coloration sometimes present; hindwing underside often with prominent angulate postmedian line.

LIFE HISTORY

OVUM Cream when first laid, later turning to pale brown; 0.6mm long, ovoid. On the flowers of the foodplant in May and June (in captivity) and again in August.

LARVA (Figs 32 l,m) *First and second instars*. Pale yellow; dorsal and subdorsal lines brown or reddish brown. *Third and fourth instars*. Thick set, slightly dorsally compressed, tapering at both ends; 20mm when full grown; head and thoracic legs brown; ground colour pale brown, buff or yellow; on each of the five central segments a broad, forward-pointing brown, reddish brown or purplish brown chevron with an oval, brownish pink centre; dorsal line fine and of a similar colour; dorsal chevrons sometimes extend laterally to spiracular region forming dark oblique stripes beneath which are further narrow russet bands; spiracular line sinuate, bold, cream or pale yellow following the course of the lateral stripes; tip of anal segment brown.

First brood feeds on hawthorn (*Crataegus monogyna*) in June. Prout (1907b) cites Klos finding larvae in Germany on blackthorn (*Prunus spinosa*), hogweed (*Heracleum sphondylium*) and gentian (*Gentiana* spp.) during June and July. Henwood (1996) records finding a larva on sallow (*Salix cinerea*) in Devon in June, 1993. In captivity, it has been reared on cow parsley (*Anthriscus sylvestris*) and angelica (*Angelica sylvestris*). Second brood on the flowers of goldenrod (*Solidago virgaurea*) and common ragwort (*Senecio jacobaea*) in September and October. In captivity, it will accept Canadian goldenrod (*Solidago canadensis*).

PUPA Length 7-8mm; wings yellowish green; abdomen yellowish red, both with fine brown streaks; thorax yellowish green with small brown flecks and outlined in brown. Unmarked individuals have been noted in captivity. In an earthen cocoon. July and August, and again from October to late April. Haggett (1968a) states that some pupae of the first brood do not hatch until the following spring.

FLIGHT PERIOD AND HABITAT Bivoltine, from late April to early June, and again in July and August. The two peaks appear to be the second week of May and the last week of July. The precise habitat requirements are uncertain, though this species may occur wherever its larval foodplants are established. The adults are stated to fly over the foodplants at dusk (Tutt, 1906c), and at night come readily to light.

DISTRIBUTION (Map 37) Locally widespread throughout the British Isles with the exception of the Outer Hebrides, the Orkneys and Shetlands.

COLLECTING AND REARING Larvae can be found by beating ragwort and goldenrod. The bold markings, and especially the sinuate spiracular line of the final instar larva, readily distinguish it from those of *E. absinthiata*, *E. expallidata* and *E. satyrata*, with which it is regularly found. Females oviposit readily in captivity. A straightforward species to rear in the usual way.

Eupithecia abbreviata Stephens, 1829   Brindled Pug
= *nebulata* (Haworth, 1809) nec (Scopoli, 1763)
= *subfasciata* Stephens, 1829
= *abbreviaria* Doubleday, 1849

Pl.3, figs 13,14; Pl.6, fig.9; Pl.8, fig.2; Pl.D, fig.1

HISTORY First recorded in Britain by Stephens (1829). The immature stages were first described by Crewe (1860o;1861k).

IMAGO *Characteristic features* (Figs 6g,8r,9a). Wingspan *c.*24mm; forewing broad, elongate, apex pointed, discal spot linear, postmedian fascia biangulate, postmedian line with small basally-pointing dark triangles, those near dorsum elongate and usually conspicuous; hindwing discal spot usually obsolete, termen concave, apex pointed. Melanic form f. *nigra* (Pl.3, fig.14; Pl.6, fig.9) with intensely black discal spot and nervures.

Genitalia. Figs 12j,18a,29a ♂; 24h ♀

Variation. There is a great deal of variation in the ground colour of this species. Many Scottish specimens appear to be very conspicuously marked against a pale ground colour, and this form reaches its extreme in ab. *striata* Lempke (1951) which is pale grey with darker nervures and very distinct discal spot and transverse lines. Although most common in Scotland, these pale forms occur throughout Britain and are sometimes frequent in RIS light-trap catches from Yarner Wood National Nature Reserve in Devon. There are many dark forms. Those frequent in northern England are often suffused with matt black or brownish black, with a conspicuous pale subterminal line. It is often quoted that the completely melanic form is f. *hirschkei* Bastelberger (e.g. Agassiz *et al.,* 1981). However, this form was originally described as having a dark grey rather than yellow-brown ground colour, with inconspicuous markings (Bastelberger, 1908). The truly melanic form *nigra* is described by Cockayne (1953) from material bred by A. L. Goodson at Tring in 1945. It is black with the markings almost obliterated. The coppery black form, so common in many parts of Britain and usually called *hirschkei*, is therefore more accurately attributed to f. *nigra*. A further (unnamed) form has the area between the base and the ante-

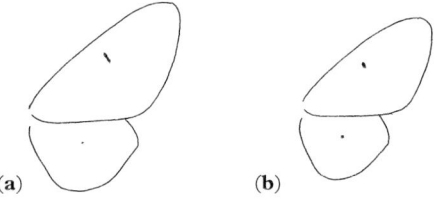

Text figure 9 Wing shapes and patterns (Similar species)
(**a**) *E. abbreviata* (**b**) *E. dodoneata* (p.119)

median lines of the forewing a darker shade than the rest. This occurs more or less regularly in populations throughout Britain.

Similar species (See Figs 8,9, pp. 69,117; Plates 6,8)
E. dodoneata (Figs 8s,9b): usually smaller and paler; forewing relatively broader, discal spot oval and more prominent, area basad of median line often darker than rest of wing, especially when worn, small pale patch often present distad of discal spot; hindwing discal spot more conspicuous.

Melanic forms of *E. trisignaria, E. vulgata, E. tripunctaria, E. subfuscata, E. virgaureata* and *E. lariciata* are similar to f. *nigra* but forewing nervures not darker than rest of wing; hindwing termen not concave, apex not pointed.

LIFE HISTORY

OVUM Creamy white when first laid, later turning to pale brown; 0.7mm; long, ovoid. On unopened leaf-buds or newly emerging leaves of the foodplant in late April and May.

LARVA (Fig.33b) *First instar.* Long and thin; pale greyish green or pale brown; dorsal markings faint. *Second and third instars.* Pale brown to yellowish buff; dorsal markings a series of anterior-pointing chevrons, brown or olive green, largest on the five central segments, reducing in size towards both the head and tail; dorsal chevrons connected by interrupted subdorsal lines of similar colour, making each resemble a capital W. *Fourth instar.* Long and fairly robust, tapering slightly towards the head; 19mm when full-grown; head brownish yellow with two dark dorsal stripes; thoracic legs brownish yellow; integument covered with tiny tubercles and hairs, giving a rough texture; dorsal markings usually obliterate most of the yellow, orange or buff ground colour; segmental divisions reddish or pink. Immediately prior to pupation the larva turns pinkish white. A larva recorded as feeding on bird's-nest orchid (*Neottia nidus-avis*) near Canford, Dorset, in 1916 was stated to be uniform orange-citrine in colour (Curtis, 1934b).

Feeds on pedunculate oak (*Quercus robur*), sessile oak (*Q. petraea*) and hawthorn (*Crataegus monogyna*). It has a strong preference for flower-buds and catkins. J. N. Greatorex-Davies (pers.comm.) records larvae on southern beech (*Nothofagus* sp.). The identification of the resulting moths was confirmed by Riley. June and July.

PUPA Length 6.5–7.5mm; bright reddish brown with paler wings, cremaster deep red. In a tough cocoon amongst ground debris. July to April.

FLIGHT PERIOD AND HABITAT Univoltine, mainly in April and May, though it has been recorded from the last week of February to the middle of June with occasional later records. The peak of emergence is around the middle of April. It is found in the vicinity of the foodplant, and occasionally at rest on the trunks of oak and hawthorn during the day; it is often abundant in large oak-woods. It flies naturally at dusk and through the night, when it comes to light – often in large numbers.

DISTRIBUTION (Map 38) Widespread and locally abundant throughout the British Isles with the exception of the Outer Hebrides, Orkneys and Shetlands.

COLLECTING AND REARING Larvae can be beaten easily from oak and hawthorn but are sometimes heavily parasitized. Females oviposit readily in captivity, and this is perhaps a more efficient method of rearing this species. Easy to rear in the standard way.

Eupithecia dodoneata Guenée, 1857   Oak-tree Pug
= *dodonearia* Morris, 1861

Pl.3, figs 15,16; Pl.6, fig.13; Pl.7, fig.4; Pl.8, fig.14; Pl.D, fig.2

HISTORY First recorded in Britain by Doubleday (1856) from a specimen he sent to be examined and named by A. Guenée. Larva was first described by Crewe (1862a).

IMAGO *Characteristic features* (Figs 1m,8s,9b). Wingspan *c.*20mm; forewing broad, discal spot oval, postmedian fascia biangulate, postmedian line with small basally-pointing dark triangles, area basad of median line darker than rest of wing, especially in worn individuals, small pale patch often present distad of discal spot; hindwing discal spot minute, termen concave, apex pointed.

Genitalia. Figs 18c,29b ♂; 24i ♀

Variation. Specimens from Devon and Cornwall appear to have a slightly paler ground colour and hence more conspicuous markings than those from elsewhere. Apart from this there is little recorded variation other than general darkening of the forewings. Ab. *approximata* Lempke (1947;1951) has the two transverse lines bordering the median area of the forewing close together. Richardson & Mere (1958) refer to a specimen caught on the Isles of Scilly in 1958 which had conspicuous red markings. This specimen cannot be traced and no formal description was published. G. M. Haggett (pers.comm.) has reared forms from Norfolk which are suffused to varying degrees with black.

Similar species (See Figs 8,9, pp. 69,117; Plates 6,7,8)

E. abbreviata (Figs 8r,9a): usually larger; forewing relatively narrower and more elongate, discal spot linear and less prominent, area basad of median line not usually darker than rest of wing, pale patch distad of discal spot absent; hindwing discal spot less conspicuous or absent.

E. tenuiata, E. denotata, E. subfuscata and *E. virgaureata*: can resemble uncommon melanochroic form but hindwing termen not concave, apex not pointed.

LIFE HISTORY

OVUM Cream when laid, later turning to dark yellowish brown; 0.5mm; ovoid. On the flowers and leaves of the foodplant in May and early June.

LARVA (Fig.33c) *First instar.* Light yellowish brown, yellowish green or pale orange; dorsal and subdorsal lines fine, dull olive or brown. *Second instar.* As first instar, but subdorsal lines more conspicuous. *Third and fourth instars.* Stout, tapering slightly towards the head; 17mm when full grown; head and thoracic legs similar in colour to rest of body; integument covered with white tubercles and hairs giving a rough texture; ground colour varies from green or pale ochreous red to deep orange-brown with all intermediate shades; dorsal markings an anterior-pointing chevron on each of the five central segments, dark brown or deep olive; dorsal line similar in colour, continuous, broadening at posterior and anterior segments, replacing the arrowheads; subdorsal line similar in colour, linking dorsal chevrons; spiracular line either dull olive or creamy yellow; between subdorsal and spiracular lines a series of oblique pale yellow stripes; occasionally dorsal chevrons absent, leaving only a dark dorsal line; segmental divisions yellow.

The larva of this species resembles that of *E. abbreviata*, but in that of the latter the segmental divisions are yellowish pink or reddish, the dorsal line is less pronounced, especially in the posterior section, and the ground colour is usually much lighter.

Feeds several weeks later than those of *E. abbreviata* on the flowers and leaves of hawthorn (*Crataegus monogyna*), pedunculate oak (*Quercus robur*) and holm oak (*Q. ilex*). Haggett (1992) states that in captivity the calyces of hawthorn fruits are eaten in preference to the leaves. Late June to early August.

PUPA Length 5.5–5.6mm; thorax and abdomen brownish or dark red; wings paler. In a cocoon in soil or under the bark of the tree on which the larva fed. August to early May.

FLIGHT PERIOD AND HABITAT Univoltine, from mid-April to the third week of June with a peak during the third week of May. It inhabits woodland and scrubland where one or more of the larval foodplants is established. Flies from early dusk and throughout the night, when both sexes come freely to light. Adults, if disturbed from trees during the day, fly directly to the ground and are then very difficult to find (Tutt, 1906d).

DISTRIBUTION (Map 39) Widespread and locally common in England and Wales as far north as Cheshire, west Lancashire and Yorkshire. Absent from Scotland. Locally distributed in Ireland and the Channel Islands.

COLLECTING AND REARING Larvae can be beaten easily from hawthorns rich in fruits but are sometimes heavily parasitized. Captive females oviposit freely and this is probably a more efficient means of obtaining stock. Straightforward to rear in the usual way.

Eupithecia massiliata Millière & Dardoin, 1866   Epping Pug
= *peyerimhoffata* Millière, 1870

Pl.3, fig.25

HISTORY The first specimen found in the British Isles was a female caught at mercury vapour light by T. Green at Epping, Essex (V.C. 18) on 2.iv.2002. The second, a male, was caught in an outbuilding at the same locality on 13.iv.2002.

The specimens were sent to B. Goodey for determination, and photographs of the moths and of their dissected genitalia were forwarded to V. Mironov in Russia who identified them as *Eupithecia massiliata* Millière & Dardoin. During April 2003, despite continuous monitoring of the area, no further specimens were found (Goodey, 2003).

The origin of these moths is the subject of speculation. *E. massiliata* occurs in Spain, Portugal and southern France; the British Isles are, therefore, a great distance from its natural range. The finder occasionally imports unfumigated cork from Portugal, which frequently hosts live ants and other insects. As the larvae of *E. massiliata* feed on several species of oak, including cork oak (*Quercus suber*), there is strong circumstantial evidence of accidental importation by this means. Goodey considers that the source of the Epping pair was Portugal and their presence in Britain fortuitous. The species' apparent need for a Mediterranean climate makes it unlikely that *E. massiliata* will colonize south-east England.

IMAGO *Characteristic features.* Wingspan *c.* 19mm; forewing ample, costa arched, apex pointed, discal spot oval and conspicuous, sinuate subterminal line pale, ending in a conspicuous pale tornal spot, postmedian line with a series of diffuse dark arrowhead-shaped marks, termen with a series of black interneural lines; hindwing termen concave with a series of black interneural lines, discal spot inconspicuous or absent, subterminal line sinuate and pale with pale tornal spot.

Genitalia. Figured by Goodey (2003).

Similar species
Flight period, size and the concave termen of the hindwing make confusion most likely with *E. dodoneata* (see Fig.1m). However, in *E. massiliata* the forewing discal spot is larger and more elongate and the pale area distad of the discal spot is absent. *E. dodoneata* lacks the black interneural lines of both fore- and hindwings.

LIFE HISTORY

OVUM Undescribed.

LARVA Yellowish, with open, chevron marks dorsally from middle segments towards the head (Spuler, 1900). Feeds during May and June on flowers and leaves of various species of oak including cork oak (*Quercus suber*), evergreen oak (*Q. ilex*) and *Q. coccifera* (Mironov, 2003).

PUPA Undescribed. Overwinters in this stage, often several times, emerging March or April (Culot, 1919).

FLIGHT PERIOD AND HABITAT March and to June in the vicinity of oaks.

Eupithecia pusillata ([Denis & Schiffermüller], 1775)

subsp. *pusillata* ([Denis & Schiffermüller], 1775)   Juniper Pug
= *laevigata* sensu Haworth, 1809
= *sobrinata* (Hübner, 1817)
= *sobrinaria* Boisduval, 1840
= *scotica* Dietze, 1910

subsp. *anglicata* Herrich-Schäffer, 1863   Kentish Tamarisk Pug
= *ultimaria* sensu Westwood, 1854
= *stevensata* Webb, 1896

Pl.3, figs 19–22; Pl.5, figs 14,15; Pl.6, fig.4; Pl.8, fig.15; Pl.D, fig.3

HISTORY First recorded in Britain by Haworth (1809) as *Phalaena laevigata*. The larva was first described by Crewe (1859k).

Subspecies *anglicata* was first recorded by Westwood (1851) as *ultimaria* from specimens caught at Dover by S. Stevens.

IMAGO *Characteristic features* (Figs 6f,8p,10). Wingspan c.20mm; forewing discal spot conspicuous, elongate, sometimes edged distally with white, postmedian fascia and line biangulate, postmedian line with small basally-pointing dark triangles, those near dorsum elongate and often prominent, median line usually conspicuous between costa and discal spot, often extends across whole wing, in worn individuals remnants of discal spot and median line form a basally-pointing T (Fig.10); hindwing discal spot absent or obsolete. Dark specimens retain vestiges of normal markings.

Genitalia Figs 12i,18e,29c ♂; 25d ♀

Variation. Subspecies *anglicata* Herrich-Schäffer (1863) (= *stevensata* Webb), the Kentish Tamarisk Pug (Pl.2, fig.14; Pl.5, fig.15), is a very pale taxon known formerly from the maritime chalk-cliffs around Dover, Kent (Westwood, 1851; Webb 1881; Stevens, 1882) and the Isle of Wight (Mutch, 1905), but it has not been recorded since 1915 and is now presumably extinct. However, a specimen of *E. pusillata* in the national collection, from Aviemore, Inverness-shire, is very similar to this subspecies. Herrich-Schäffer described this taxon in 1861 but did not actually name it until two years later (Herrich-Schäffer, 1861;1863). A form erroneously accorded subspecific status is f. *scotica* Dietze (1910;1913) (Pl.3, fig.21; Pl.5, fig.14). Specimens of this form have extensive white scaling, particularly in the median area of the forewing, which gives a paler, more mottled

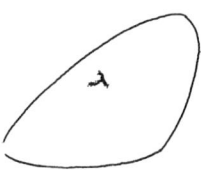

Text figure 10

Worn individual of *E. pusillata* showing characteristic T marking in the forewing discal area

appearance. From individuals recorded in the RIS light-traps, it appears that this and the nominotypical form are sympatric in some localities. Pale forms reach their extreme in ab. *achromata* Dannehl (1927), which is whitish grey with the normal markings considerably reduced or obsolete. However, the present authors are not aware of this aberration having been recorded in Britain. Differences in colour-intensity of the median area of the forewing are sometimes frequent. In f. *confluens* (Dietze 1910;1913) (= ab. *conjuncta* Dietze, 1910;1913), this area is extremely dark, and in f. *nigrofasciata* Dietz (1910;1913) (Pl.3, fig.20; Pl.6, fig.4) it is entirely black. Similar to these is ab. *rittichi* Dioszeghy (1929–30), which has the same basic form of variation but with a more reddish brown appearance with reddish brown median fascia and dark brown costal and postmedian maculation. A form with the median area paler and more sharply marked than the basal and distal areas is referrable to f. *expressaria* Herrich-Schäffer (1848). Small, dark and weakly marked individuals sometimes occur in large isolated populations. These are described as ab. [= f.] *luneburgensis* by Dietze (1910;1913), and are possibly the result of underfed larvae. F. *impuncta* Lempke (1947;1951) has the discal spot of the forewing absent. More recently, a dark form (Pl.3, fig.19; Pl.8, fig.15) was discovered at Teesdale, Co. Durham, which has very dark, brownish black forewing, a small conspicuous white tornal spot and intense black remnants of the usual markings (Agassiz et al., 1981). Similar specimens were also bred from Bedfordshire larvae in 1987 (Riley, pers.obs.).

Similar species
Melanic forms of *E. trisignaria*, *E. vulgata*, *E. tripunctaria*, *E. subfuscata*, *E. virgaureata* and *E. lariciata* are similar to dark forms of *E. pusillata* but forewings without vestiges of normal fasciae and lines.

LIFE HISTORY

OVUM Glossy yellow, later turning to dull leaden blue; 0.6mm; ovoid. Laid in small batches between the needles of the foodplant from July to early September, remaining in this stage until the following March or April.

LARVA (Figs 33d,e) *First instar*. Pale yellow with no discernible markings. *Second and subsequent instars*. Slender, tapering slighting towards the head; 17mm when full-grown; head and thoracic legs brown or green; ground colour variable, from pale buff to deep reddish brown and from pale yellowish to deep grass-green; dorsal markings also very variable, e.g. blotches of various shapes including chevrons, rectangles, a single, fine, dark green or brown dorsal line, or sometimes entirely absent; spiracular line undulate, conspicuous, pale yellow or cream; ventral line whitish; spiracular and ventral lines are consistently present.

Feeds on juniper (*Juniperus communis*) and cultivated varieties of *Juniperus*, *Thuja* and *Chamaecyparis* (Skinner, 1984). Subsp. *anglicata* Herrich-Schäffer was stated by Newman & Leeds (1913) and others to feed on tamarisk (*Tamarix gallica*), but there appears to be no evidence to support this. Late March to early June.

PUPA Length 6.5–7.5mm; thorax and abdomen usually greenish brown; wings bright green. In a slight cocoon on the ground or amongst the needles of the foodplant; from four to six weeks between early June and late July.

FLIGHT PERIOD AND HABITAT Univoltine, from the end of June to late September with a peak around the third week of July. This species can be abundant where the foodplant is common. It is also able to maintain huge colonies on apparently isolated specimens of juniper, despite often heavy mortality through parasitism. Sometimes common in gardens where cultivated conifers are grown. Stated to fly around the foodplant during hot sunny afternoons (Agassiz et al., 1981). Flies strongly at dusk and through the night when it is attracted to light.

DISTRIBUTION (Map 40) Locally common throughout England, Wales and Scotland as far north as the Orkneys. Not recorded from the Isle of Man, the Outer Hebrides or the Shetlands. Locally common in Ireland; the Channel Islands.

COLLECTING AND REARING Larvae of several instars can sometimes be beaten in huge numbers from juniper bushes but are often heavily parasitized. A female can be induced to oviposit in captivity but this species does not usually do so freely (see Crewe, 1861b). Fresh sprigs of the foodplant are essential as a stimulus, and eggs will normally only be laid thereon. The ova should be kept cold (or outdoors) over the winter until they are due to hatch. Larvae are easy to rear in captivity but they appear to be very touch-sensitive, biting at anything which strays too close. Such bites usually prove fatal to other larvae, so even slight overcrowding should be avoided. It is perhaps safest to rear them in individual boxes. Juniper quickly desiccates once cut and, as the larvae eat only fresh young needles, a constant supply of the foodplant is needed close to hand.

Eupithecia phoeniceata (Rambur, 1834)   Cypress Pug
Pl.3, fig.17; Pl.4, fig.13; Pl.D, fig.4

HISTORY First recorded in the British Isles by Leech (1966) from specimens caught at Brelade's Bay, Jersey, in September 1956. A further specimen was caught in Guernsey by C. J. Shayer on 21 August 1958 but misidentified as *E. oxycedrata* Rambur. The first mainland capture was made at Penzance, Cornwall, on 11 September 1959 and recorded by de Worms & Messenger (1960). The early stages were first described by Haggett (1968c).

IMAGO *Characteristic features*. Wingspan *c.*20mm; forewing clearly and characteristically marked.
  Genitalia. Figs 12k,18b,29d ♂; 25b ♀
  Variation. Varies very little. Two specimens of a form with a dark median area to the forewing have been recorded: the first, caught at Brockenhurst, Hampshire, on 24 September 1989 by H. G. M. Middleton, is figured by BENHS (1990); the second was caught at Trinity, Jersey, on 21 August 1990.

Similar species
None.

LIFE HISTORY

OVUM Yellow when first laid, later peachy pink; 0.6mm; ovoid. On the leaf-tips of the foodplant, usually during August and September.

LARVA (Fig.33f) *First instar.* Pale purplish brown; segmental divisions broad, pale green. *Subsequent instars.* Elongate, not tapering; 17mm when full grown; head largely similar in colour to body; thoracic legs pale grey with red tips; posterior prolegs dark green or greyish brown; anterior pair with red tips; integument smooth, glossy, deeply indented at segmental divisions; usually bright green; dorsal markings brown and situated at the segmental divisions, giving the larva an alternate green and brown banded appearance; some forms with white segmental divisions and a very dark brown dorsal line with small black spots on the central segments; a brown form has ground colour light purplish; dorsal markings horseshoe-shaped, dark brown. Both green and brown forms sometimes with dark brown and white lateral spots; in all forms anal flap has a triangular red marking with a dark centre. Highly procryptic; the green form closely resembles the living leaves of the foodplant and the brown form those which have died.

Feeds on the young leaf-tips of Monterey Cypress (*Cupressus macrocarpa*), Leyland cypress (×*Cupressocyparis leylandii*), Lawson's cypress (*Chamaecyparis lawsoniana*) and Phoenician juniper (*Juniperus phoenicea*). September to March.

PUPA Length 7–8mm; thorax and abdomen yellowish brown; wings green. In captivity, pupation takes place in a fairly tough cocoon in ground debris or amongst the leaves of the foodplant. Usually from late February to August, though some larvae pupate before the end of the winter.

FLIGHT PERIOD AND HABITAT Univoltine, usually from August to early October with a peak during the first week of September. However, Pickering (1980) states that adults had been caught as early as May at a locality in Sussex. Inhabits gardens, parkland, plantations or wherever the foodplant grows, though it is presently restricted to southern England and is most common on the south coast. During the day it can be disturbed easily from the foodplant, and at night both sexes come readily to light.

DISTRIBUTION (Map 41) Locally common in England south of a line from Somerset to Essex, with a single record from Warwickshire. In Wales, it has been recorded only from Glamorgan. In Ireland it has been recorded from the east of Co. Cork. Absent from Scotland.

COLLECTING AND REARING Since its relatively recent discovery in Britain, this species has quickly extended its range, possibly assisted by the horticultural popularity of its larval foodplant. Captive females oviposit freely in captivity if supplied with a sprig of cypress on which to lay. Larvae are easy to obtain by beating the foodplant, though mature trees are preferred. The larvae will eat only the young growth at the tips of the leaves and quickly exhaust their supply, so fresh food must be given frequently.

Eupithecia ultimaria Boisduval, 1840   Channel Islands Pug
Pl.3, fig. 18; Pl.7, fig.2; Pl.D, fig.5

HISTORY The first record of this species in the British Isles was of a single male caught in the RIS light-trap operating at St. Martin's, Guernsey, on 20/21 August (Riley, 1985a). Larvae were subsequently found in Guernsey in 1985 on tamarisk (*Tamarix gallica*) (Peet, 1988). The first mainland specimen was caught by A. King in a garden at Bishop's Stortford, Hertfordshire, in June 1989 and was identified by A. M. Riley (Riley, 1991). Further single adults were caught by J. R. Langmaid in June and July 1995 at Southsea, Hampshire, and larvae were found in August 1995 on tamarisk at nearby Hayling Island by Langmaid and S. M. Palmer (Langmaid, 1996). Previously unidentified specimens from West Sussex, caught at Walberton in 1990 and 1995 and Littlehampton in 1992, were found in the collection of J. Radford. The species is now known to be established in the Isle of Wight (Knill-Jones, 1997), and at several localities on the coasts of Hampshire and West Sussex.

IMAGO *Characteristic features* (Fig.1l) Wingspan *c.*18mm; forewing narrow, costa and termen rounded, postmedian fascia biangulate and prominent, discal spot characteristically linear, fringes of both wings chequered; hindwing discal spot linear.

Genitalia. Figs 17b,28n ♂; 24a

Similar species (See Fig. 1, p. 35; Plate 7)
E. tenuiata (Fig.1b): forewing broader, postmedian fascia much less conspicuous, discal spot oval; hindwing discal spot oval.

E. distinctaria (Fig.1h): forewing broader, postmedian fascia geniculate and less conspicuous, discal spot elongate, not linear; hindwing discal spot oval.

E. indigata (Fig.1i): forewing costa straighter, apex more pointed, generally ochreous-tinged, postmedian fascia inconspicuous or absent, discal spot elongate or oval, not linear; hindwing discal spot oval.

LIFE HISTORY (details taken from continental literature unless otherwise stated)
OVUM Undescribed.

LARVA Full-grown, 20mm, slender; colour variable, usually green but sometimes whitish, yellowish or brownish; lateral markings sometimes absent but usually a series of oblique yellowish, brown or purplish dashes, one per segment; dorsal markings sometimes absent but usually W-shaped, purplish; head, prothoracic and anal plates concolorous with the body (Langmaid, 1996).

Feeds on tamarisk (*Tamarix gallica*). August and September
PUPA Undescribed. Overwinters in this stage; September to May.
FLIGHT PERIOD AND HABITAT Bivoltine. May and June, with a partial second emergence in late August and September. Found in coastal localities and possibly in gardens inland where tamarisk is grown.

DISTRIBUTION (Map 42) Presently known only from Hertfordshire, East and West Sussex, south-east Hampshire, the Isle of Wight, and Guernsey, Channel Islands.

Eupithecia lariciata (Freyer, 1842) Larch Pug
Pl.3, figs 23,24; Pl.6, fig.8; Pl.8, fig.4; Pl.D, fig.6

HISTORY First caught in Britain by T. Eedle of Hackney in 1862 (no locality cited) and recorded by Doubleday (1864). Larvae were first described by Crewe (1864b;1865a).

IMAGO *Characteristic features* (Figs 5k,8d) Wingspan *c.*22mm; metathorax with white spot; forewing apex pointed, distinctly light and dark fasciae giving overall striped appearance, all fasciae and striae strongly geniculate, discal spot conspicuous; hindwing discal spot not usually prominent. Melanic form with distinct forewing discal spot.

Genitalia. Figs 18d,29e ♂; 25c ♀

Variation. Dietze (1910;1913) and others named many aberrations of this species which are summarized below:

Forewing strongly darkened with iron grey: ab. *ferrearia* Nitsche (Goodson, unpubl.)

Forewing darkened basad of discal spot: ab. *basifasciata* Dietze (1910;1913) (=*bifasciata* Prout, 1914).

Median area of forewing darker: ab. *mediofasciata* Dietze (1913).

Forewing with distal half of median area broadly paler: ab. *mediopallens* Dietze (1910;1913).

Median area of forewing brighter; similar to that of *E. egenaria*: ab. *strigata* Dietze (1910;1913).

Markings obsolete, with only the discal spot remaining: ab. *uniformis* Dietze (1913).

Forewing median and postmedian lines closer together than in the type and united at costa; posterior part of median area suffused heavily with dark scales: ab. *virgata* Cockayne (1953).

Much more copiously marked than nominotypical form: ab. *luxuriosa* Dietze (1913).

By far the most common form is f. *nigra* Prout (1915) (Pl.3, fig.24; Pl.6, fig.8). This is uniformly sooty black with intensely black discal spots. It occurs frequently in populations throughout Britain though RIS light-trap records suggest more commonly in Wales and northern England.

Similar species (See Figs 5,8, pp. 61,69; Plates 6,8)

E. egenaria (Figs 5a,8 l): metathorax without white spot; forewing with characteristic double ante- and postmedian lines, overall appearance not distinctly striped.

E. trisignaria f. *angelicata* (Fig.8k): similar to f. *nigra* but forewing broader with more rounded apex. Examination of the male abdominal plate or the genitalia is recommended.

E. intricata (Figs 5d,8b): metathorax without white spot; abdomen with dark sub-basal band, unicolourous, not paler ventrally; forewing costa straighter, dark median ante- and postmedian lines more prominent, especially at costa.

E. satyrata (Figs 5g,8j): metathorax without white spot; forewing narrower, more uniform, not striped in overall appearance, discal spot smaller; hindwing discal spot smaller or absent.

E. vulgata (Figs 5b,8e): some pale forms similar but methathorax without white spot; forewing usually tinged ochreous, ante- and median striae and fasciae less prominent, postmedian fascia biangulate, discal spot smaller or absent. Melanic f. *atropicta* without conspicuous forewing discal spot.

E. tripunctaria (Figs 5f,8h): fore- and hindwings with conspicuous and prominent white tornal spot; forewing without overall striped appearance. Examination of the male abdominal plate or the genitalia is necessary to separate the melanic forms.

E. denotata subsp. *jasioneata* (Figs 5i,8f): metathorax without white spot; forewing postmedian fascia biangulate, antemedian fascia curved.

E. subfuscata (Figs 5l,8g): metathorax without white spot; forewing apex less pointed, tornal spot more conspicuous, pale fasciae less conspicuous, giving more uniform appearance. Examination of the male abdominal plate or the genitalia is necessary to separate the melanic forms.

E. virgaureata (Figs 5l,8g): forewing apex more rounded, tornal spot more conspicuous, pale fasciae less conspicuous giving more uniform overall appearance. Examination of the male abdominal plate or the genitalia is necessary to separate the melanic forms.

E. pusillata (Fig.8p): dark forms similar to f. *nigra* but forewing narrower, retaining vestiges of normal fasciae and lines.

E. abbreviata f. *nigra* (Fig.8r): similar to f. *nigra* but forewing more elongate, nervures intensely black; hindwing termen concave, apex pointed.

LIFE HISTORY

OVUM Greenish white, later turning yellow; 0.7mm; ovoid. Laid on the needles of the foodplant during June.

LARVA (Fig.33g) *First instar.* Integument translucent, pale yellowish green or brownish buff; dorsal line darker. *Second instar.* Pale green or brownish buff; dorsal line darker green or brown; subdorsal and spiracular lines pale yellow; a red triangular mark at the posterior end of the dorsal line; segmental divisions yellow. *Third and fourth instars.* Long and slender, tapering towards the head; 18mm when full grown; head and thoracic legs pale green or brown; integument smooth, grass-green, ventrally greenish white; dorsal line green, flanked by fine cream lines; subdorsal line dark brown; spiracular line yellowish white.

Alternative form light brown, ventrally yellowish or brownish white; dorsal and subdorsal lines dark brown; spiracular line greenish yellow. In both forms posterior tip of dorsal line sometimes reddish.

Feeds on the needles of European larch (*Larix decidua*). G. M. Haggett (pers.comm.) states that larvae have also been found on Japanese larch (*L. kaempferi*) and hybrid larch (*L.* ×*eurolepis* (= *L.* ×*marschlinsii*). Late June to September.

PUPA Length 7.5–8.5mm; abdomen and thorax greenish yellow or brown; wings sometimes dark green; segmental divisions reddish; cremaster bright pale chestnut. In a cocoon in soil. From late August to May.

FLIGHT PERIOD AND HABITAT Usually from late May to the end of June but specimens have been recorded regularly in RIS light-traps from the beginning of May to the end of September with occasional individuals even later. These records suggest a partial second emergence (Riley, 1986b;1987). Larvae resulting from this second emergence have not so far been found. *E. lariciata* is most commonly found in larch plantations but is also occasionally associated with single larch trees in mature gardens or parks. It should therefore be expected wherever the foodplant grows. Although it flies naturally at night and both sexes are fairly common at light, during the day it can be disturbed into flight by jarring the larch boughs with a stick.

DISTRIBUTION (Map 43) Locally common throughout the British Isles but not recorded from Orkney.

COLLECTING AND REARING Larvae can be collected easily by beating larch but are sometimes heavily parasitized. Captive females oviposit readily. An easy species to rear in the normal way but, as cut larch desiccates quickly, a nearby supply of the foodplant is advantageous.

Eupithecia tantillaria Boisduval, 1840 Dwarf Pug
= *subumbrata* sensu Hübner, 1799
= *pusillata* sensu auctt.
= *piceata* Prout, 1914

Pl.3, fig.26; Pl.5, fig.11; Pl.D, fig.7

HISTORY This species may be that included by Stephens (1829) as '*Ge. pusillata* ... Fabr.; ... Hüb...; the small grey Pug'. *Ge. pusillata* appears again in his later work (1831) but the larva is stated to feed on 'the common juniper and birch' and the adult to emerge in June, which conflicts with the known life-history of *E. tantillaria*. Stephens' records are therefore very doubtful and the first definite British record is probably that of Doubleday (1856) who records specimens of this species caught in Devon by S. Wood as *E. pusillata* Hübner. The larva was first described by Crewe (1861e;1862a).

IMAGO *Characteristic features.* Wingspan *c.*20mm; abdomen with characteristic pale basal and dark sub-basal band (Fig.4b); forewing costa arched, fasciae prominent and curved, discal spot oval and prominent; hindwing discal spot conspicuous, postmedian line geniculate.

Genitalia. Figs 12 l,19a,29f ♂; 25a ♀

Variation. Variation is unusual, consisting mainly of differences in the intensity of the ground colour which result in greater or lesser enhancement of the markings. Two pale aberrations have been described: ab. *pallida* Lempke (1951) lacks the usual dark margins to fore- and hindwings and has whitish grey forewing with sharp markings. Ab. *piceata* Prout (1915) has a lighter ground colour, sometimes with a greenish or reddish tinge. Ab. *nigricata* Vorbrodt (Goodson, unpubl.) is blackish grey and sharply marked and ab. *mediopallens* Silbernagel (1943) has the median area of the forewing paler, contrasting more with the darker borders.

Similar species (See Fig. 4, p. 49; Plate 5)

E. irriguata (Fig.4c): forewing costal markings forming complete bar, striae in median area inconspicuous or absent, sub-basal fascia conspicuous; hindwing postmedian line less prominent; abdomen without pale basal and dark sub-basal band.

LIFE HISTORY

OVUM Glossy cream, later turning orange; 0.6mm; elongate ovoid. Laid on the needles of the foodplant during May and June.

LARVA (Fig.33h) *First and second instar.* Pale reddish buff or green; dorsal line darker, broadening slightly at the centre of each segment. *Third instar.* Orange, brick-red or greyish green; dorsal line dark brown or blackish; subdorsal line similarly-coloured, fine; spiracular line pale yellow; segmental divisions red. *Fourth instar.* Long, slender and tapering towards the head; 16mm when full grown; integument smooth, velvety; colour and markings similar to those of previous instars but darker and bolder.

Recorded on the needles mainly of Norway spruce (*Picea abies*), but also of European larch (*Larix decidua*), Douglas fir (*Pseudotsuga menziesii*), giant fir (*Abies grandis*), Scots pine (*Pinus sylvestris*), Sitka spruce (*Picea sitchensis*), western red-cedar (*Thuja plicata*), western hemlock-spruce (*Tsuga heterophylla*), Austrian pine (*Pinus nigra*) and also Lawson's cypress (*Chamaecyparis lawsoniana*) (Hatcher, 1989). July and August.

PUPA Length 6.5–7.5mm; pale brownish yellow; wings bordered dorsally with two black spots and ventrally with a black line (Crewe, 1861e); a jet-black prothoracic belt which is divided by a narrow streak of brown (G. M. Haggett, pers.comm.). In a cocoon spun between foliage (G. M. Haggett, pers.comm.) or in soil. Late August to May.

FLIGHT PERIOD AND HABITAT Univoltine, from the end of April to early July, with a peak during the first week of June. Commonest in spruce plantations. During the day it can be disturbed from spruce trees and, at night, both sexes come freely to light.

DISTRIBUTION (Map 44) Locally common throughout the British Isles though apparently absent from the Inner and Outer Hebrides and the northern islands of Scotland.

COLLECTING AND REARING The larvae can be obtained easily by beating spruce, and captive females oviposit readily. A straightforward species to rear, though larch should be avoided as a cut foodplant as it is prone to desiccate quickly. Larvae should not be crowded as they can be cannibalistic.

Chloroclystis v-ata (Haworth, 1809) V-Pug
= *coronata* (Hübner, 1813)
= *coronaria* (Doubleday, 1849)

Pl.3, figs 27,28; Pl.4, fig.15; Pl.5, fig.5; Pl.D, figs 8,9

HISTORY First recorded in Britain by Haworth (1809). The larva was originally described by Crewe (1861k) under the name *Eupithecia coronata*.

IMAGO *Characteristic features*. Wingspan *c*.20mm; forewing ground colour green or yellowish green, costal half of postmedian line forming bold characteristic black dentate marking.

Genitalia. Figs 12m,19c,29g ♂; 25e ♀

Variation. The markings vary very little. Ab. *bistrigata* (Lempke, 1951) has the central area of the forewing bordered on both sides by a complete dark line. The ground colour varies in shade from yellowish to bluish green.

Similar species
None.

LIFE HISTORY

OVUM Glossy yellow; 0.5mm; ovoid. Laid on the flowers of the foodplant, in southern localities from late April to August in one or two generations; in northern Britain usually only in June and July.

LARVA (Fig.33i) *First and second instars*. Creamy white or very pale green; dorsal and subdorsal lines fine, pale green or brown. *Third and fourth instars*. Short and stout, 10mm when full grown; head and thoracic legs similar colour to body; integument smooth with sparse, fine whitish hairs, deep, clear yellowish green, bright yellow or pink, varying to match the shade of the flowers on which it feeds; dorsal markings on five central segments a series of three anterior pointing dark brownish red, pink or green elongated triangles joined posteriorly by a white transverse line extending laterally as fine stripes or flecks; segmental divisions pink; subdorsal line fine, similar in colour to dorsal markings. Wilson (1880) states that occasionally the dorsal markings are absent and also describes a form in which the lateral and ventral sufaces are suffused with red. When feeding, the larva often adopts a curled posture in which the thoracic legs and anal claspers are almost touching.

Widely polyphagous on the flowers of herbaceous plants and trees including goldenrod (*Solidago virgaurea*), yarrow (*Achillea millefolium*), mugwort (*Artemisia vulgaris*), hemp-agrimony (*Eupatorium cannabinum*), angelica (*Angelica sylvestris*), traveller's-joy (*Clematis vitalba*), bramble (*Rubus fruticosus*), apple (*Malus domestica*), pear (*Pyrus communis*) and hawthorn (*Crataegus monogyna*). In southern Britain from late May to October in two protracted generations; in the north usually from July to September.

PUPA Length 5–6mm; uniform golden yellowish brown; thorax and wings with fine black spots; wings noticeably ribbed. In northern Britain where the species is univoltine, from September to June; in bivoltine populations, June and July (often amongst spun leaves of the foodplant), and again from September to April, usually in a cocoon in soil. The adult sometimes overwinters apparently fully developed inside the pupal skin (Riley, 1990a).

FLIGHT PERIOD AND HABITAT Univoltine in the north, in June and July; bivoltine in the south, from late April to June and again in July and August. These flight periods can sometimes be difficult to separate as both are somewhat protracted and vary according to the weather. Further detailed study is desirable to clarify the phenology of this species. It is very generally distributed and occurs in most lowland habitats. At dusk it can be found flying around traveller's-joy, and at night both sexes come to light in small numbers.

DISTRIBUTION (Map 45) Widespread and common throughout England and Wales. In Scotland it appears to be restricted to the lowland areas of the south. Widely distributed in Ireland. Common in the Channel Islands.

COLLECTING AND REARING Larvae are beaten easily from flowers of the foodplant. It is usually very common on *Clematis* but is often heavily parasitized. Both sexes are frequent at light and females oviposit readily. A straightforward species to rear in captivity and larvae will change from one foodplant to another without difficulty. They are not usually cannibalistic but Jager (1887) states that half-a-dozen he reared ate a large noctuid larva. Emergence is difficult to predict. Agassiz *et. al.* (1981) cite cases of second generation adults emerging in November and December whilst other first generation larvae did not emerge until the following spring. Prout (1908) records a third generation in September reared on bramble.

Pasiphila chloerata (Mabille, 1870) — Sloe Pug
Pl.3, fig.29; Pl.5, fig.3; Pl.D, fig.11

HISTORY A relative newcomer to the British list, the first published record of this species was that by Pelham-Clinton (1972) based on larvae found by him at Effingham, Surrey, in April 1971. However, it was later discovered that several specimens had been reared in 1944 from larvae collected by C. G. M. de Worms at Salisbury, Wiltshire, which were erroneously identified as *Chloroclystis* (now *Pasiphila*) *rectangulata* (L.) (de Worms, 1978). The life history was first described by Pelham-Clinton (1972).

IMAGO *Characteristic features* (Fig.11a). Wingspan *c.*20mm; forewing tinged green or bronze, postmedian fascia curved, postmedian line not dentate near costa; hindwing postmedian fascia curved or smoothly geniculate, underside with postmedian fascia absent or obsolete.

Genitalia (Figs 19d,29h ♂; 26c ♀). The male abdominal plate is very similar to those of *P. rectangulata* and *P. debiliata* and takes the form of two fine, lightly sclerotized prongs. In *P. rectangulata* and *P. debiliata* these are usually long and crossed; those of *P. debiliata* are also less robust in appearance. In *P. chloerata* the prongs are relatively shorter and do not usually cross. Although these characters are useful guides, it must be stressed they are not completely reliable and it is therefore unwise to use the male abdominal plates of these three species as the sole means of identification.

Variation. The melanic forms so often found in its close relative *P. rectangulata* do not occur in this species, though Dietze (1910;1913) records ab. *hadenata* which is much darker than the type and ab. *nigrofasciata* in which the median area of the wings is darkened.

Similar species (Fig. 11; see also Plate 5)
P. rectangulata (Figs 11d,e,f): forewing postmedian line dentate near costa, underside postmedian line angulate or strongly geniculate; hindwing underside postmedian line angulate.

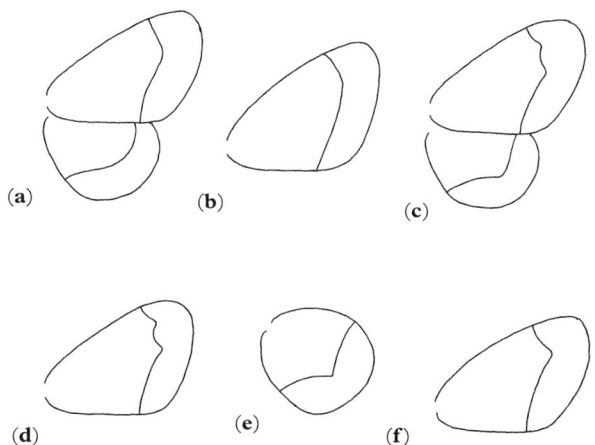

Text figure 11 Wing shapes and patterns (Similar species)

(**a**) *Pasiphila chloerata* (p. 132) (**b**) *P. debiliata*, forewing underside (p. 137)
(**c**) *P. debiliata* (**d**) *P. rectangulata*, forewing upperside (p. 135)
(**e**) *P. rectangulata*, hindwing underside (**f**) *P. rectangulata*, forewing underside

P. debiliata (Figs 11b,c): ground colour very pale; forewing postmedian line dentate near costa; hindwing postmedian line angulate, particularly conspicuous on underside.

LIFE HISTORY

OVUM White when first laid, later turning to pale red then, just prior to emergence, black; 0.6mm, ovoid. In crevices in the bark of the foodplant during June, overwintering in this stage and hatching the following March.

LARVA (Fig.33j) *First and second instars.* Pure white or greenish white; dorsal line red, very faint. *Third and fourth instars.* Short and grub like, tapering sharply to head and tail; 10mm when full-grown; head pale brown or deep yellow; thoracic legs white; thoracic plate brown; integument smooth but wrinkled into folds along the flanks, white or greenish white; dorsal markings a series of carmine-red thoracic spots, decreasing in size from the head to nearly half way along the body, continuing as a thin red line on the posterior segments; subdorsal lines faint, grey. In early instars the larvae bore into unopened flower-buds. Later they often spin the petals together, feeding for preference on the unripe seeds, often leaving the empty calyx and stamens.

Feeds on Blackthorn (*Prunus spinosa*). March to early May.

PUPA Length 6–7mm; at first whitish green turning to deep honey-brown. It occasionally pupates on the foodplant amongst the opening leaves but usually in a fairly tough cocoon amongst ground litter. It remains in this stage for only two or three weeks during late April or May.

FLIGHT PERIOD AND HABITAT Univoltine, from late May to mid-July with a peak during the last week of June. A single, very late individual was caught in an RIS light trap at Harpenden, Hertfordshire, on 29 July 1984 (Riley, 1985c). Rarely seen during the daytime, at night it comes readily to light. It inhabits blackthorn thickets and hedges and scrubland where blackthorn grows. Although Haggett (1980b) states that isolated and small clumps of blackthorn are rarely used, the present authors find that this is not the case.

DISTRIBUTION (Map 46) Locally common in most of England as far north as Yorkshire and Westmorland, it appears absent from much of the South-West. In Wales, it has been recorded only from Glamorgan in the south and Denbighshire and Flintshire in the north. It has not been found in Scotland or Ireland.

COLLECTING AND REARING The females are reluctant to lay eggs in captivity, but wild larvae are easily beaten from blackthorn. There is no distinction in the age or size of bush or tree preferred by the moth for oviposition. Larvae can be found on trees of all ages, either in hedgerows or individual specimens. The timing of blackthorn flowering is geographically variable and depends greatly on weather conditions. As the flowering period is so short, a constant check must be kept on the development of the blossom. The debris which falls on to the beating tray or sheet should be retained and either examined in detail at home or kept in a box or cage for up to six weeks. The larvae often cling to this debris and are extremely well camouflaged amongst the flowers and easily overlooked. They develop rapidly on blackthorn blossom but, because they do not eat the petals, they must be given what appears to be an excess of flowers.

Pasiphila rectangulata (Linnaeus, 1758)   Green Pug
= *viridulata* (Hufnagel, 1767)
= *nigrosericeata* (Haworth, 1809)
= *sericeata* (Haworth, 1809)
= *subaerata* (Hübner, 1817)
= *rectangularia* (Boisduval, 1840)

Pl.3, figs 31–34; Pl.5, figs 1,2; Pl.6, figs 10,11; Pl.D, figs 11,12

HISTORY First recorded in Britain by Haworth (1809) as three species: *Phalaena rectangulata* (Green Pug), *P. nigrosericeata* (Black Silk Pug) and *P. sericeata* (Satin Pug). The larva was first fully described by Crewe (1860h;1861k).

IMAGO *Characteristic features* (Figs 6h,8q,11d,e,f). Wingspan *c*.20mm; forewing green or greenish-tinged, postmedian line dentate near costa, underside postmedian line angulate or strongly geniculate; hindwing underside postmedian line angulate.
   Genitalia. Figs 19e,29i ♂; 26a ♀
   Variation. This species is extremely variable in colour and markings. Pale varieties, in which the green is replaced by ochre-yellow, are referrable to f. *ochrea* Derenne (1932), and those in which the green is replaced by grey are described by Lempke (1951) as ab. *grisescens*. The markings are much reduced in ab. *subaerata* Hübner (1813), leaving only the ante- and postmedian lines developed. These lines are highly strengthened in ab. *bistrigata* Dietze (1913). F. *joannisata* Culot (1919) is bright green with a dark median band. In ab. *effusa* Cockayne (1953) this band is narrowed, clearly bounded by the ante- and postmedian lines, and the discal spot is situated in the latter. The outer margins are suffused with darker scales. Several dark taxa are described and these are summarized as follows: black increased, particularly in the median area, f. *cydoniata* Borkhausen (1794); ground colour light brown instead of green, ab. *brunneata* Dannehl (1927); almost entirely black with remnants of green crosslines, f. *nigrosericeata* Haworth (1809) (Pl.3, fig.34; Pl.6, fig.11); completely black, f. *anthrax* Dietze (1910) (Pl.3, fig.33; Pl.6, fig.10). The latter two forms often outnumber the type in some British localities. Ab. *mediospolinata* Schwingenschuss (1954) has the postmedian line beginning considerably basad of its normal position on the costa and then acutely angled. Ab. *pilcheri* Haggett (1989) is an extraordinary aberration in which all the wings are pale dusky cream. The nervures and costa are paler; discal spots blackish and prominent; postmedian lines undulate and black; outer margins heavily darkened. Carrington (1885b) suggests that adults reared from crab apple are usually greener and more distinctly marked than those from cultivated trees.

Similar species (See Fig. 11, p. 133; Plate 5)
P. chloerata (Fig.11a): forewing postmedian line smoothly curved, not dentate near costa; hindwing postmedian fascia curved, underside postmedian line absent or obsolete.

P. debiliata (Fig.11b): ground colour much paler; forewing underside postmedian line curved.

LIFE HISTORY

OVUM Creamy white when first laid, later red and immediately prior to overwintering, black; 0.6mm, ovoid. Laid in cracks and crevices of the bark of the foodplant. Late June to April.

LARVA (Fig.33k) *First and second instars.* White or pale yellow; dorsal line faint, red, sometimes absent. *Third and fourth instars.* Short and stumpy, not tapering to either end; 12mm when full grown; head and thoracic legs brown; integument smooth, waxy and translucent, revealing the gut, covered with fine hairs, cream, pale yellow or pale green; dorsal line usually red, rarely green, running the whole length of the body, occasionally broken into a series of dashes, sometimes absent from the central segments.

Feeds on apple (*Malus domestica*), crab apple (*M. sylvestris*), pear (*Pyrus communis*), wild pear (*P. pyraster*), blackthorn (*Prunus spinosa*), wild cherry (*P. avium*) and possibly other fruit trees. First-instar larvae bore into flower-buds and later feed on the opened flowers, sometimes spinning the petals together. They make no attempt to eject their droppings, resulting in untidy accumulations of frass which indicate tenanted flowers. Late April and May.

PUPA Length 7–8mm; at first honey-brown; later thorax and abdomen golden brown; wings green. In a cocoon spun amongst ground litter or occasionally in flowers still attached to the foodplant. This stage lasts three or four weeks. Late May and June.

FLIGHT PERIOD AND HABITAT Univoltine, recorded from late May to the end of August. However, the main flight period is from June to early August with a peak during the first week of July. It is found in orchards, gardens, hedgerows, scrubland and woodland and sometimes swarms around apple trees at dusk. It is often readily attracted to light.

DISTRIBUTION (Map 47) Locally common throughout the British Isles but not recorded from the Outer Hebrides, and the northern isles.

COLLECTING AND REARING It is very difficult to induce oviposition in captive females, though they may lay if supplied with pieces of bark on which eggs can be oviposited, as in the wild. Larvae are very easy to collect. Apple or pear trees should be beaten when in full blossom and the flowers which fall on to the tray searched carefully for the larvae. It is perhaps wisest to retain all the debris as the larvae are difficult to spot and many may be missed unless a thorough search is undertaken. If this debris is kept for six weeks or so in conditions which are not too dry, but not so damp as to cause mould, adults should emerge satisfactorily. Larvae which are noticed on the beating tray and separated can be reared in the usual way, though what may appear to be an over-abundance of flowers must be supplied as the petals are not usually eaten. It is good practice to change the food completely each day and remove any larvae which have begun to pupate.

By collecting larvae from many different localities, an impressive selection of adult varieties can be bred; however, within each locality only limited variation usually occurs.

Pasiphila debiliata (Hübner, 1817)   Bilberry Pug
= *nigropunctata* (Chant & Bentley, 1833)
= *debiliaria* (Boisduval, 1840)

Pl.3, fig.30; Pl.5, fig.4; Pl.D, fig.13

HISTORY First recorded in Britain as *Eupithecia nigropunctata* by Chant & Bentley (1833) who discovered the moth at Spitchwick, Devon, in June 1832. The larva was first described by Crewe (1863f;1865a) after much debate regarding possible larval foodplants (Campbell, 1862; Crewe, 1862q,m).

IMAGO *Characteristic features* (Figs 11b,c). Wingspan *c.*20mm; wings very pale, tinged lime-green in fresh specimens; forewing postmedian line dentate near costa, underside postmedian line curved; hindwing postmedian line angulate, particularly conspicuous on underside.

Genitalia. Figs 19f,29j ♂; 26b ♀

Variation. Variation in colour and markings is uncommon. F. *nigropunctata* Chant has the ante- and postmedian lines replaced by rows of spots, and f. *albescens* Cockayne (1953) is very pale with only pale grey, barely discernible markings. In ab. *grisescens* Dietze (1913) the ground colour is silver-grey, without the usual greenish tinge, and in ab. *obscurevirescens* Lempke (1951) the ground colour is dark green. Ab. *mediofasciata* Dietze (1913) has the median area of the forewing darkened.

Similar species (See Fig. 11, p. 133; Plate 5)

P. chloerata (Fig.11a): much darker, forewing postmedian line not dentate near costa; hindwing postmedian fascia curved, postmedian line absent or obsolete.

P. rectangulata (Fig.11f): much darker, often bright green; forewing underside postmedian line angulate or strongly geniculate.

LIFE HISTORY

OVUM Brown; 0.6mm; roundish ovoid. Laid on the twigs of the foodplant during June and July. Overwinters in this stage until late April of the following year.

LARVA (Fig.33 l) *First and second instars.* Head black; body pale, dirty greenish white; markings absent. *Third and fourth instars.* Short and stumpy, tapering slightly towards the head and more so at the posterior end; 14mm when full grown; integument smooth, translucent, faintly revealing the gut, sparsely scattered with tubercles and hairs, dirty greenish yellow; dorsal line dark green; spiracular line dull yellow.

Feeds on Bilberry (*Vaccinium myrtillus*). In the early instars the larva feeds inside the flowers, eating the stamens and pistils. In the final instar the entire flower is eaten. It feeds mainly at night, spinning together one or two leaves in which it hides during the day. April and May.

PUPA Length 7–8mm; thorax green; wings pale yellowish green; abdomen pale yellowish green; cremaster red. Late May to early June in a cocoon amongst ground-debris.

FLIGHT PERIOD AND HABITAT Univoltine, from June to early August with a peak during the first week of July. It is found in open woodland where the ground flora is dominated by bilberry. Adults fly over the foodplant in the early evening and at night are attracted to light.

DISTRIBUTION (Map 48) Locally common south-west of a line from Staffordshire to West Sussex, including Wales. Elsewhere recorded sporadically, with no known breeding colonies. In Scotland, known only from Aberdeenshire. Local in Ireland.

COLLECTING AND REARING Ova can be obtained from captive females only by caging them on mature bilberry plants. Wild larvae can be found by carefully searching bilberry plants for spun leaves which are usually conspicuous for their content of dark frass. These should be cut off and allowed to drop straight into a container for later examination. They should not be handled in the field as the larvae will immediately vacate the chamber and fall to the ground. A large number of spinnings must be collected as a fresh one is made after each feeding session, resulting in many old habitations occupied only by frass. Larvae can be expected in approximately 15 per cent of all spinnings collected. In captivity, larger larvae will eat the leaves of the foodplant and Skinner (1984) states that leaves are also eaten in the wild. Provided a ready supply of bilberry is available, this is a straightforward species to rear.

Gymnoscelis rufifasciata (Haworth, 1809)   Double-striped Pug

= *bistrigata* (Haworth, 1809) nec (Borkhausen, 1790)
= *pumilata* (Hübner, 1813)
= *strobilata* sensu Stephens, 1829
= *recictaria* (Boisduval, 1840)
= *globulariata* Millière, 1861

Pl.3, figs 35,36; Pl.4, figs 16,17; Pl.D, figs 14,15

HISTORY First recorded in Britain by Haworth (1809) as *Phalaena bistrigata* (Double Striped Pug) and *P. rufifasciata* (Red Barred Pug). He gives his own garden (possibly in Chelsea, London) as a locality for this species. Although the larva was known as early as 1854 (Crewe, 1854), it was not described until seven years later (Crewe, 1861a,k).

IMAGO *Characteristic features.* Wingspan c.18mm; abdomen with dark sub-basal band; forewing costa and termen straight, apex pointed, discal spot absent, costal portion of postmedian line characteristically prominent, black, with small basally-pointing dark triangles, antemedian fascia dark and prominent, area basad of antemedian fascia often unicolorous dark, marginal band broken one-quarter distance from costa by pale patch between postmedian and subterminal line; hindwing termen concave; postmedian line usually with small basally-pointing triangular marks. In fresh specimens, forewing fasciae distinctively reddish.

Genitalia. Figs 12n,19b,29k ♂; 26d

Variation. Herrich-Schäffer (1848) describes a dwarf form *parvularia*. Most superficial variation concerns general differences in the intensity of the ground colour and the definition of the cross lines. Dark forms are represented by ab. *nigrofasciata* Deitze (1910) in which the basal half of the median area is blackened, and ab. *tenebrata* Dietze (1910) which is almost wholly dark, retaining only fragments of the normal ground colour. Paler forms are more common and often result in more clearly defined markings. Ab. *albescens* Lempke (1951) has the basal and median area of the forewing whitish. Similar to this is ab. *obsolescens* Richardson (1952) in which the basal two-thirds of the wings are almost devoid of markings, leaving the postmedian line and terminal markings strongly pronounced. Plant (1989) describes ab. *mediopallens* which has a broad whitish median band on the forewing. The type was caught in the Wyre Forest, Shropshire, on 18 June 1984 and is illustrated in colour by BENHS (1989). Ab. *bucovinata* Hormuzaki (Goodson, unpubl.) has the whole ground colour ash-grey with pure white, sharply defined cross-lines, while f. *tempestivata* Zeller (1847) is generally greyer in colour and less reddish. Increased development of the dark cross-lines has been described under the forms *nigrostriata* Dietze (1910), in which the ante- and postmedian lines stand out as dark stripes, *incertata* Millière (1875), in which the numerous fine, dark, transverse markings are more coalesced into broader dentate bands, and *contrastata* Lempke (1951), in which the dark markings are normal in form but blackish brown and highly defined. Ab. *puncta* Lempke (1947;1951) has a conspicuous discal spot on each forewing.

Similar species
None.

LIFE HISTORY

OVUM Cream when first laid, later turning to bright yellow; 0.6mm; ovoid. On the flower-stalks of the foodplant. March to early May and again in July and August.

LARVA (Fig.33m) *First instar.* Cream; dorsal line well defined, dark green, red or brown. *Second, third and fourth instars.* Short and stout, tapering towards the head; 12mm when full grown; head and thoracic legs usually yellowish brown; ground colour varies greatly, matching that of the flowers on which it feeds; dorsal marking on each of the six central segments dark brown to purplish red or dark green, resembling a three-pronged eel-fork, connected by a fine, similarly-coloured dorsal line; subdorsal line broad, dark brown to purplish red or dark green; spiracular line yellowish cream; markings often obliterate most of the ground colour.

Widely polyphagous on the flowers of trees, shrubs and herbaceous plants. The following are commonly recorded: hawthorn (*Crataegus monogyna*), gorse (*Ulex europaeus*), broom (*Cystisus scoparius*), bramble (*Rubus fruticosus*), rose (*Rosa* sp.), common ragwort (*Senecio jacobaea*), heather (*Calluna vulgaris*) and hemp-agrimony (*Eupatorium cannabinum*). May and June, and September and October.

PUPA Length 5–7mm; pale reddish brown. Usually in a cocoon in soil. June and July and October to March.

FLIGHT PERIOD AND HABITAT Although this species has been recorded in every month from February to November, it is usually bivoltine, flying from March to May and late June to September, with peaks in early April and late July. It is possible that records from October and November represent a third emergence. Found in many habitat types including waste ground, gardens and woodland, but it is perhaps most abundant on heathland such as that in the New Forest, Hampshire. It is easily disturbed from herbage during the day and often flies naturally in sunny conditions. It is most active from dusk onwards through the night when it is attracted to light, sometimes in very large numbers, with males usually predominant.

DISTRIBUTION (Map 49) Widespread and generally common throughout the British Isles except the Shetlands, where it has not been recorded.

COLLECTING AND REARING A straightforward species to collect. Larvae can be easily swept or beaten from the foodplant, and adult females can often be found at sallow blossom or netted at dusk, and will oviposit readily in captivity. Three generations can frequently be obtained.

Erroneous Additions to the British List

The two following species have appeared in previous literature but have subsequently been found to be misidentifications.

Eupithecia alliaria Staudinger, 1870   Isle of Man Pug
[re-identified as *E. pygmaeata* (Hübner, 1799) – Marsh Pug]

On 26 July 1945, W. S. Cowin found a female *Eupithecia* on the Isle of Man which had been caught by a rare asilid robber fly, *Epitriptus cowini*. W. H. Tams (then of the British Museum (Natural History)) identified the specimen as *E. alliaria* Staudinger (Cowin, 1946).

The larval foodplant (*Allium* spp., usually *A. flavum* on the Continent) was not present in the locality of capture and further searches for the moth were unsuccessful.

The specimen was later re-examined by D. S. Fletcher (British Museum (Natural History)) and was found, after examination of the genitalia, to be a female *E. palustraria* Doubleday (= *E. pygmaeata* (Hübner)), the Marsh Pug (de Worms, 1958). It is now in the Hope Department of Oxford University.

Eupithecia oxycedrata (Rambur, 1833)
[re-identified as *E. phoeniceata* Rambur, 1834 – Cypress Pug]

A single specimen, caught on Guernsey on 21 August 1958 by C. J. Shayer (Shayer, 1959), was identified at the British Museum (Natural History) as *E. oxycedrata* (Rambur). In 1983, it was re-examined by Peet and found to be *E. phoeniceata* Rambur (T. N. D. Peet, pers.comm.). In 1985, further examination at the Natural History Museum, confirmed this. It is believed the specimen is now in the Museum's collection (R. Long, pers.comm.).

It is interesting to note that the record of *E. phoeniceata* previously regarded as the first not only for Britain but also for the British Isles was that of de Worms and Messenger (1960) from Penzance, Cornwall (see p. 124). However, the Guernsey record precedes theirs by a year so should now be regarded as the first record for the British Isles.

The following taxa are forms which were previously given specific status but are not now regarded as being distinct species.

Eupithecia knautiata Gregson, 1874   Scabious Pug
[= syn. of *E. absinthiata* (Clerck, 1759), see p. 76]

Eupithecia knautiata was proposed as a distinct species by Gregson from larvae collected by a Mr Porter at Bulls Hill Lodges, near Bolton, Lancashire. The foodplant was stated to be field scabious (*Knautia arvensis*) (flowers and seeds) (Gregson, 1874b). Fierce debate ensued between Gregson (1875a,b), Crewe (1874a), Johnson (1875), Bird (1875) and Melvill (1875) as to the status of *E. knautiata*. Crewe (1874a) and Johnson (1875) each stated that he was familiar with this insect and considered it to be *E. minutata* ([Denis & Schiffermüller,

1775]) (=*E. absinthiata* (Clerck) f. *goossensiata* Mabille). Specimens supposedly of *E. knautiata* from Gregson's collection, currently held in the Hope Department of Oxford University, have been examined by the present authors and they support this view.

Eupithecia goossensiata Mabille, 1869   Ling Pug
[= f. of *E. absinthiata* (Clerck, 1759), see p. 76; Pl.B, fig.7]

Eupitheciata goossensiata Mabille was first recorded in Britain by Stephens (1831) and its taxonomic status has subsequently been the subject of considerable debate. Most British authors have continued to regard it as a distinct species, but some (e.g. Skinner, 1984) make the suggestion that it is merely a heathland form of *E. absinthiata*.

The argument for specific status is based on differences in the colour of the larvae and in the ground colour of the adults. Larvae of *goossensiata* which feed on heather (*Calluna vulgaris*) are generally slightly smaller than those of typical *E. absinthiata* and are usually some shade of pink; the resulting adults are likewise often smaller than those of *E. absinthiata* and greyer in colour. However, the complex is reviewed in detail by Riley (1986a) who explains that these characters can be controlled by the plants on which the larvae feed; *goossensiata*-like larvae and adults can be produced in captivity by rearing *E. absinthiata* larvae on heather. Further, *goossensiata* larvae fed on ragwort (*Senecio jacobaea*) lose their typical pink colouration and produce *E. absinthiata*-like adults. No other consistent superficial differences are apparent and structurally the two taxa appear identical. Subspecific status is not warranted as the two forms are sympatric and synchronic. Consequently, Riley concludes that *goossensiata* should be considered to be a heather-feeding heathland form of *E. absinthiata*.

Eupithecia curzoni Gregson, 1884   Curzon's Pug
[= f. of *E. satyrata* Hübner, 1813 – Satyr Pug, see p. 72]

Eupithecia curzoni was described as a distinct species by Gregson (1884a,b;1885) from material collected on the Shetland Islands. Although McArthur (1884) considered it to be a form of *E. nanata* (Hübner), it has subsequently been recognized as a form of *E. satyrata* (Hübner).

Eupithecia ultimaria (Duponchel, 1831)   Kentish Tamarisk Pug
(=*E. stevensata* Webb, 1896)
[= syn. of *E. pusillata* ([Denis & Schiffermüller]) subsp. *anglicata* Herrich-Schäffer, 1863 – Juniper or Kentish Tamarisk Pug, see p. 122]

Specimens caught near Dover, Kent in 1850 by S. Stevens (Westwood, 1851) were figured by Carrington (1881) and were misidentified by Guenée as *E. ultimaria* Duponchel (Webb, 1881; Stevens, 1882). A further specimen was caught at Freshwater, Isle of Wight, in September 1904 and recorded by Mutch (1905) as *E. stevensata* Webb. Mutch also inferred that tamarisk (*Tamarix gallica*) could be a possible foodplant but there is no evidence, other than circumstantial, to support this view. Step (1892) cites Purdey as disagreeing with Barrett, who identified specimens as *E. pusillata*. Purdey stated that the adults flew at different times and the larvae would not feed on juniper. However, an alternative foodplant is not given and there is no evidence extant to support Purdey's statements. Prout (1914) states that *E. stevensata* Webb is subsp. *anglicata* Herrich-Schäffer of *E. pusillata* and this has been confirmed by subsequent examinations of material held in the National Collection. No specimens have been caught since 1915 and this subspecies is now thought to be extinct.

The original name of *ultimaria* (Duponchel) should not be confused with *ultimaria* of Boisduval (Channel Islands Pug). Although South (1961) referred to the Kent race of *E. pusillata* as *ultimaria* Boisduval, this is erroneous and the two are quite distinct.

Eupithecia innotata (Hufnagel, 1767)   Angle-barred Pug
[British status uncertain, see p. 112)]

First listed as a British species by Stephens (1829), the status of *E. innotata* in this country has been the focus of considerable debate ever since. Haggett (1963) comprehensively reviewed the subject and concluded that supposed British *E. innotata* should not be regarded as distinct from *E. fraxinata* and questioned existing records that suggested the opposite.

There are no structural or superficial differences between *E. innotata* and *E. fraxinata*, but the opinion that the British insect is distinct from the Continental *E. innotata* is supported by their biology. On the Continent, *E. innotata* is bivoltine, the larvae of the first brood feeding on deciduous trees and shrubs such as *Fraxinus*, *Sambucus* and *Prunus*, and those of the second brood on *Artemisia*. This reliance on *Artemisia* during the second brood holds the key to the existence of *E. innotata* in this country. In Britain, *E. fraxinata* is bivoltine but, although it feeds on ash, tamarisk and sea buckthorn, none has been recorded reliably in the wild state on *Artemisia*. Furthermore, Haggett (1963) tried to rear captive larvae obtained from sea buckthorn on *Artemisia* but failed. Larvae from the same source fed readily on *Fraxinus*.

Those records, which supposedly originated from *Artemisia*-feeding larvae (and particularly those from Durham, where a colony of *E. innotata* was thought to have occurred between 1902 and 1931 (Robson, 1902; Heslop-Harrison, 1931)), were attributed by Haggett (1963) either to misidentification or the lack of any confirmed association with that plant. His arguments were well constructed and, with an absence of evidence to the contrary, the present authors have agreed with his conclusions. However, it is interesting to note that Riley has reared this species in captivity on the leaves of mugwort (*Artemisia vulgaris*).

Eupithecia tamarisciata Freyer, 1836   Tamarisk Pug
[= syn. of *E. fraxinata* Crewe, 1863 – Ash Pug, see p. 112]

First recorded as a British species from larvae found on tamarisk (*Tamarix gallica*) in Cornwall in 1905 by Mr and Mrs E. M. Holmes of Kent. These were reared by Mrs Holmes, and two adults and a drawing of the larvae were identified by Tutt (1906a;1908) as *E. tamarisciata* (Freyer). Prout (1914) disagreed with the identification and considered the specimens to be *E. fraxinata* Crewe. Current opinion suggests that all previous records of *E. tamarisciata* should be regarded as *E. fraxinata*. However, it is interesting to note that tamarisk is not listed as a larval foodplant for *E. fraxinata* by either Agassiz *et. al.* (1981) or Skinner (1984). Rather, *E. fraxinata* is divided into two 'races' – one feeding on ash in woodland and hedgerows, and the other feeding on sea buckthorn on the coast. Further to the 1905 record from Cornish tamarisk, in 1979 Prior beat a single larva, once again from tamarisk in Cornwall. The genitalia of the resulting adult (a female) were examined by Riley and found to be typical of *E. fraxinata*. A single male *E. fraxinata* was also caught at light in a dense patch of tamarisk on Guernsey in May 1989. The lack of sea buckthorn on the island and absence of ash in the vicinity suggest that this individual was associated with tamarisk (Riley, 1990c). Both Riley and Haggett (1963) have reared *E. fraxinata* in captivity on this plant. These observations support the view that *E. tamarisciata* is merely *E. fraxinata* associated with tamarisk.

Eupithecia nigrosericeata (Haworth, 1809)   Black Silk Pug
Eupithecia sericeata (Haworth, 1809)   Satin Pug
[= forms of *Pasiphila rectangulata* (Linnaeus), see p. 135]

Both the above were described as distinct species by Haworth (1809) but were later discovered to be merely dark forms of *P. rectangulata* (Linnaeus).

Eupithecia arceuthata Freyer, 1842   Freyer's Pug
[= subsp. of *E. intricata* (Zetterstedt, 1839) (Freyer, 1842), see p. 67]

Crewe (1862p) proposed this as a distinct species on the basis of adult winglength and colour. His views on this matter were not widely supported and it is now correctly regarded as one of the three subspecies of *E. intricata*.

Eupithecia castigata (Hübner, 1813)   Paisley Pug
[= syn. of *Eupithecia subfuscata* (Haworth, 1809), f. *obscurissima* Prout, 1914 – Grey Pug, see p. 91]

Originally described from specimens collected in Paisley, near Glasgow. This melanic insect caused great confusion and at different times was thought to be a form of *E. tripunctaria*, *E. satyrata*, *E. trisignaria* and *E. virgaureata*. However, Tugwell (1892) finally concluded that it was, in fact, a dark form of *E. subfuscata*.

Species cited as likely to occur in Britain

Eupithecia immundata (Lienig & Zeller, 1846)
Eupithecia actaeata (Walderdorff, 1869)

Cockayne (1952a) states that Dr S. Hoffmeyer considered it worth searching for these two species in northern England. The larval foodplant, baneberry (*Actaea spicata*) occurs in Lancashire, Yorkshire and Westmorland and, in Hoffmeyer's opinion, the British climate would be suitable for *E. immundata* and *E. actaeata*.

Between 5 and 10 September 1951, E. W. Classey and H. S. Robinson searched baneberry plants growing on limestone pavement at Colt Park, Raisghyll, and three further sites in the West Riding of Yorkshire. They were unsuccessful in locating larvae and concluded that neither pug was resident in this country (Cockayne, 1952a). However, the timing of this search was perhaps not well judged. The larvae of *E. immundata* feed in late July and early August and those of *E. acteata* from July to early September.

In a later article, Cockayne (1952b) cites a list of further localities where baneberry occurs. Although some of these are also on limestone pavement, others are in woodland. Furthermore, Agassiz *et. al.* (1981) states that baneberry forms the predominant undergrowth in some Yorkshire woodlands. As both *E. actaeata* and *E. immundata* inhabit deciduous woodland on the Continent, it would seem logical to concentrate future searches in this habitat-type.

A complete account of the biology of *E. actaeata*, with photographs of the larva, pupa and imagine, is given by Weigt (1979). The adults and larvae of both species are illustrated by Skou (1986).

Eupithecia analoga Djakonov, 1926
Eupithecia selinata Herrich-Schäffer, 1861
Eupithecia gelidata Möschler, 1860
subsp. *hyperboreata* Staudinger, 1861
Eupithecia unedonata Mabille, 1868
Eupithecia lanceata (Hübner, 1825)
Eupithecia conterminata (Lienig & Zeller, 1846)

The above species were listed, with short descriptions, by Agassiz *et. al.* (1981) as also likely to occur in the British Isles. None has yet been found and the authors are not aware of any serious attempts to find them. Their list also included *E. sinuosaria* (Eversmann) which turned up in Britain in 1991 but there is no evidence that it has yet become established (see p. 102).

5: The Genitalia

The male aedeagus of each species (Figs 13–19) is illustrated at a magnification of $c. \times 80$ and the female bursa copulatrix (Figs 20–26) at $c. \times 60$. The sclerotized plate, which is present beneath the scales of the eighth abdominal sternite of the males of each species (Figs 27–29), is drawn at $c. \times 30$. These can be examined by brushing scales from the abdomen without need for dissection.

When the name of a species in the caption is followed by an asterisk (*), the reader is referred to the text of the Descriptions of British and Irish Pug Moths in Chapter 4 for the discussion, under the heading *Similar species*, on how to separate it from other closely allied species. For ease of identification, the orientation of dissected genitalia should match that of those illustrated in the Figures.

Those species which have atypical and immediately recognizable valvae are illustrated at $c. \times 15$ (Fig.12).

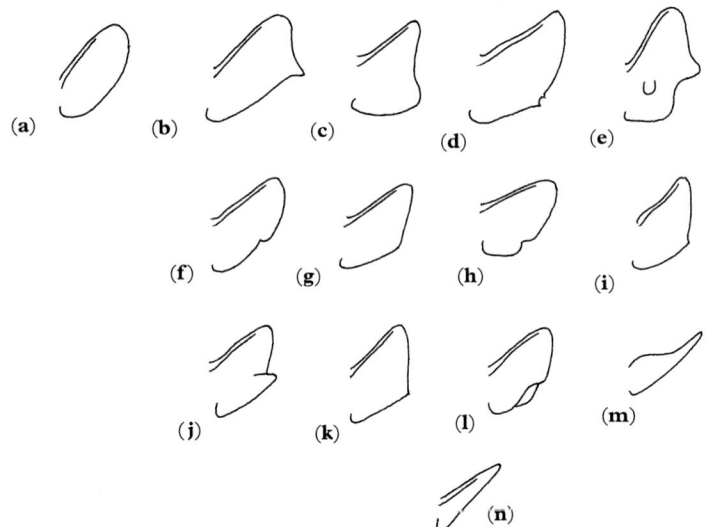

Text figure 12 Eupitheciine species with distinctively-shaped valvae

(**a**) A typical eupitheciine valva (**b**) *E. insigniata* (p. 52) (**c**) *E. exiguata* (p. 51)
(**d**) *E. egenaria* (p. 60) (**e**) *E. centaureata* (p. 63) (**f**) *E. trisignaria* (p. 64)
(**g**) *E. denotata* (p. 89) (**h**) *E. distinctaria* (p. 103) (**i**) *E. pusillata* (p. 122)
(**j**) *E. abbreviata* (p. 117) (**k**) *E. phoeniceata* (p. 124) (**l**) *E. tantillaria* (p. 130)
(**m**) *C. v-ata* (p. 131) (**n**) *G. rufifasciata* (p. 139)

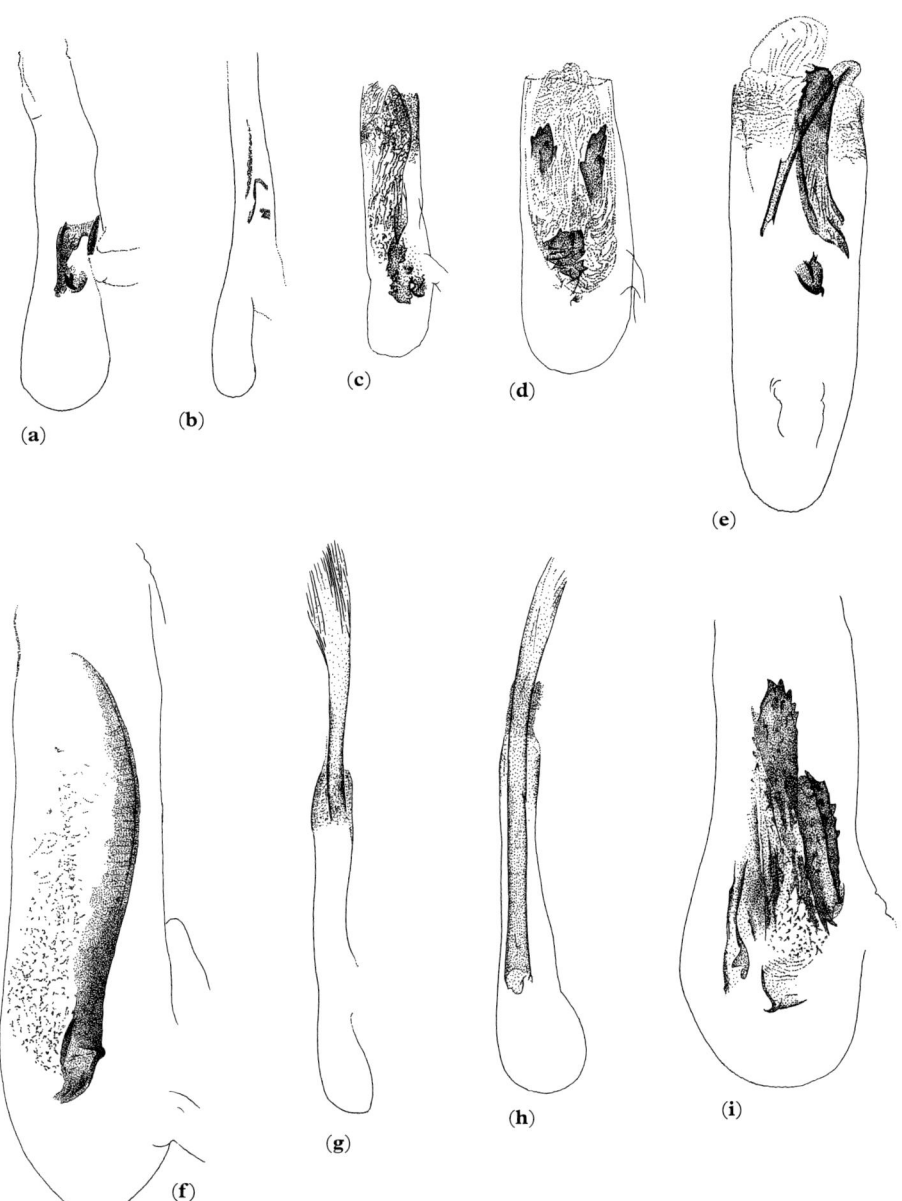

Text figure 13 Male aedeagus

(a) *E. tenuiata* (p. 33) (b) *E. inturbata* (p. 37) (c) *E. haworthiata* (p. 39)
(d) *E. irriguata* (p. 49) (e) *E. plumbeolata* (p. 41) (f) *E. abietaria* (p. 43)
(g) *E. linariata* (p. 45) (h) *E. pulchellata* (p. 47) (i) *E. exiguata* (p. 51)

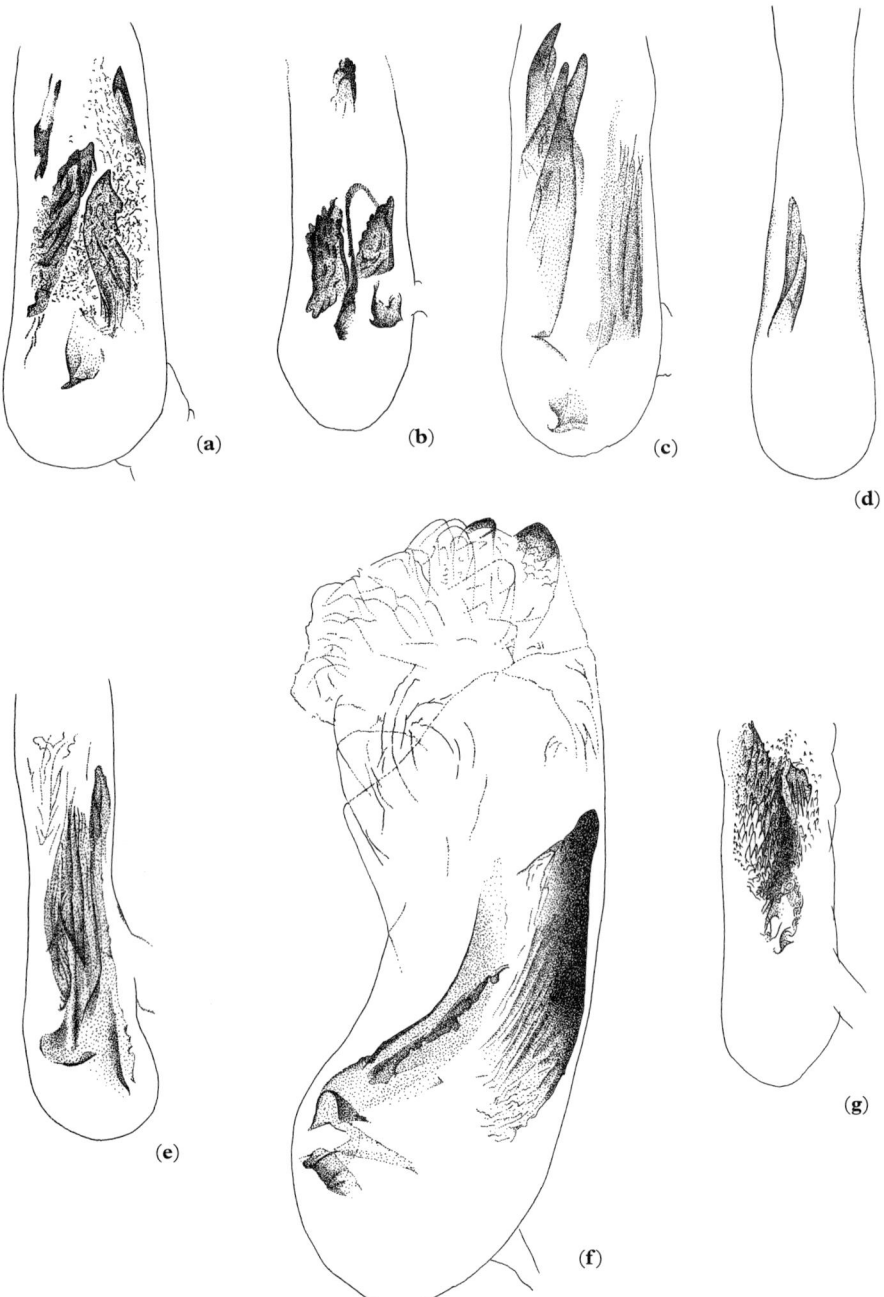

Text figure 14 Male aedeagus

(**a**) *E. insigniata* (p. 52) (**b**) *E. valerianata* (p. 54) (**c**) *E. pygmaeata* (p. 55)
(**d**) *E. venosata* (p. 58) (**e**) *E. centaureata* (p. 63) (**f**) *E. egenaria* (p. 60)
(**g**) *E. trisignaria* (p. 64)

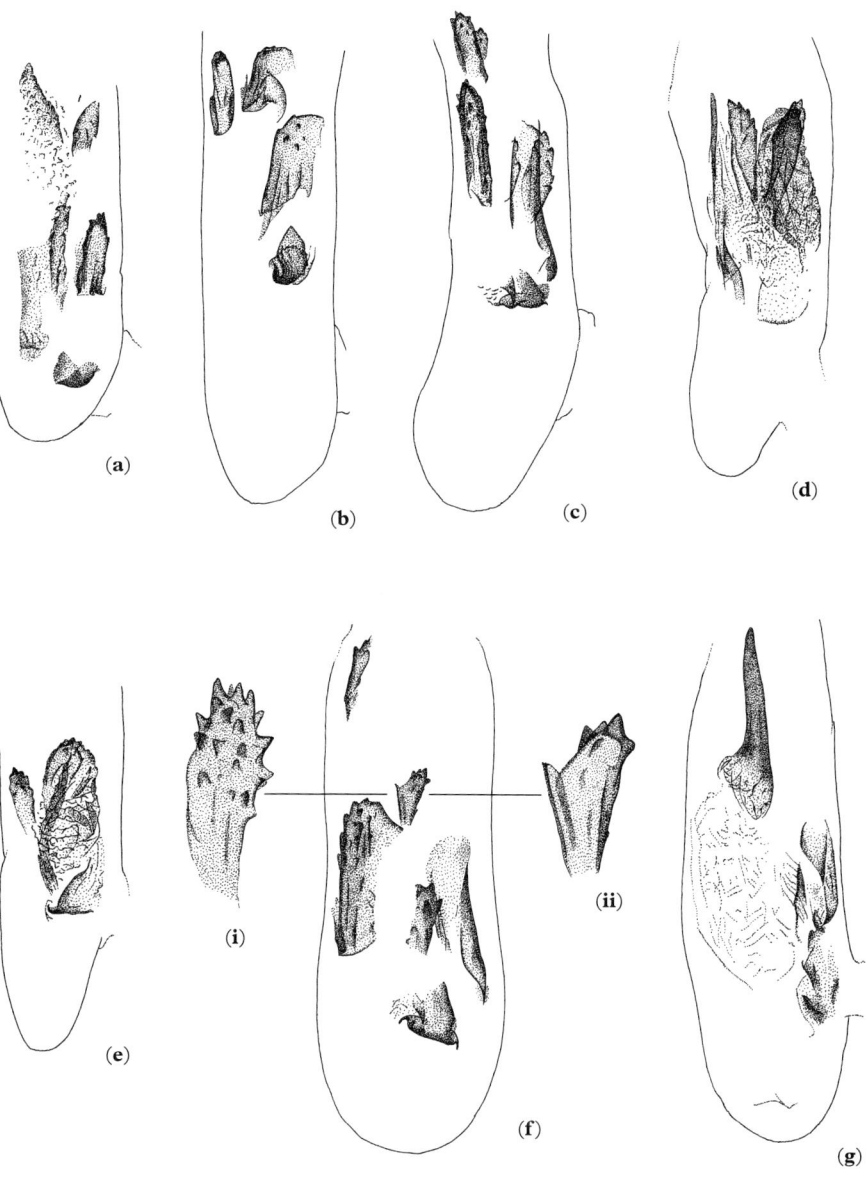

Text figure 15 Male aedeagus

(**a**) *E. intricata* (p. 68) (**b**) *E. cauchiata* (p. 76) (**c**) *E. satyrata* (p. 72)
(**d**) *E. assimiliata* (p. 81) (**e**) *E. vulgata* (p. 83) (**f**) (i) *E. expallidata* (p. 79)
(ii) *E. absinthiata* (p. 77) (**g**) *E. tripunctaria* (p. 86)

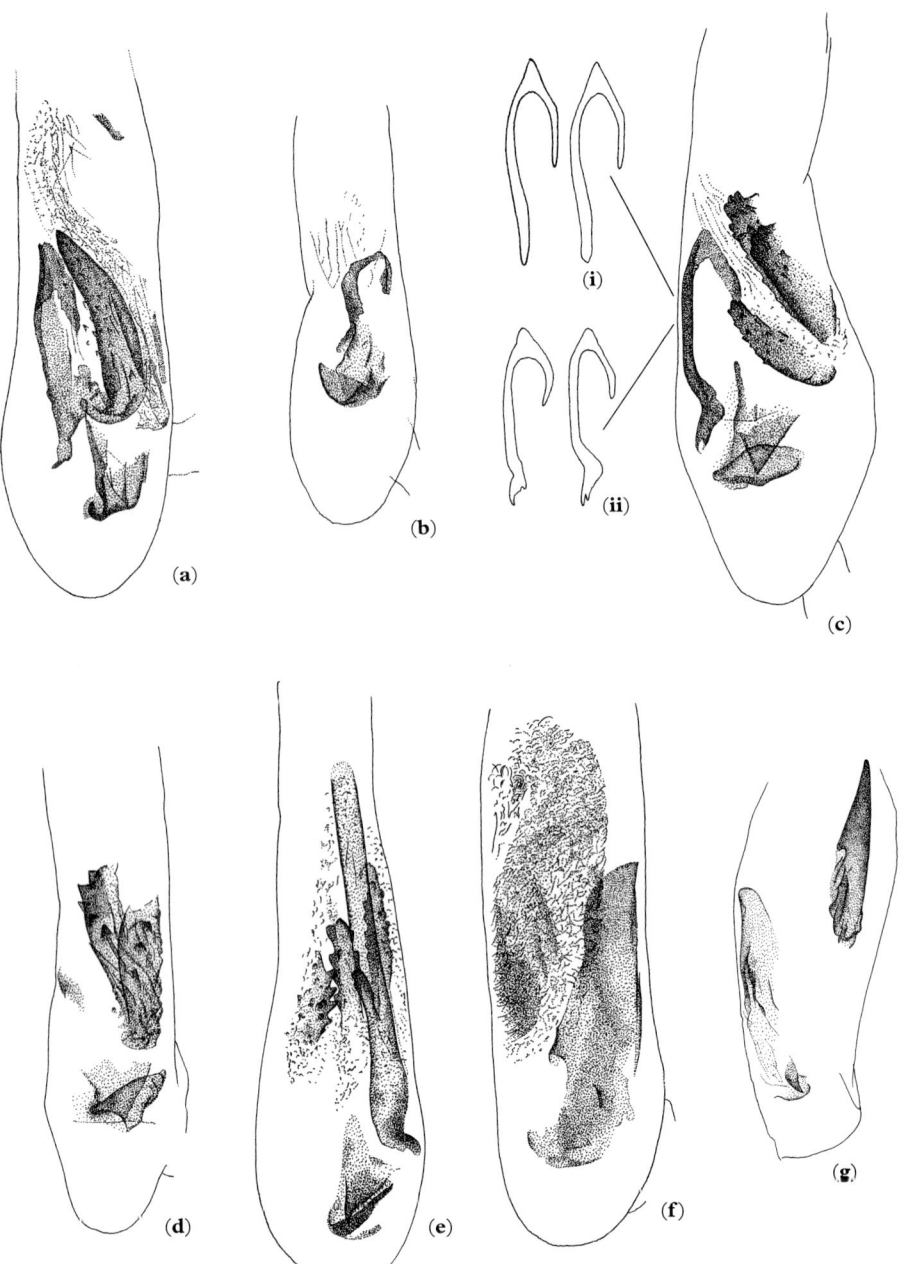

Text figure 16 Male aedeagus

(a) *E. denotata* (p. 89) (b) *E. subfuscata* (p. 91) (c) (i) *E. succenturiata* (p. 96)
(c) (ii) *E. icterata* (p. 94) (d) *E. subumbrata* (p. 98) (e) *E. millefoliata* (p. 99)
(f) *E. simpliciata* (p. 101) (g) *E. sinuosaria* (p. 103)

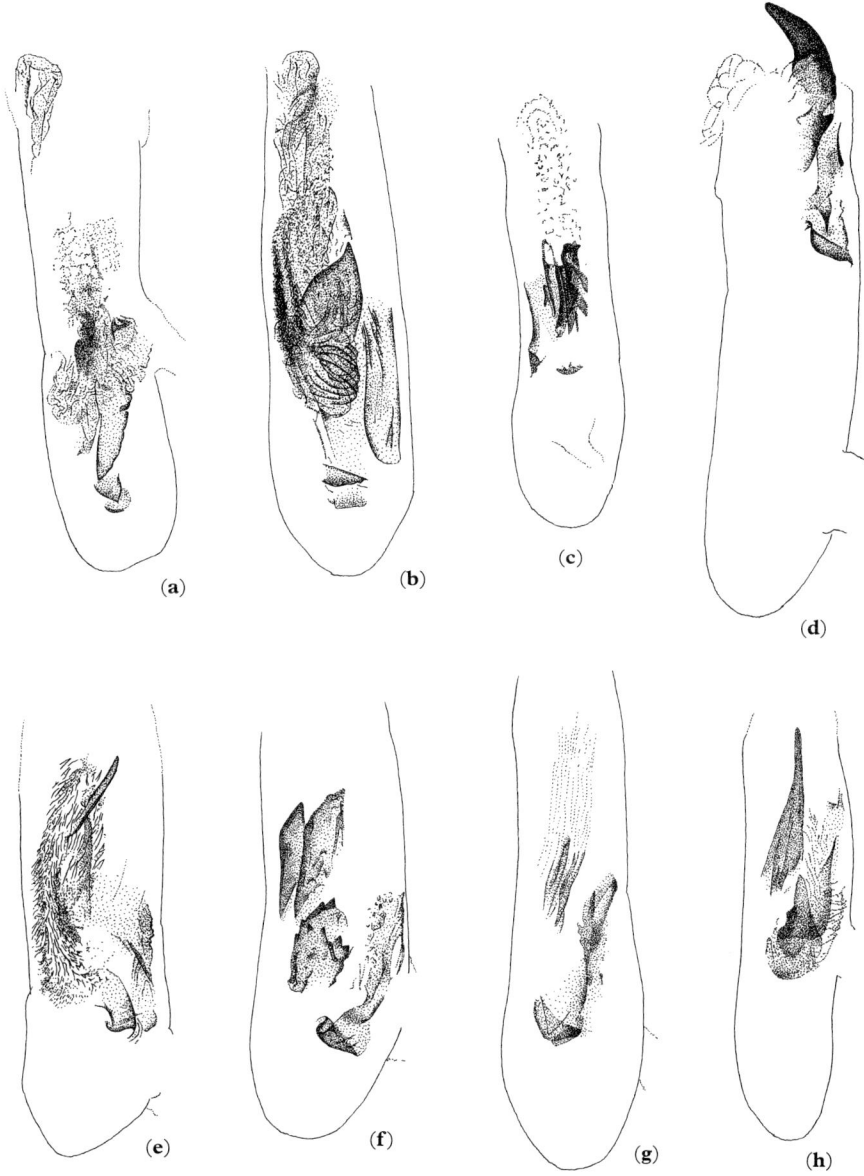

Text figure 17 Male aedeagus

(a) *E. distinctaria* (p. 103) (b) *E. ultimaria* (p. 126) (c) *E. indigata* (p. 105)
(d) *E. pimpinellata* (p. 106) (e) *E. nanata* (p. 108) (f) *E. extensaria* (p. 110)
(g) *E. fraxinata* (p. 112) (h) *E. virgaureata* (p. 114)

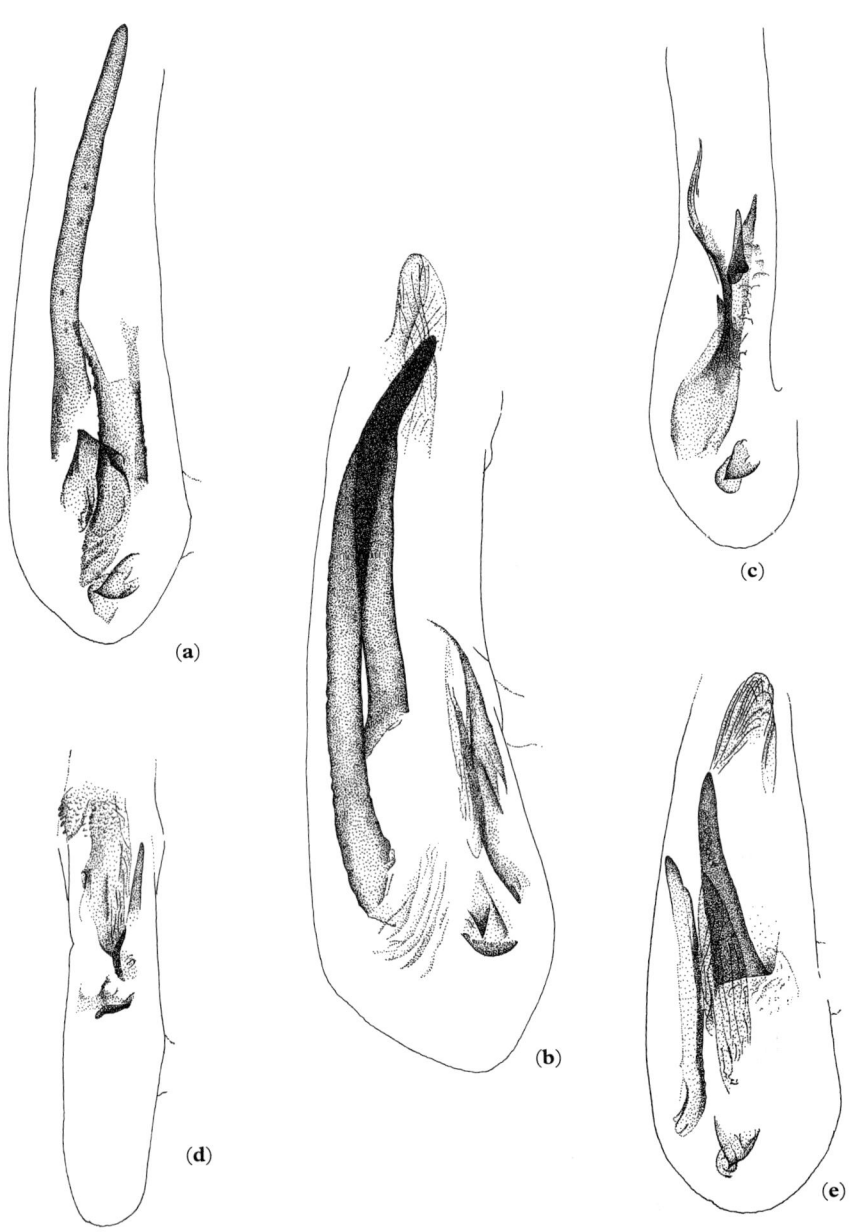

Text figure 18 Male aedeagus

(a) *E. abbreviata* (p. 117) (b) *E. phoeniceata* (p. 124) (c) *E. dodoneata* (p. 119)
(d) *E. lariciata* (p. 127) (e) *E. pusillata* (p. 122)

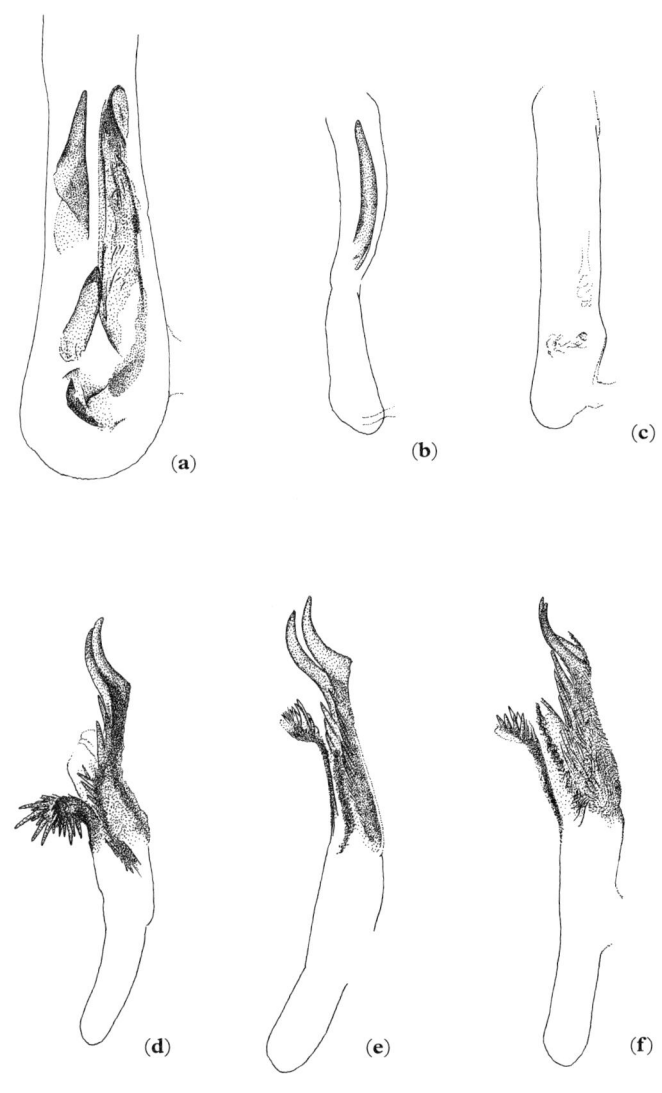

Text figure 19 Male aedeagus

(a) *E. tantillaria* (p. 130) (b) *G. rufifasciata* (p. 139) (c) *C. v-ata* (p. 131)
(d) *P. chloerata* (p. 133) (e) *P. rectangulata* (p. 135) (f) *P. debiliata* (p. 137)

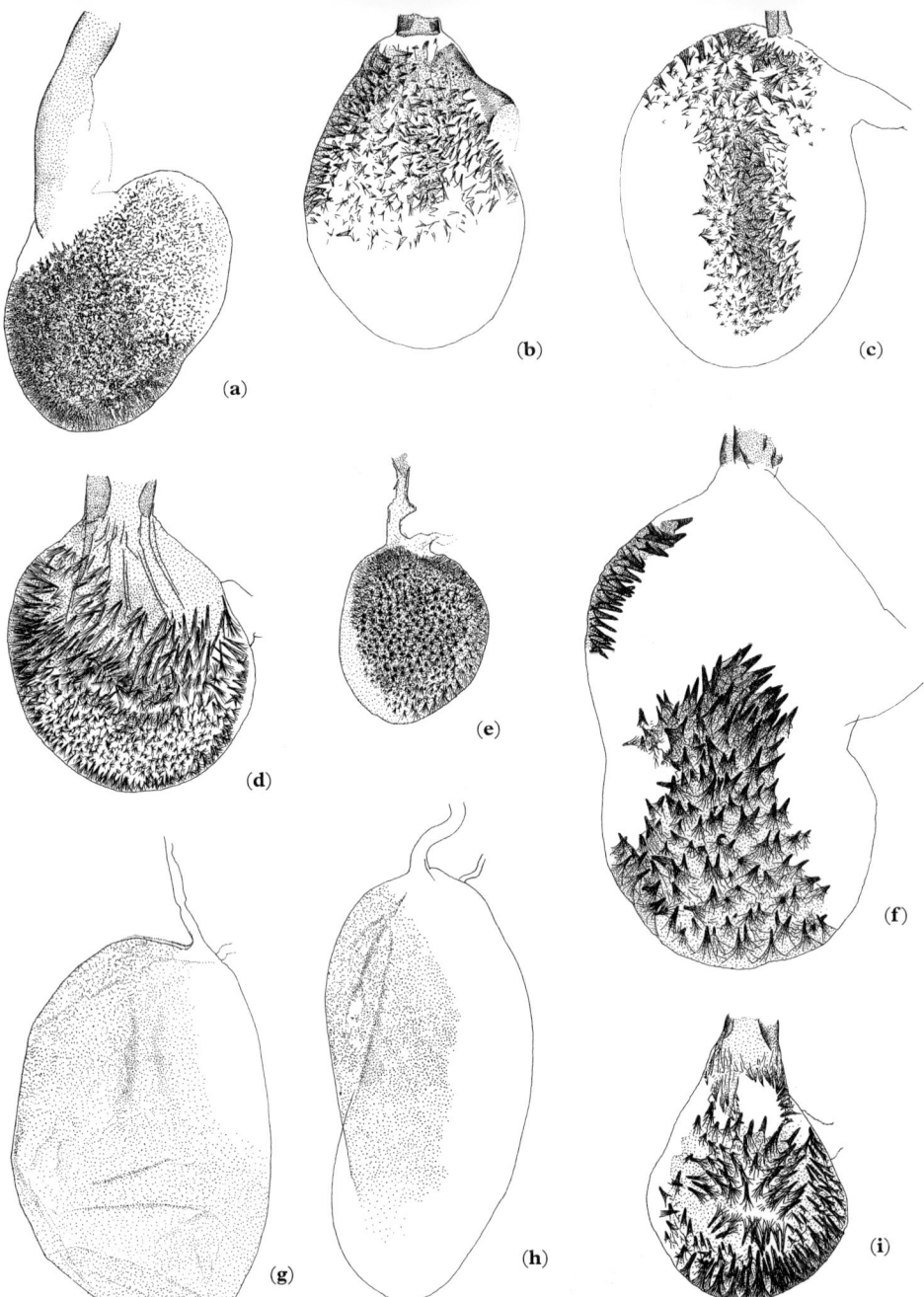

Text figure 20 Female bursa copulatrix

(a) *E. tenuiata* (p. 33) (b) *E. plumbeolata* (p. 41) (c) *E. haworthiata* (p. 39)
(d) *E. irriguata* (p. 49) (e) *E. inturbata* (p. 37) (f) *E. abietaria* (p. 43)
(g) *E. linariata* (p. 45) (h) *E. pulchellata* (p. 47) (i) *E. valerianata* (p. 54)

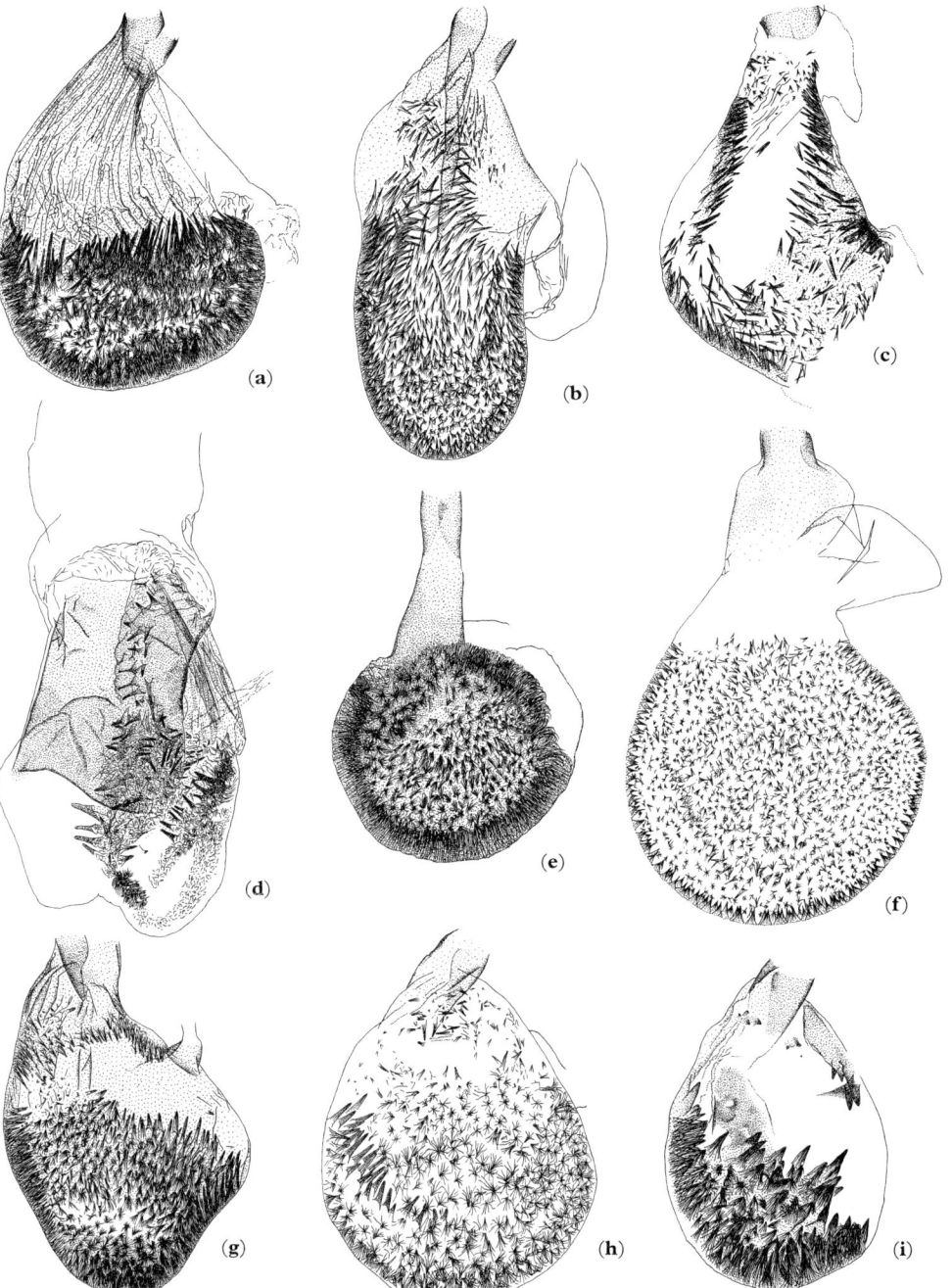

Text figure 21 Female bursa copulatrix

(a) *E. exiguata* (p. 51) (b) *E. insigniata* (p. 52) (c) *E. pygmaeata* (p. 55)
(d) *E. egenaria* (p. 60) (e) *E. venosata* (p. 58) (f) *E. centaureata* (p. 63)
(g) *E. trisignaria* (p. 64) (h) *E. satyrata* (p. 72) (i) *E. intricata* (p. 68)

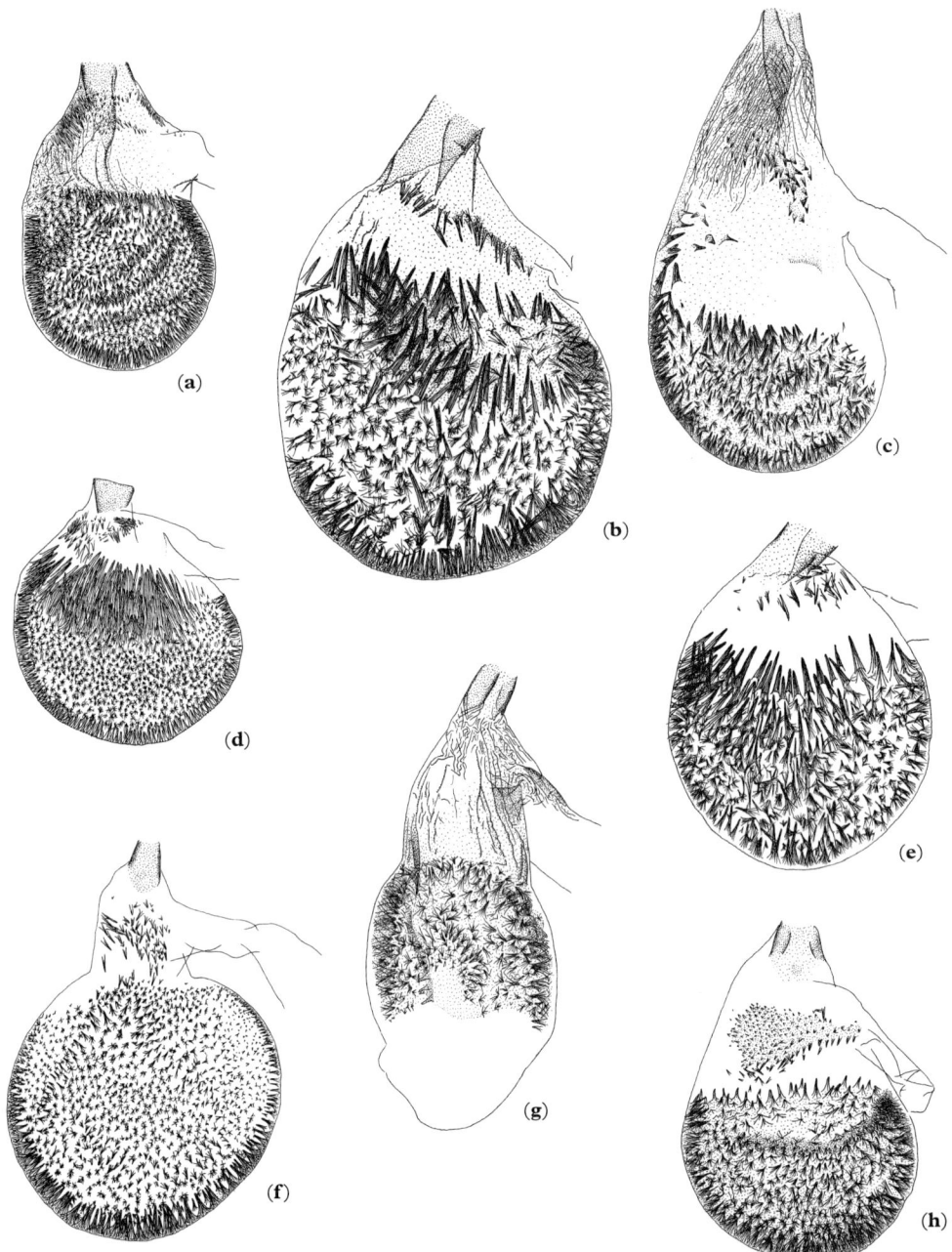

Text figure 22 Female bursa copulatrix

(**a**) *E. assimilata* (p. 81) (**b**) *E. absinthiata/ expallidata*★ (pp. 77, 79) (**c**) *E. denotata* (p. 89)
(**d**) *E. vulgata* (p. 83) (**e**) *E. cauchiata* (p. 76) (**f**) *E. subfuscata* (p. 91)
(**g**) *E. tripunctaria* (p. 86) (**h**) *E. subumbrata* (p. 98)

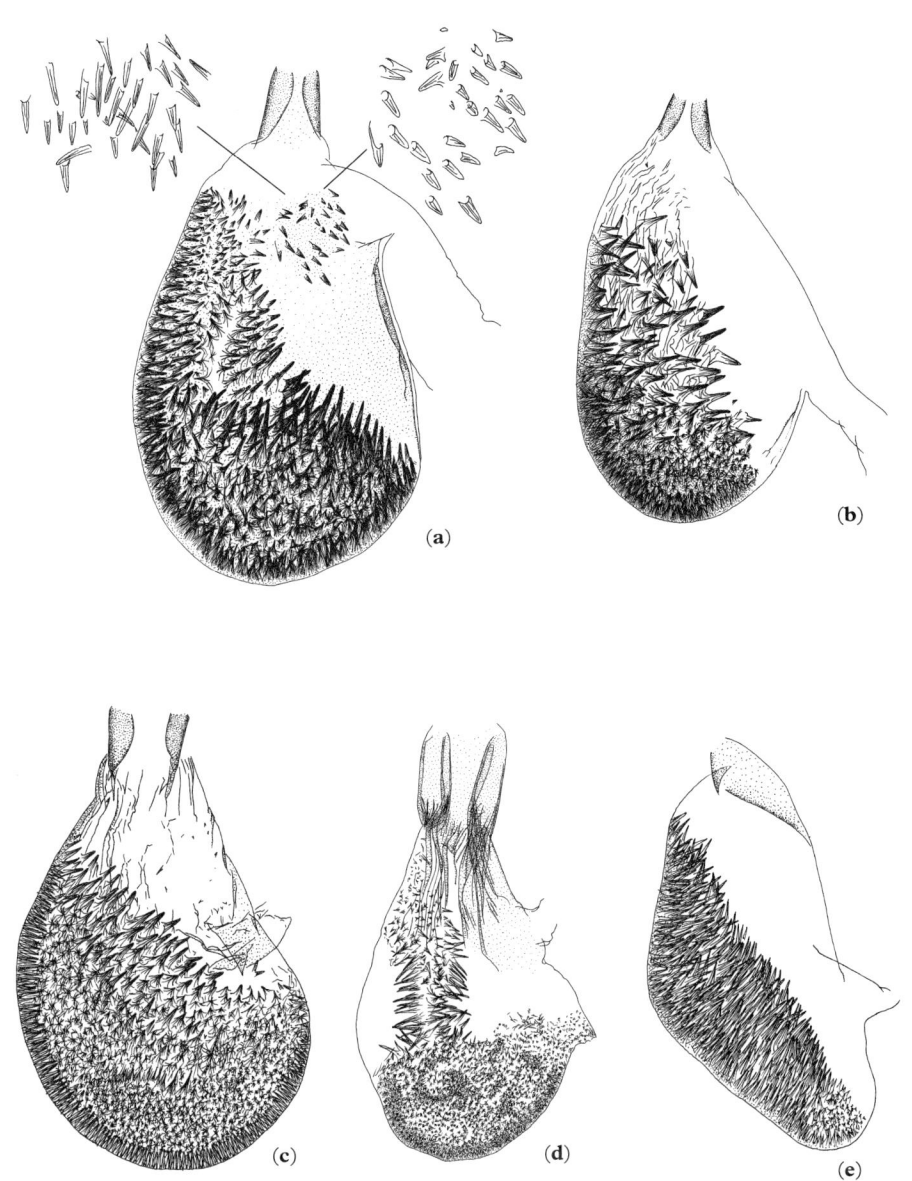

Text figure 23 Female bursa copulatrix

(**a**) (i) *E. succenturiata* (p. 96) (ii) *E. icterata* (p. 94) (**b**) *E. millefoliata* (p. 99)
(**c**) *E. simpliciata* (p. 101) (**d**) *E. sinuosaria* (p. 103) (**e**) *E. distinctaria* (p. 103)

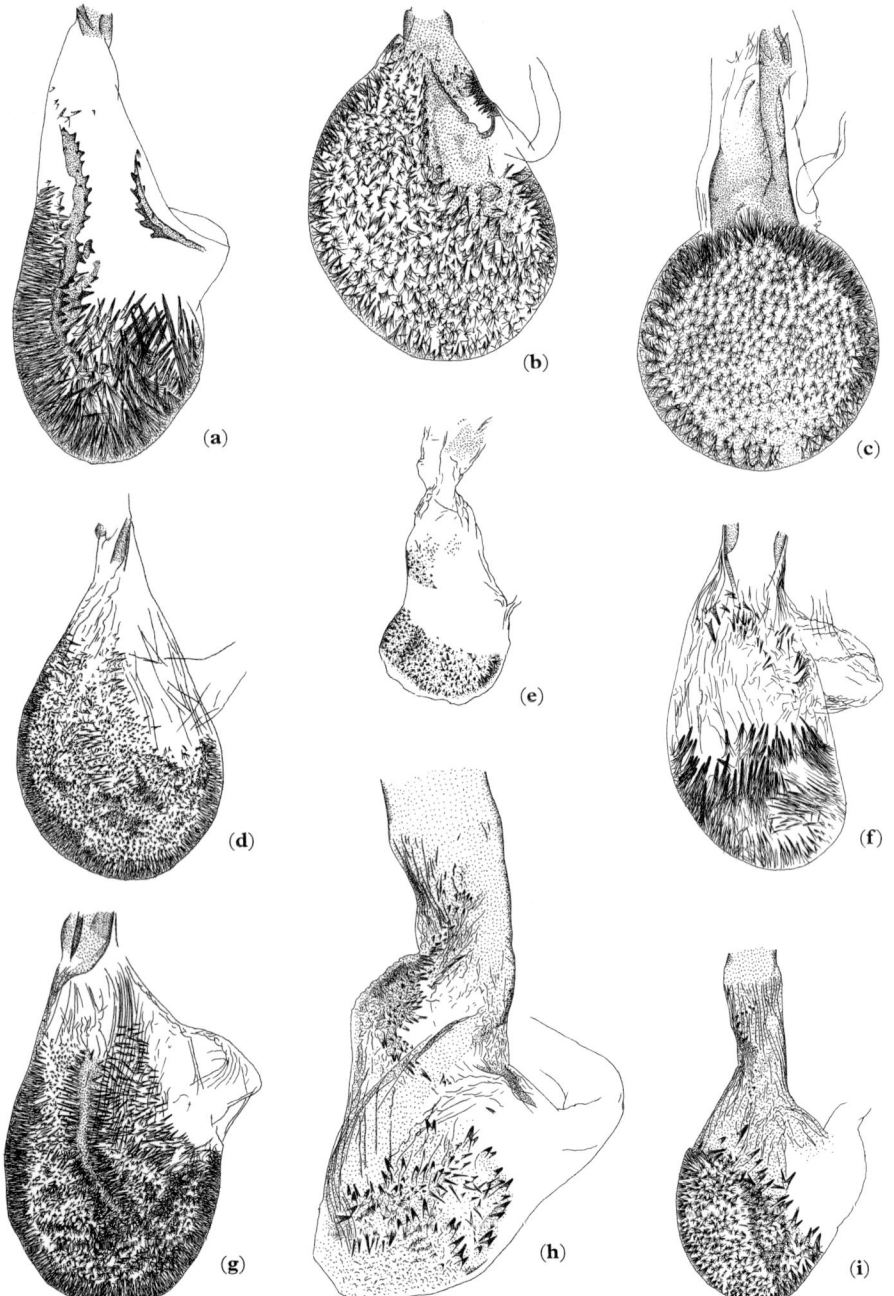

Text figure 24 Female bursa copulatrix

(a) *E. ultimaria* (p. 126) (b) *E. indigata* (p. 105) (c) *E. pimpinellata* (p. 106)
(d) *E. nanata* (p. 108) (e) *E. virgaureata* (p. 114) (f) *E. extensaria* (p. 110)
(g) *E. fraxinata* (p. 112) (h) *E. abbreviata* (p. 117) (i) *E. dodoneata* (p. 119)

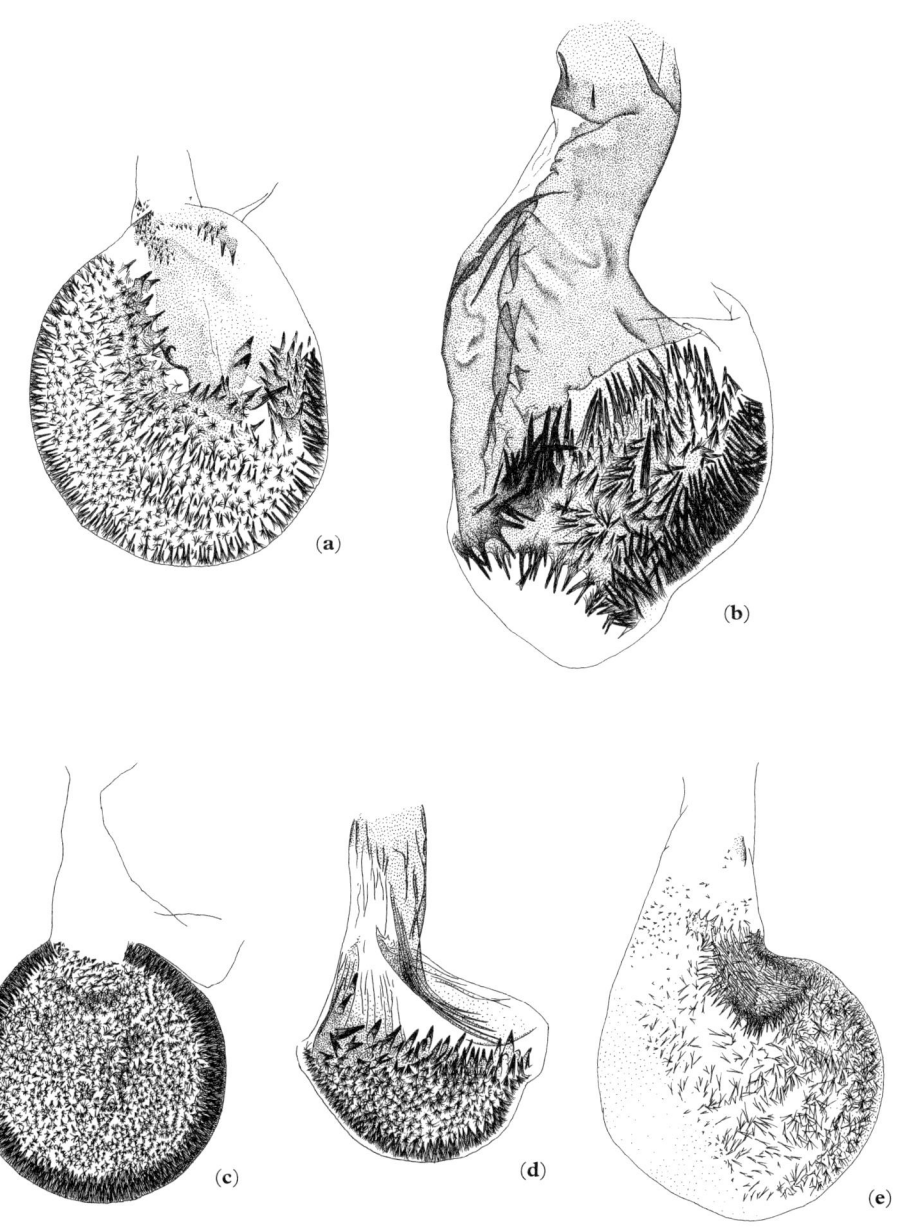

Text figure 25 Female bursa copulatrix

(a) *E. tantillaria* (p. 130) (b) *E. phoeniceata* (p. 124) (c) *E. lariciata* (p. 127)
(d) *E. pusillata* (p. 122) (e) *C. v-ata* (p. 131)

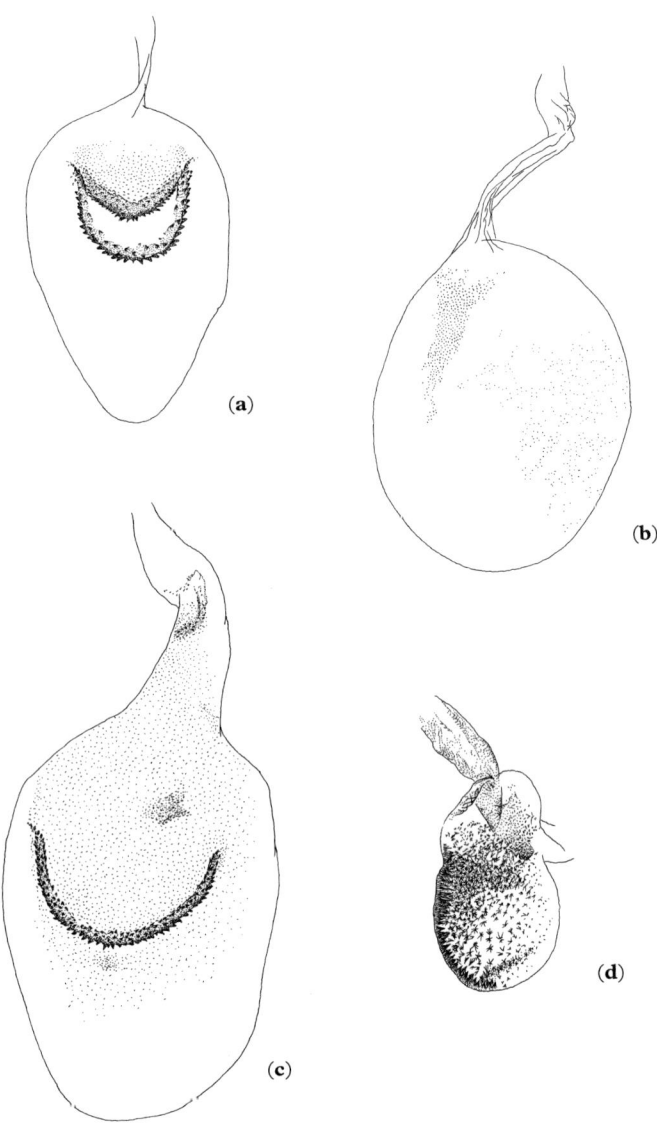

Text figure 26 Female bursa copulatrix

(**a**) *P. rectangulata* (p. 135) (**b**) *P. debiliata* (p. 137) (**c**) *P. chloerata* (p. 133)
(**d**) *P. rufifasciata* (p. 139)

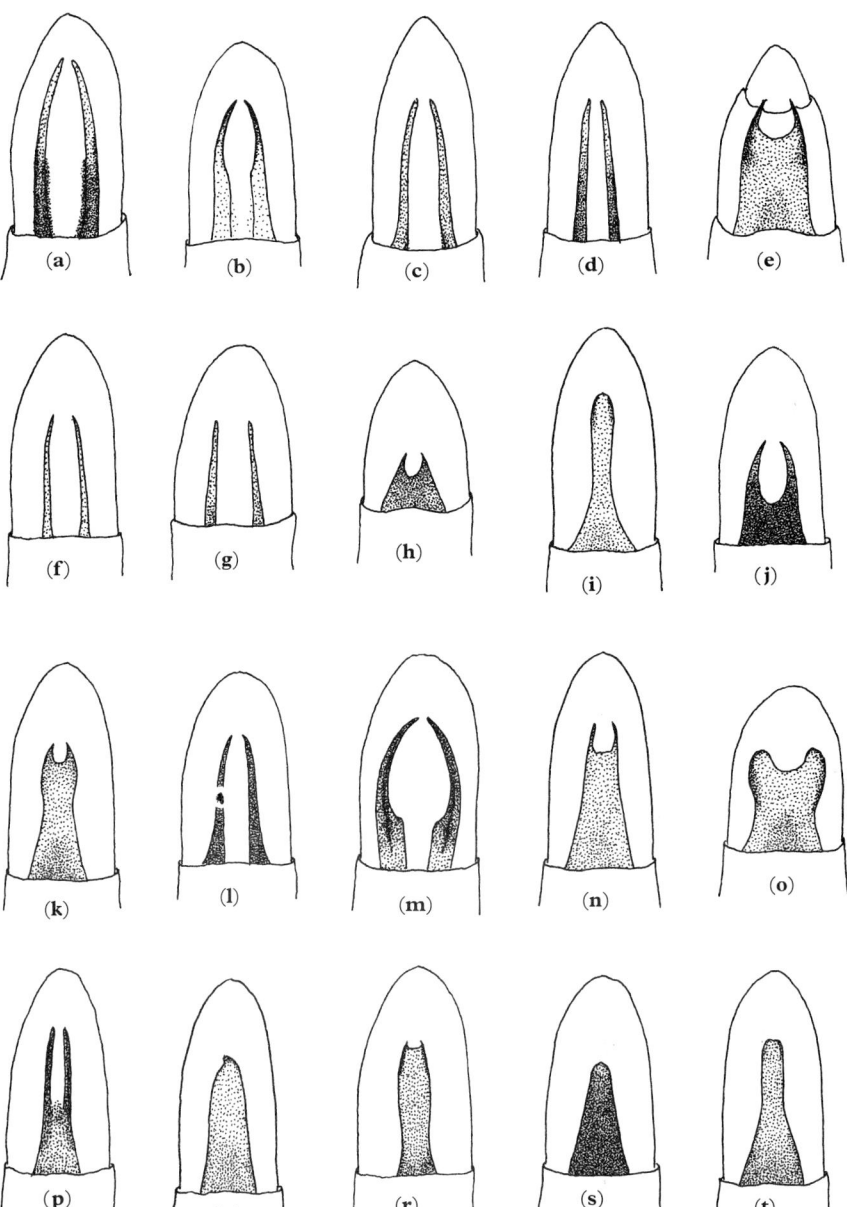

Text figure 27 Male abdominal plate

(a) *E. tenuiata* (p. 33) (b) *E. inturbata* (p. 37) (c) *E. haworthiata* (p. 39)
(d) *E. plumbeolata* (p. 41) (e) *E. abietaria* (p. 43) (f) f. *E. linariata*★ (p. 45)
(g) *E. pulchellata*★ (p. 47) (h) *E. irriguata* (p. 49) (i) *E. exiguata* (p. 51)
(j) *E. insigniata* (p. 52) (k) *E. valerianata* (p. 54) (l) *E. pygmaeata* (p. 55)
(m) *E. venosata* (p. 58) (n) *E. egenaria* (p. 60) (o) *E. centaureata* (p. 63)
(p) *E. trisignaria* (p. 64) (q) *E. intricata* (p. 68) (r) *E. cauchiata* (p. 76)
(s) *E. satyrata* (p. 72) (t) *E. absinthiata*★ (p. 77)

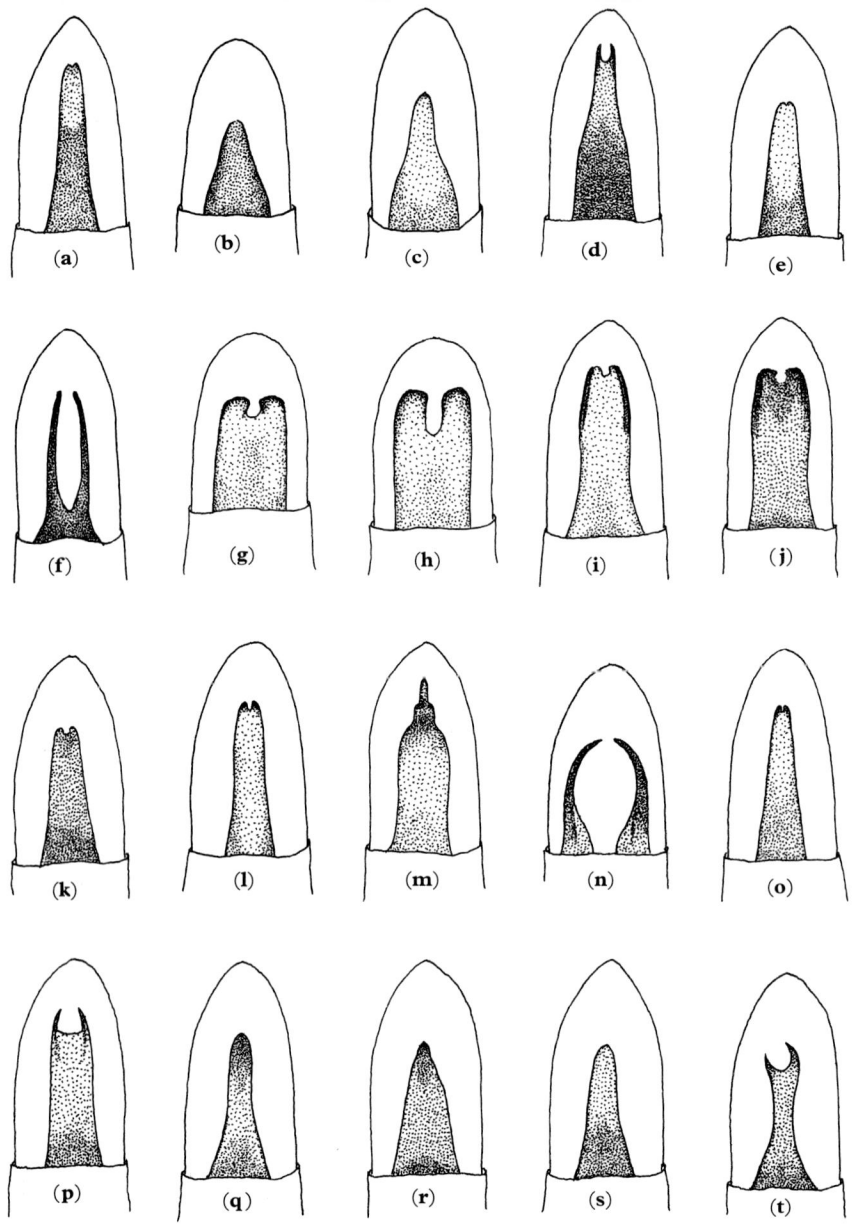

Text figure 28 Male abdominal plate

(a) *E. expallidata** (p. 79) (b) *E. assimilata* (p. 81) (c) *E. vulgata* (p. 83)
(d) *E. tripunctaria* (p. 86) (e) *E. denotata* (p. 89) (f) *E. subfuscata* (p. 91)
(g) *E. icterata* (p. 94) (h) *E. succenturiata* (p. 96) (i) *E. subumbrata* (p. 98)
(j) *E. sinuosaria* (p. 103) (k) *E. millefoliata* (p. 99) (l) *E. simpliciata* (p. 101)
(m) *E. distinctaria* (p. 103) (n) *E. ultimaria* (p. 126) (o) *E. indigata* (p. 105)
(p) *E. pimpinellata* (p. 106) (q) *E. nanata* (p. 108) (r) *E. extensaria* (p. 110)
(s) *E. fraxinata* (p. 112) (t) *E. virgaureata* (p. 114)

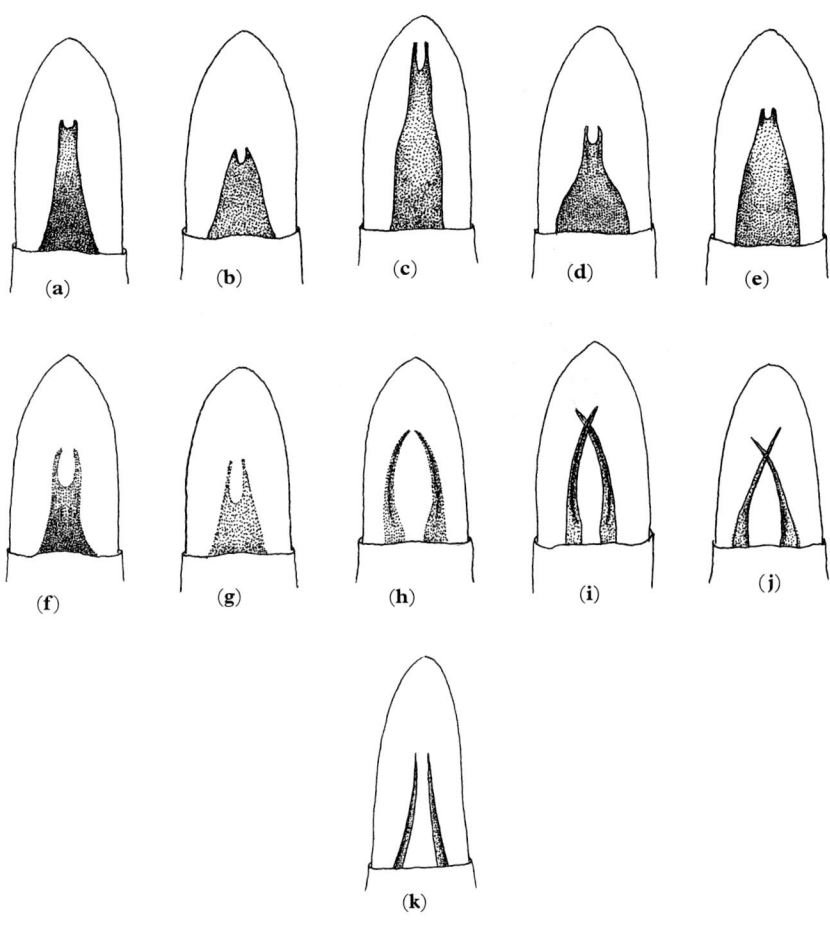

Text figure 29 Male abdominal plate

(a) *E. abbreviata* (p. 117) (b) *E. dodoneata* (p. 119) (c) *E. pusillata* (p. 122)
(d) *E. phoeniceata* (p. 124) (e) *E. lariciata* (p. 127) (f) *E. tantillaria* (p. 130)
(g) *C. v-ata* (p. 131) (h) *P. chloerata* (p. 133) (i) *P. rectangulata* (p. 135)
(j) *P. debiliata* (p. 137) (k) *G. rufifasciata* (p. 139)

6: Illustrations of the larvae

Final-instar larvae are illustrated (Figs 30–33) mainly from the dorsal aspect as most can be identified from their dorsal markings. Where lateral markings are important diagnostic features, a lateral view is also given. In species where the dorsal markings are extremely variable, examples of the variation on a central segment are shown. The actual size of the fully-grown larva is indicated by the line to the left of each illustration.

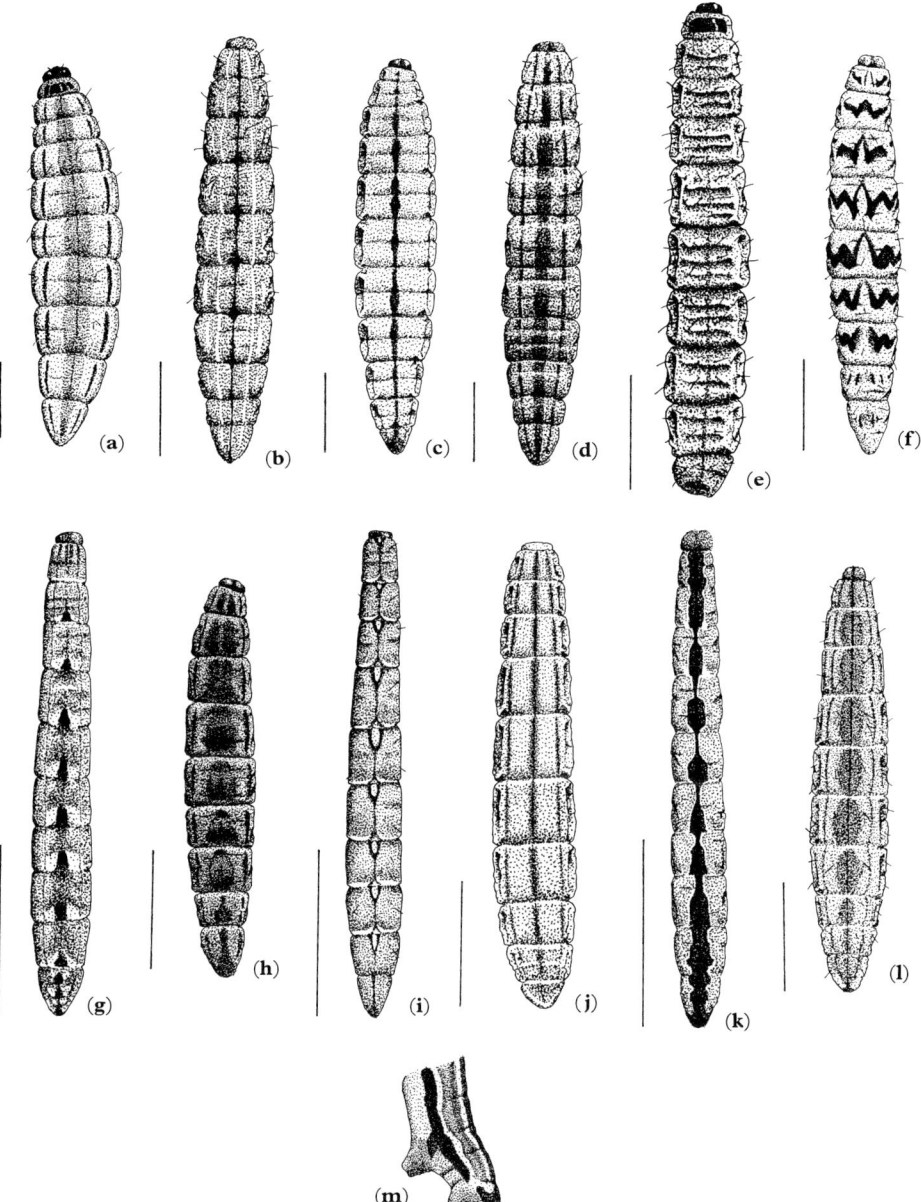

Text figure 30 Larvae

(**a**) *E. tenuiata* (p. 34) (**b**) *E. inturbata* (p. 38) (**c**) *E. haworthiata* (p. 40)
(**d**) *E. plumbeolata* (p. 42) (**e**) *E. abietaria* (p. 44) (**f**) *E. linariata* (p. 46)
(**g**) *E. irriguata* (p. 50) (**h**) *E. pulchellata* (p. 48) (**i**) *E. exiguata* (p. 51)
(**j**) *E. valerianata* (p. 55) (**k**) *E. insigniata* (p. 52) (**l**) *E. pygmaeata* (p. 56)
(**m**) *E. insigniata*, posterior lateral view (p. 52)

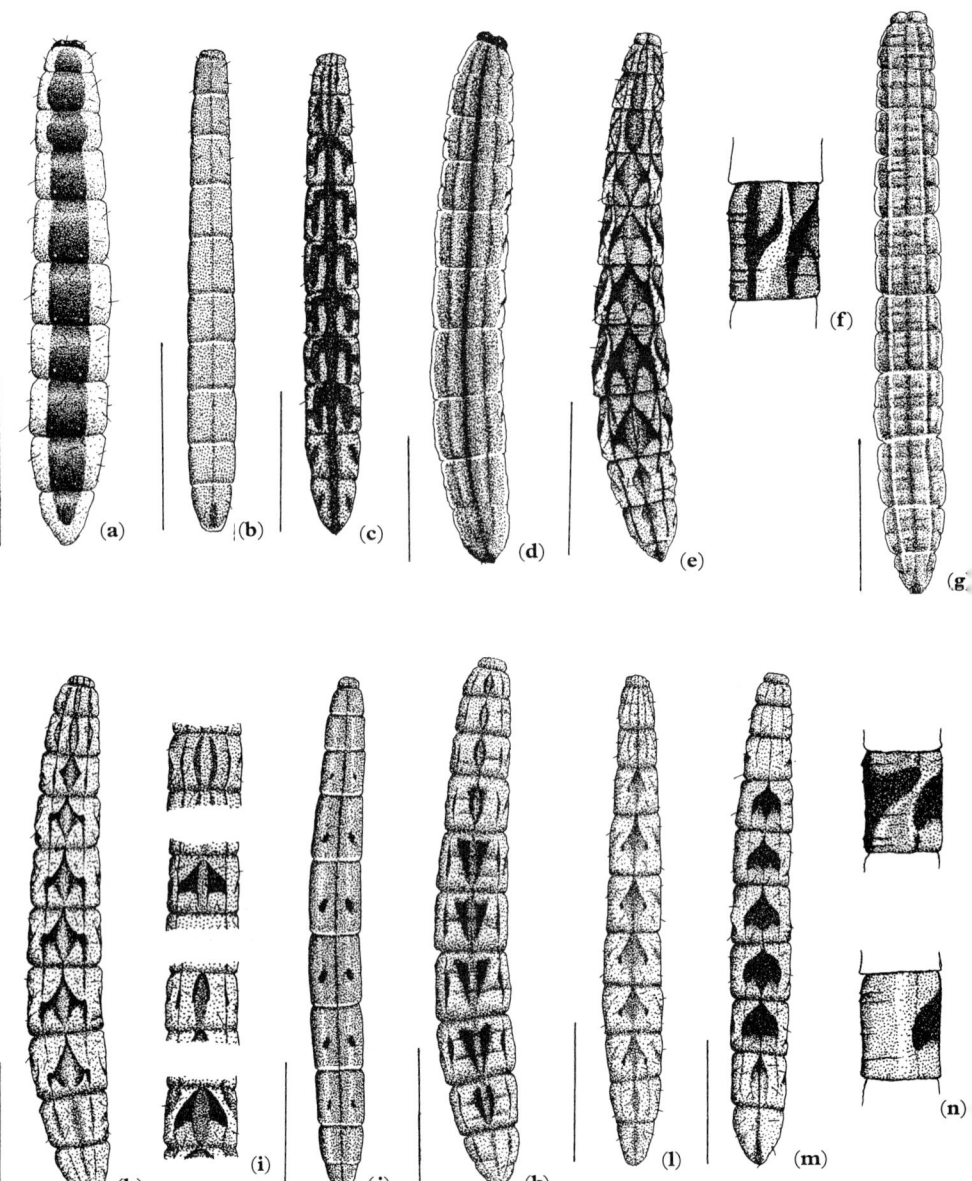

Text figure 31 Larvae

(**a**) *E. venosata* (p. 59) (**b**) *E. egenaria* (p. 61) (**c**) *E. centaureata* (p. 63) (**d**) *E. trisignaria* (p. 66) (**e**) *E. satyrata* (p. 74) (**f**) *E. satyrata*, lateral view of central segment (p. 74) (**g**) *E. intricata* (p. 70) (**h**) *E. absinthiata* (p. 78) (**i**) *E. absinthiata*, variation in dorsal markings (p. 78) (**j**) *E. assimilata* (p. 82) (**k**) *E. expallidata* (p. 800) (**l**) *E. vulgata* (p. 85) (**m**) *E. tripunctaria* (p. 87) (**n**) *E. tripunctaria*, lateral view of central segment showing variation (p. 87)

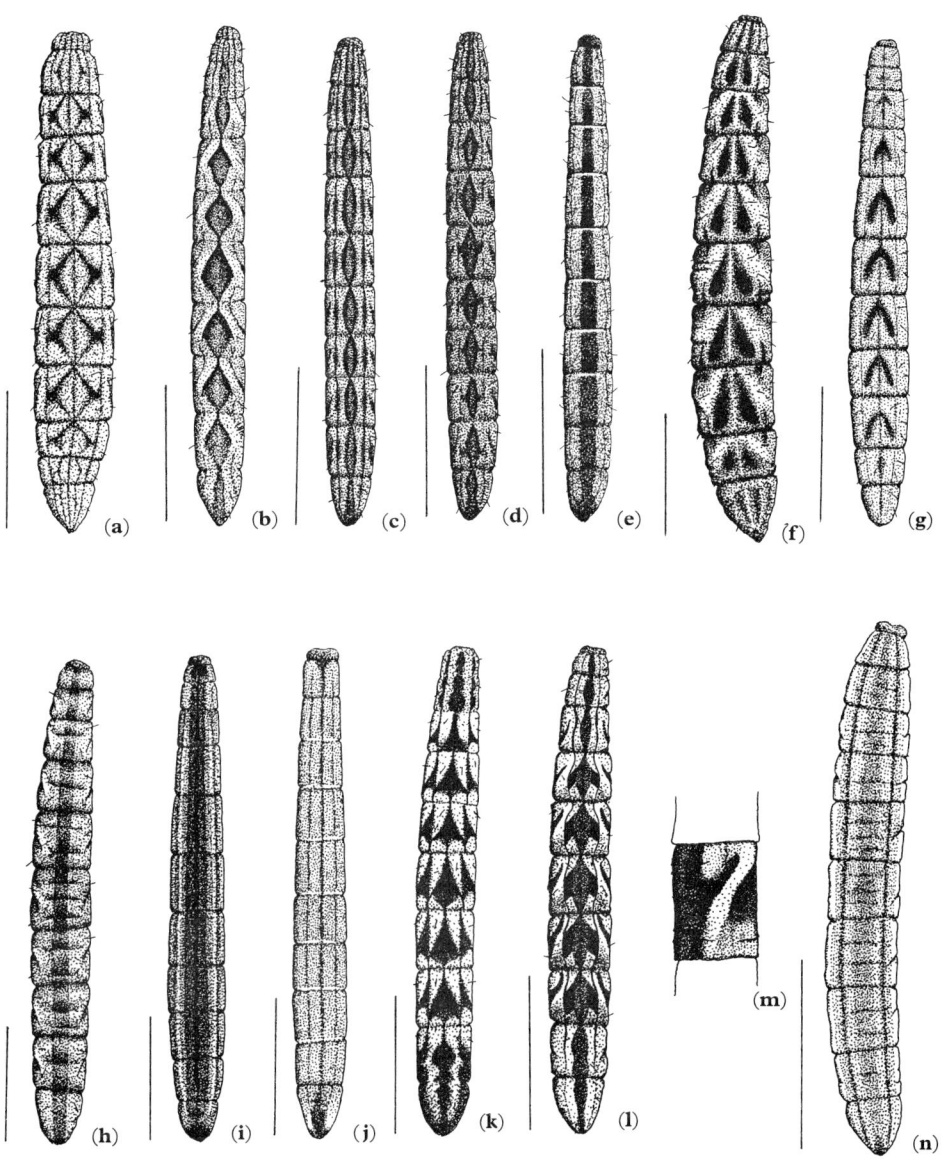

Text figure 32 Larvae

(**a**) *E. denotata* (p. 90) (**b**) *E. subfuscata* (p. 93) (**c**) *E. icterata* (p. 95)
(**d**) *E. succenturiata* (p. 97) (**e**) *E. subumbrata* (p. 98) (**f**) *E. millefoliata* (p. 100)
(**g**) *E. simpliciata* (p. 101) (**h**) *E. distinctaria* (p. 104) (**i**) *E. indigata* (p. 105)
(**j**) *E. pimpinellata* (p. 107) (**k**) *E. nanata* (p. 109) (**l**) *E. virgaureata* (p. 116)
(**m**) *E. virgaureata*, lateral view of central segment (p. 116) (**n**) *E. extensaria* (p. 111)

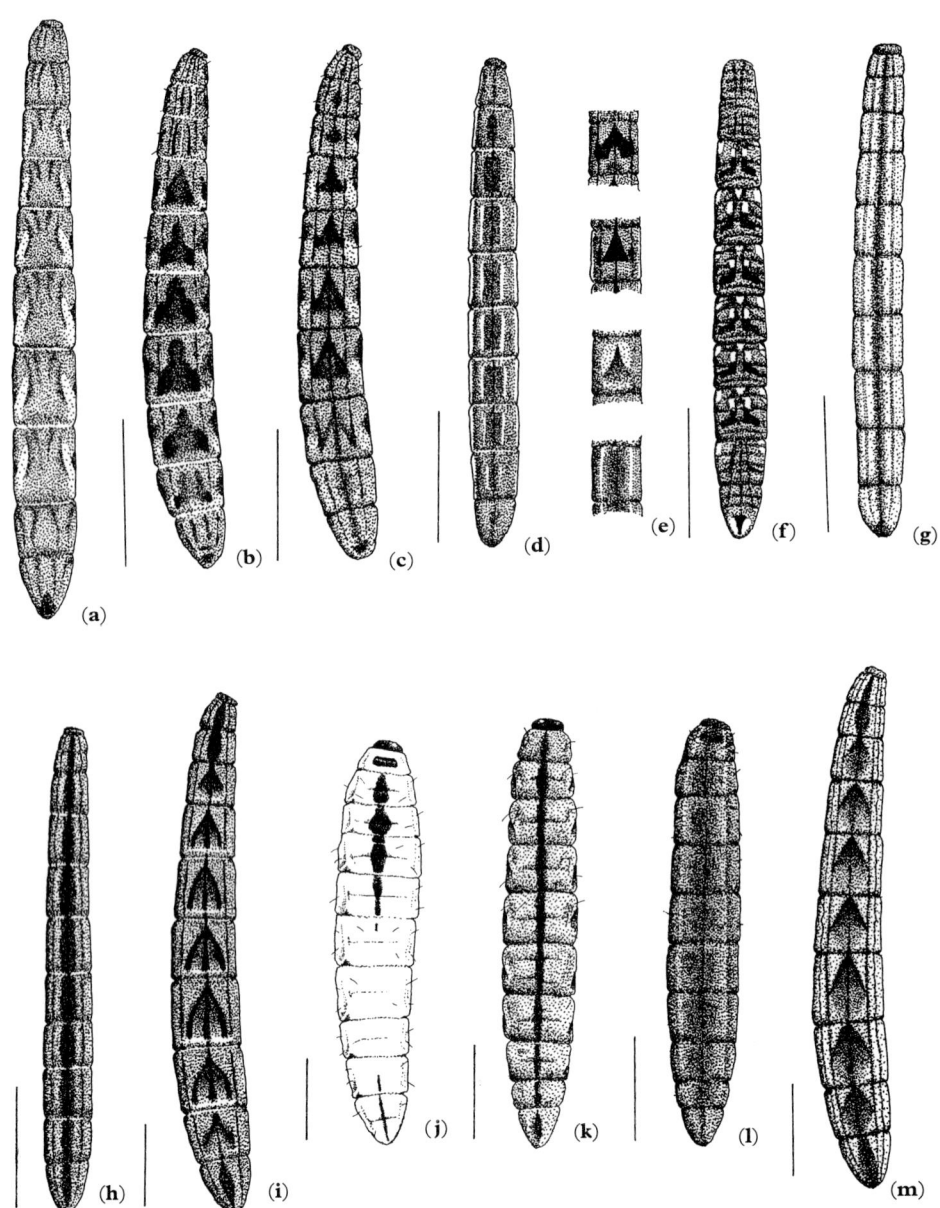

Text figure 33 Larvae

(**a**) *E. fraxinata* (p. 112) (**b**) *E. abbreviata* (p. 117) (**c**) *E. dodoneata* (p. 119)
(**d**) *E. pusillata* (p. 123) (**e**) *E. pusillata*, variation in dorsal markings (p. 123)
(**f**) *E. phoeniceata* (p. 125) (**g**) *E. lariciata* (p. 128) (**h**) *E. tantillaria* (p. 130)
(**i**) *C. v-ata* (p. 131) (**j**) *P. chloerata* (p. 134) (**k**) *P. rectangulata* (p. 136)
(**l**) *P. debiliata* (p. 137) (**m**) *G. rufifasciata* (p. 139)

7: Vice-county distribution maps

Key to Vice-counties

England and Wales

1. West Cornwall (with Scilly)
2. East Cornwall
3. South Devon
4. North Devon
5. South Somerset
6. North Somerset
7. North Wiltshire
8. South Wiltshire
9. Dorset
10. Isle of Wight
11. South Hampshire
12. North Hampshire
13. West Sussex
14. East Sussex
15. East Kent
16. West Kent
17. Surrey
18. South Essex
19. North Essex
20. Hertfordshire
21. Middlesex
22. Berkshire
23. Oxfordshire
24. Buckinghamshire
25. East Suffolk
26. West Suffolk
27. East Norfolk
28. West Norfolk
29. Cambridgeshire
30. Bedfordshire
31. Huntingdonshire
32. Northamptonshire
33. East Gloucestershire
34. West Gloucestershire
35. Monmouthshire
36. Herefordshire
37. Worcestershire
38. Warwickshire
39. Staffordshire
40. Shropshire (Salop)
41. Glamorgan
42. Breconshire
43. Radnorshire
44. Carmarthenshire
45. Pembrokeshire
46. Cardiganshire
47. Montgomeryshire
48. Merionethshire
49. Caernarvonshire
50. Denbighshire
51. Flintshire
52. Anglesey
53. South Lincolnshire
54. North Lincolnshire
55. Leicestershire (with Rutland)
56. Nottinghamshire
57. Derbyshire
58. Cheshire
59. South Lancashire
60. North Lancashire
61. South-east Yorkshire
62. North-east Yorkshire
63. South-west Yorkshire
64. Mid-west Yorkshire
65. North-west Yorkshire
66. Durham
67. South Northumberland
68. North Northumberland (Cheviot)
69. Westmorland with North Lancashire
70. Cumberland
71. Isle of Man

Scotland

72. Dumfries-shire
73. Kirkcudbrightshire
74. Wigtownshire
75. Ayrshire
76. Renfrewshire
77. Lanarkshire
78. Peebleshire
79. Selkirkshire
80. Roxburghshire
81. Berwickshire
82. East Lothian (Haddington)
83. Midlothian (Edinburgh)
84. West Lothian (Linlithgow)
85. Fifeshire (with Kinross)
86. Stirlingshire
87. West Perthshire (with Clackmannan)
88. Mid Perthshire
89. East Perthshire
90. Angus (Forfar)
91. Kincardineshire
92. South Aberdeenshire
93. North Aberdeenshire
94. Banffshire
95. Moray (Elgin)
96. East Inverness-shire (with Nairn)
97. West Inverness-shire
98. Argyll Main
99. Dunbartonshire
100. Clyde Isles
101. Kintyre
102. South Ebudes
103. Mid Ebudes
104. North Ebudes
105. West Ross
106. East Ross
107. East Sutherland
108. West Sutherland
109. Caithness
110. Outer Hebrides
111. Orkney Islands
112. Shetland Islands (Zetland)

113. Channel Islands

Ireland

H.1 South Kerry
H.2 North Kerry
H.3 West Cork
H.4 Mid Cork
H.5 East Cork
H.6 Waterford
H.7 South Tipperary
H.8 Limerick
H.9 Clare
H.10 North Tipperary
H.11 Kilkenny
H.12 Wexford
H.13 Carlow
H.14 Leix (Queen's County)
H.15 South-east Galway
H.16 West Galway
H.17 North-east Galway
H.18 Offaly (King's County)
H.19 Kildare
H.20 Wicklow
H.21 Dublin
H.22 Meath
H.23 West Meath
H.24 Longford
H.25 Roscommon
H.26 East Mayo
H.27 West Mayo
H.28 Sligo
H.29 Leitrim
H.30 Cavan
H.31 Louth
H.32 Monaghan
H.33 Fermanagh
H.34 East Donegal
H.35 West Donegal
H.36 Tyrone
H.37 Armagh
H.38 Down
H.39 Antrim
H.40 Londonderry

Key to symbols

● Generally distributed or widespread
● Not generally distributed
○ Uncertain status

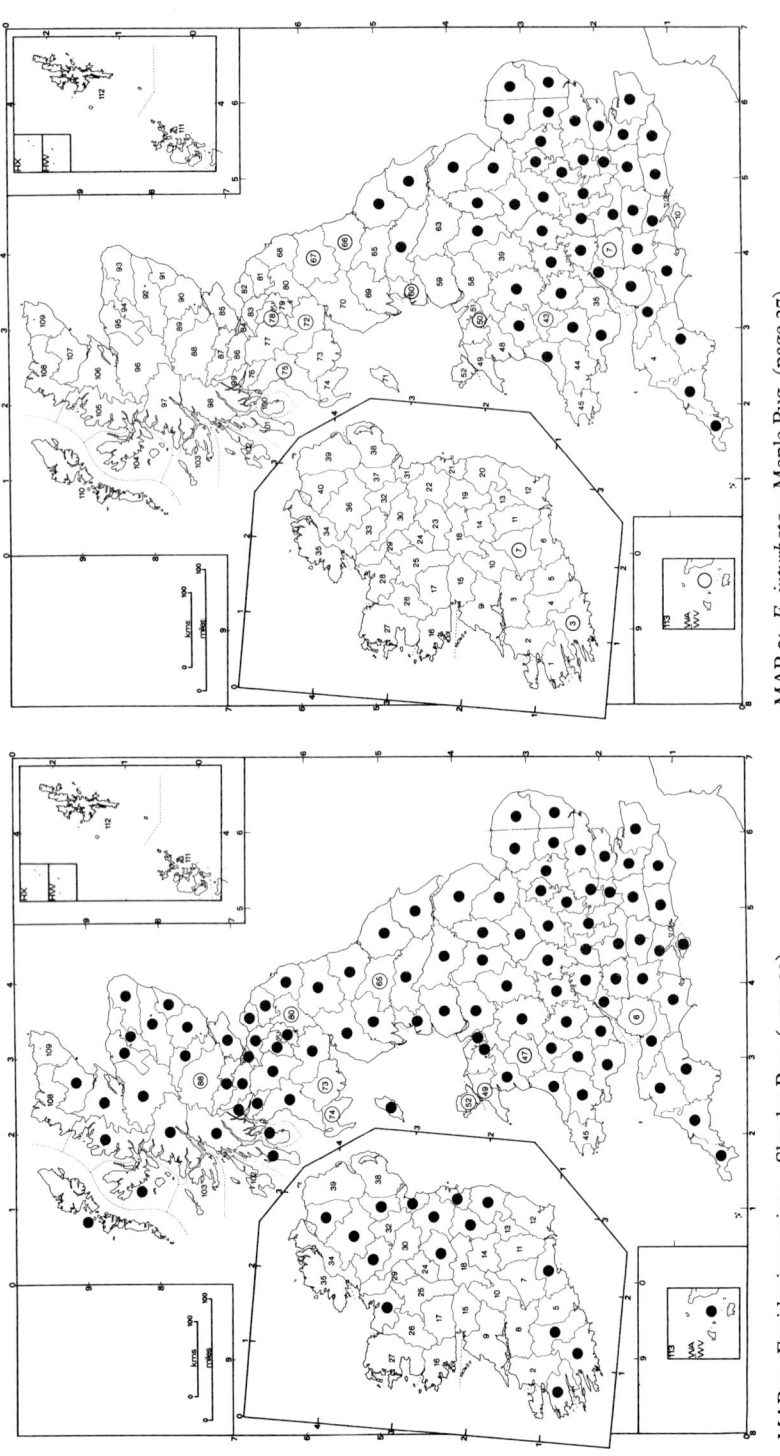

MAP 2: *E. inturbata* – Maple Pug (page 37)

MAP 1: *Eupithecia tenuiata* – Slender Pug (page 33)

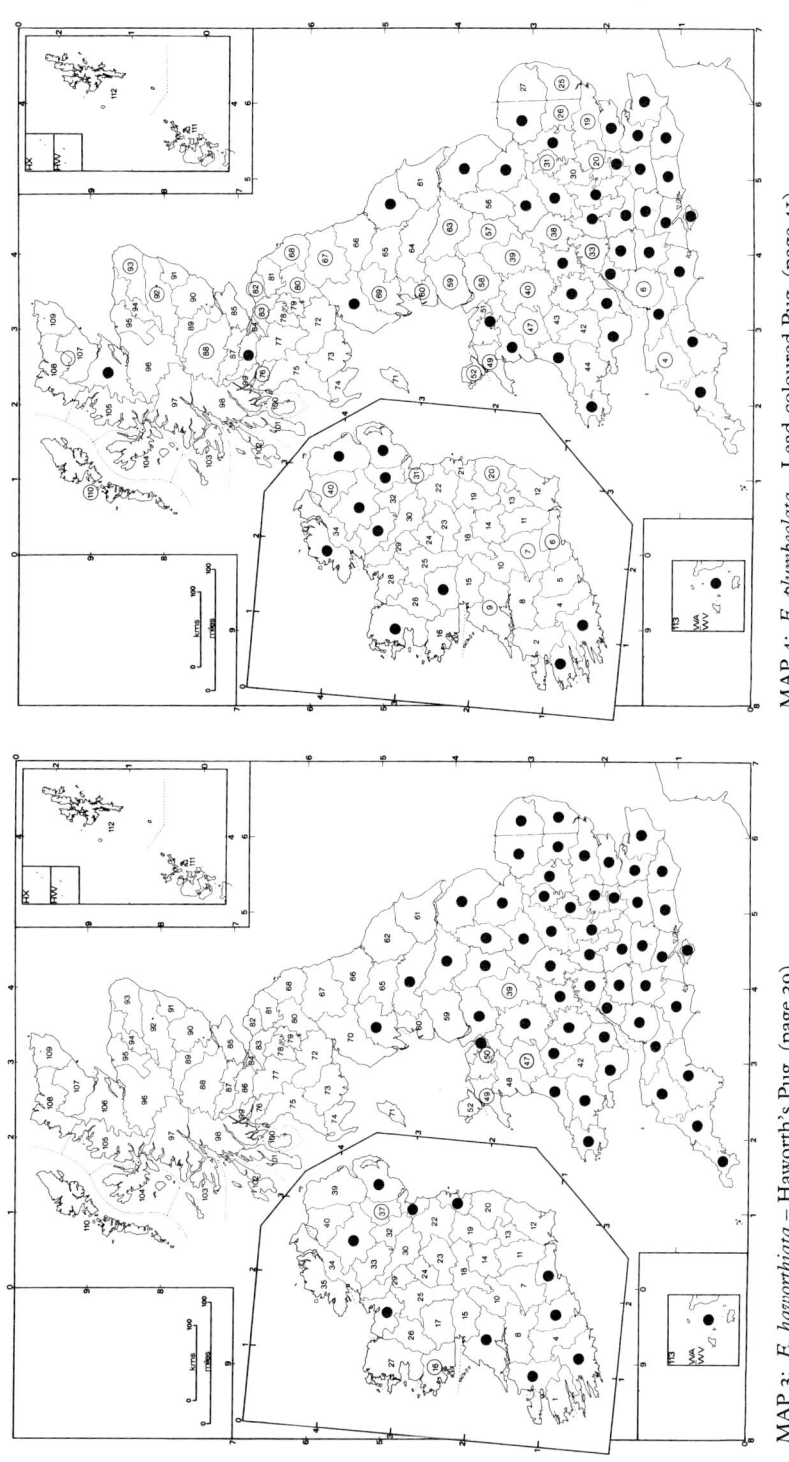

MAP 3: *E. haworthiata* – Haworth's Pug (page 39)

MAP 4: *E. plumbeolata* – Lead-coloured Pug (page 41)

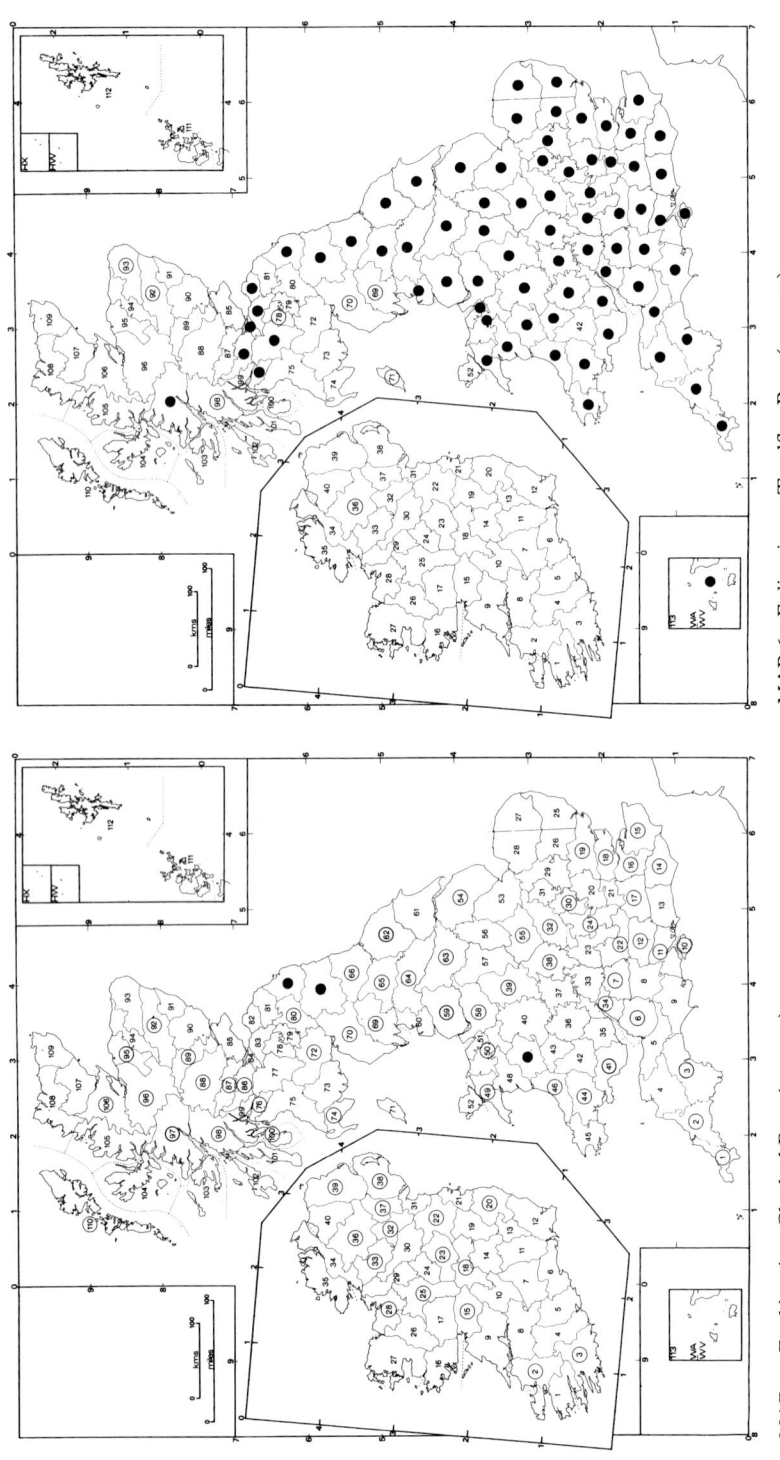

MAP 5: *E. abietaria* – Cloaked Pug (page 43)

MAP 6: *E. linariata* – Toadflax Pug (page 45)

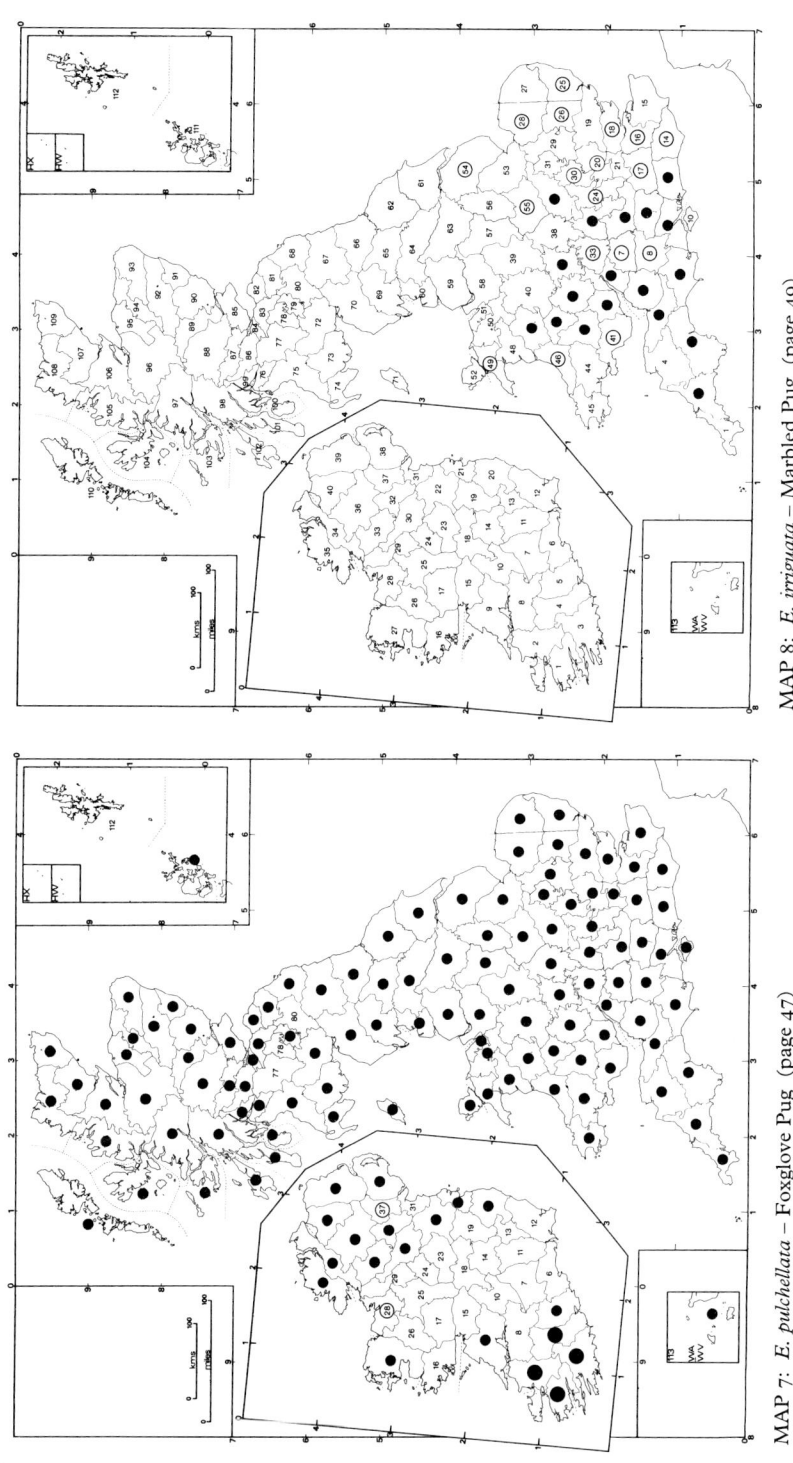

MAP 7: *E. pulchellata* – Foxglove Pug (page 47)

MAP 8: *E. irriguata* – Marbled Pug (page 49)

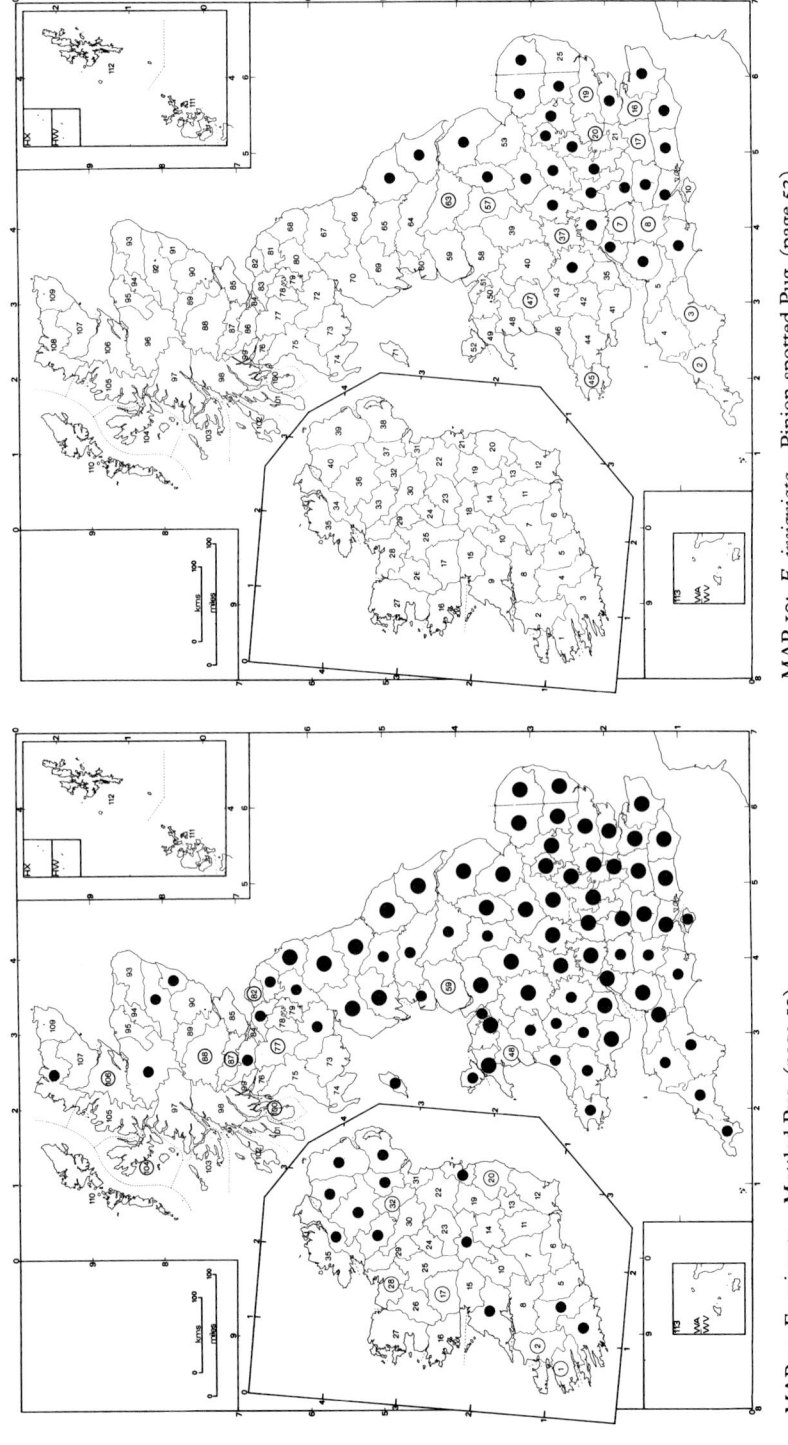

MAP 9: *E. exiguata* – Mottled Pug (page 50)

MAP 10: *E. insigniata* – Pinion-spotted Pug (page 52)

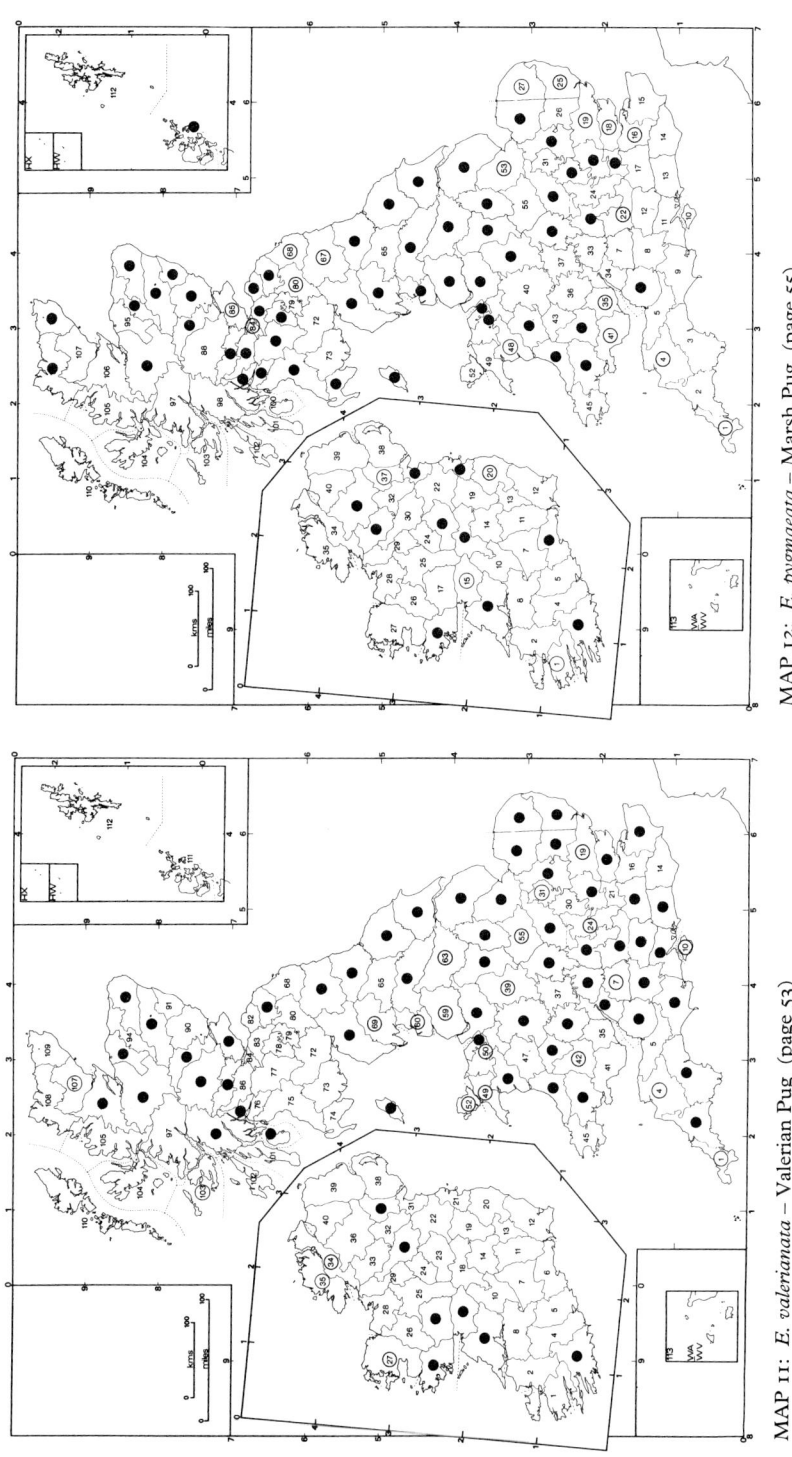

MAP 11: *E. valerianata* – Valerian Pug (page 53)

MAP 12: *E. pygmaeata* – Marsh Pug (page 55)

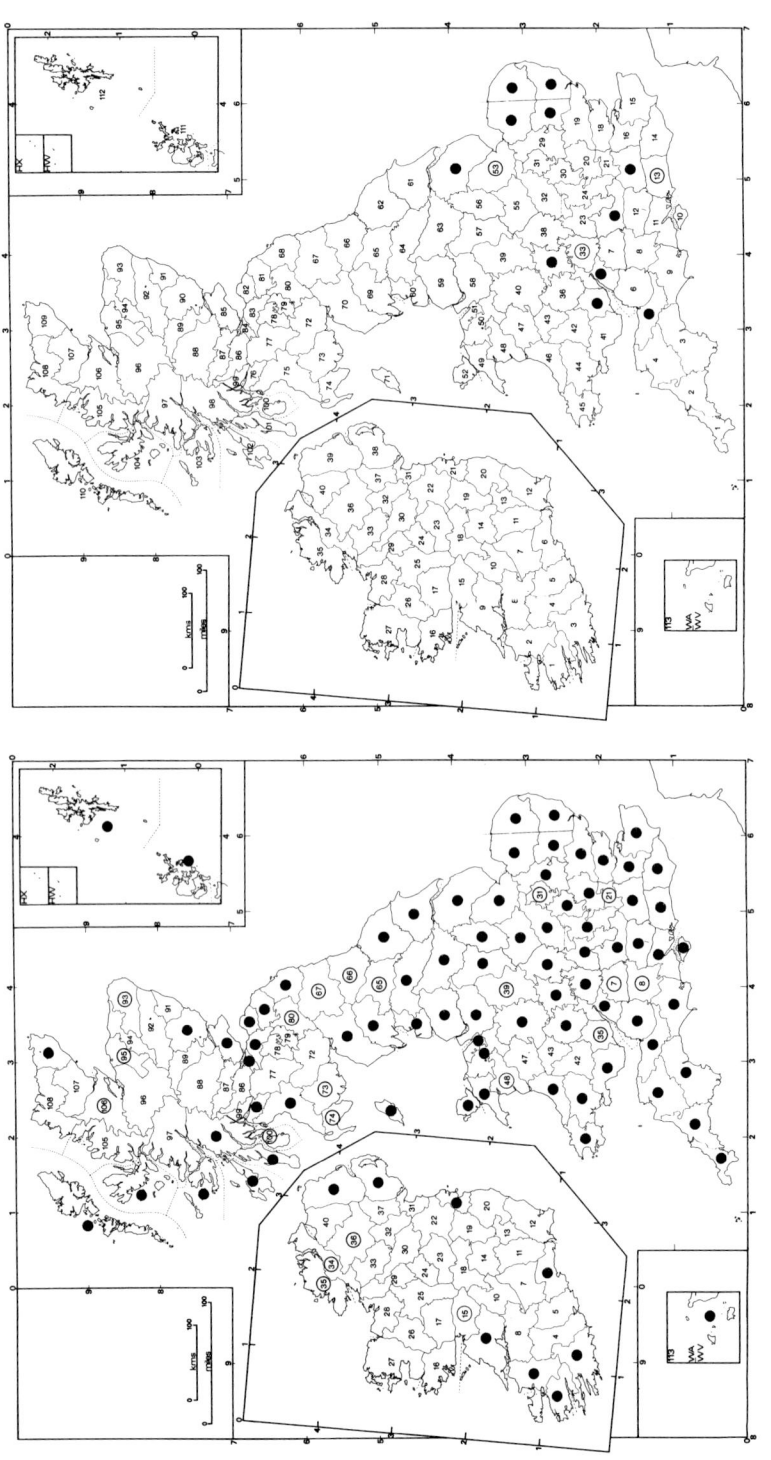

MAP 13: *E. venosata* – Netted Pug (page 57)

MAP 14: *E. egenaria* – Pauper Pug, Fletcher's Pug (page 59)

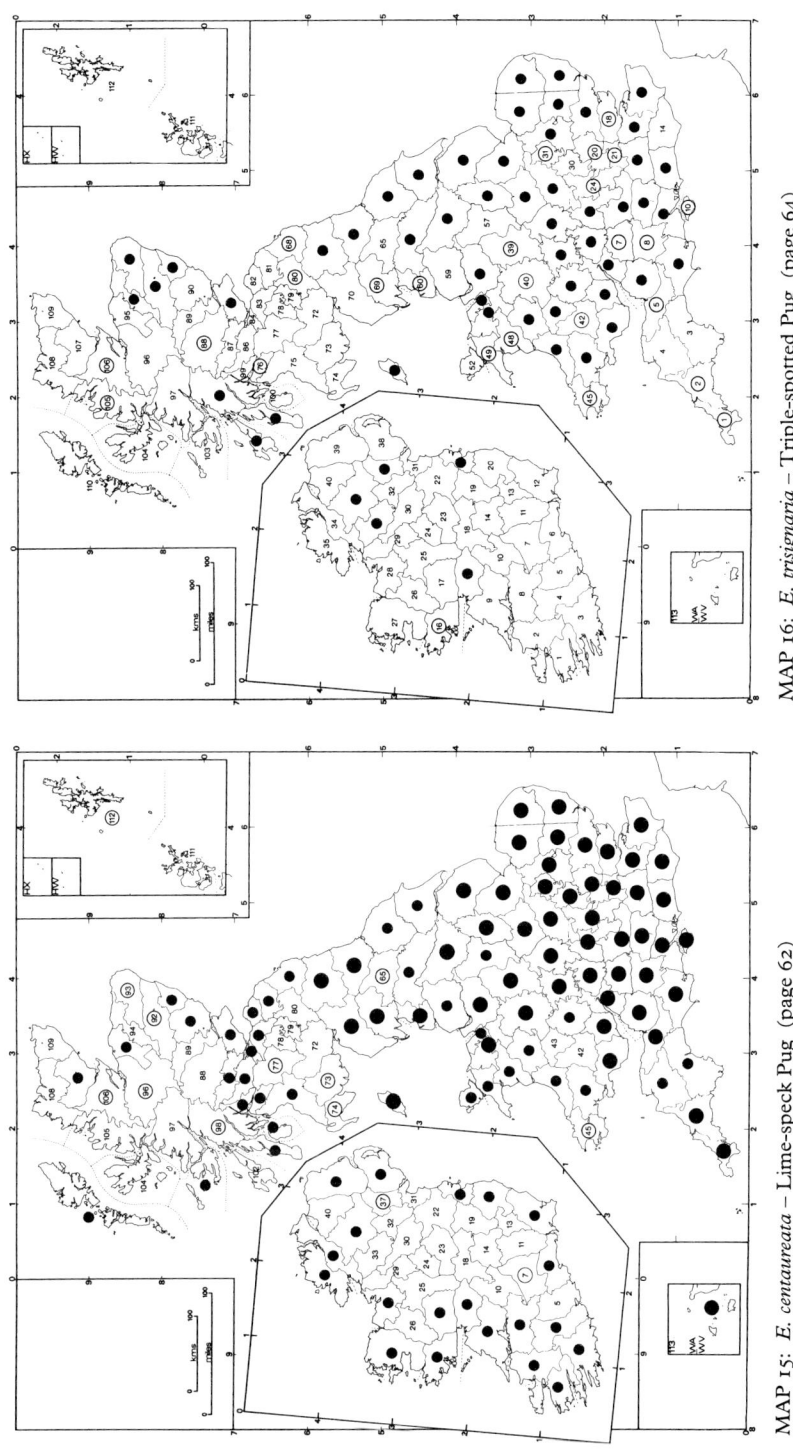

MAP 16: *E. trisignaria* – Triple-spotted Pug (page 64)

MAP 15: *E. centaureata* – Lime-speck Pug (page 62)

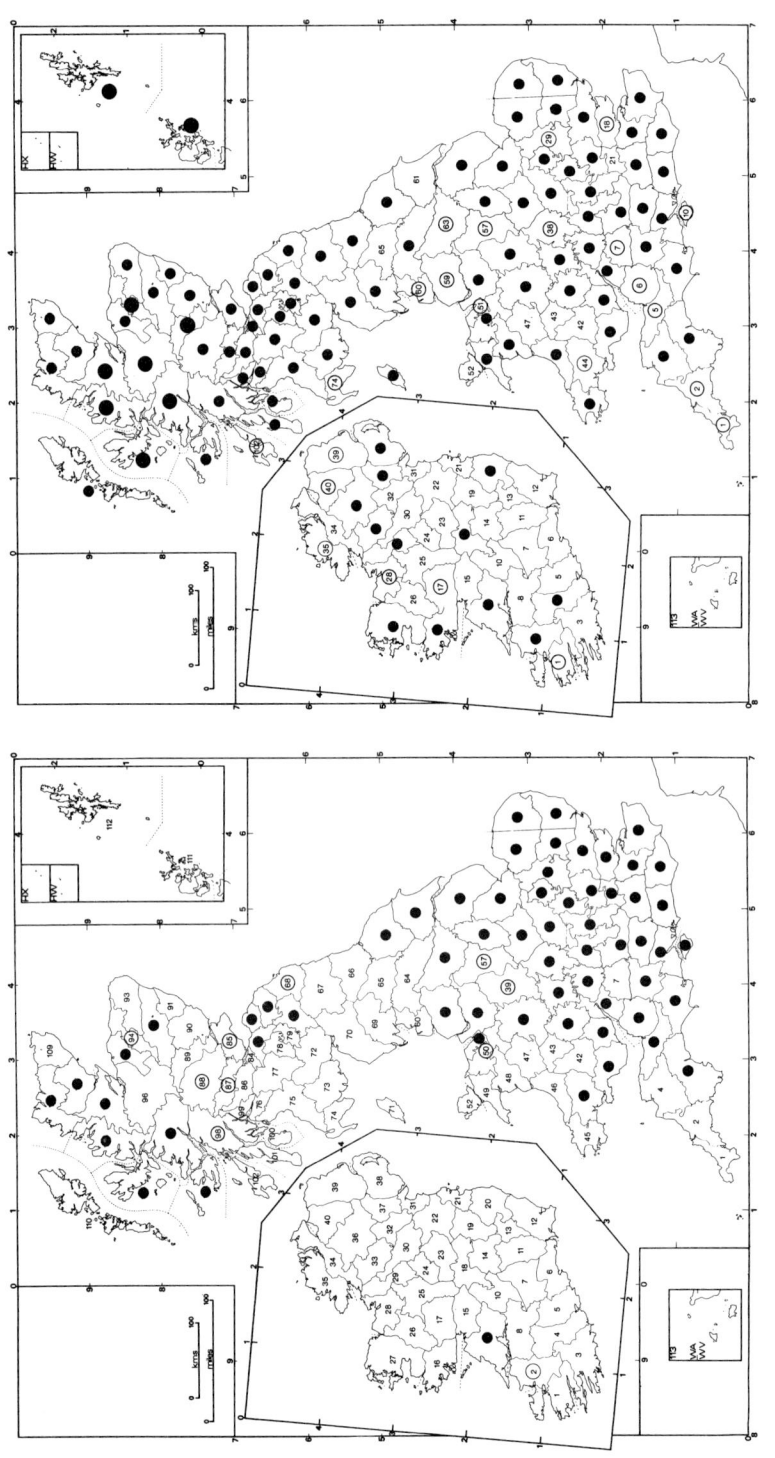

MAP 18: *E. satyrata* – Satyr Pug (page 72)

MAP 17: *E. intricata* – Freyer's Pug (page 67)

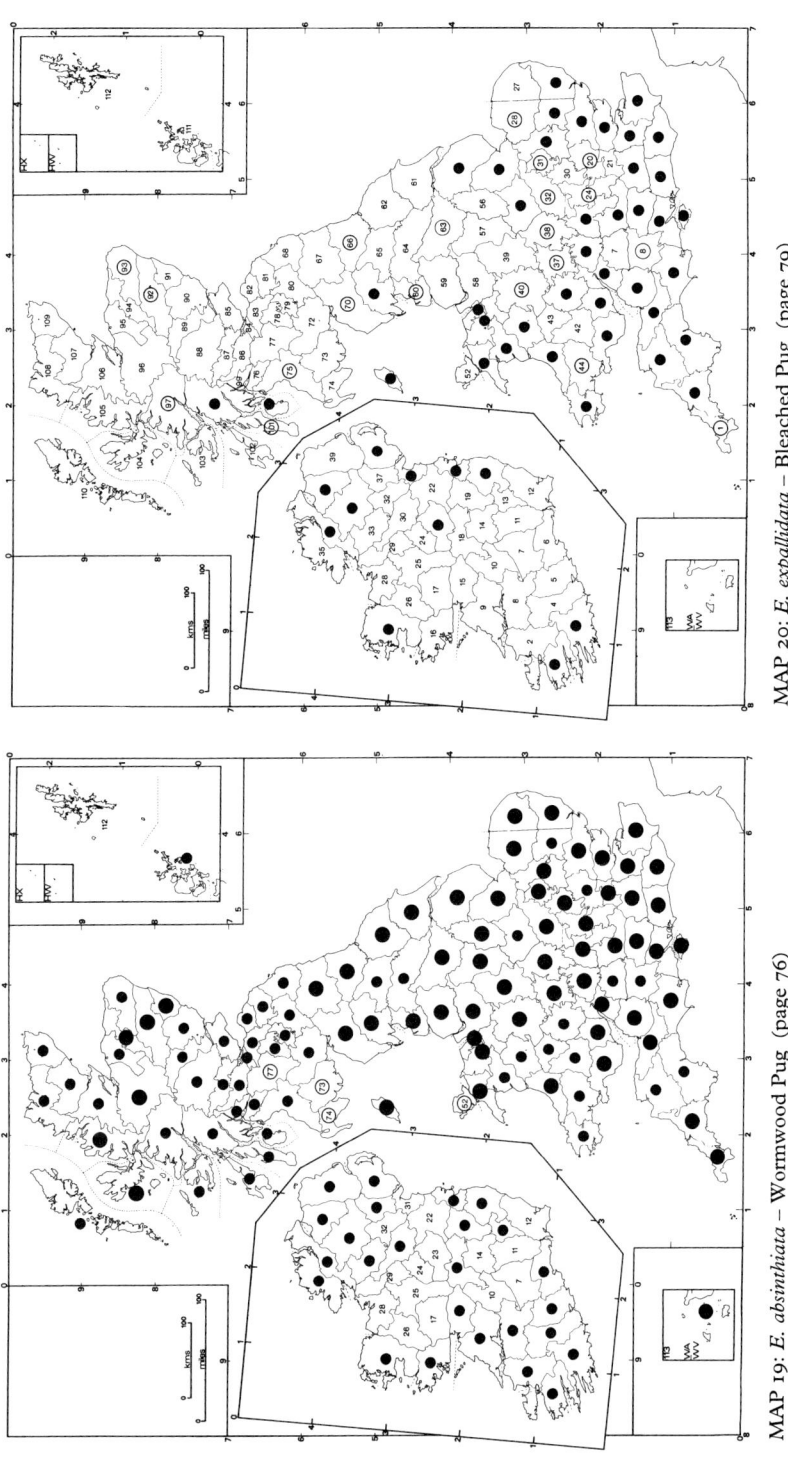

MAP 20: *E. expallidata* – Bleached Pug (page 79)

MAP 19: *E. absinthiata* – Wormwood Pug (page 76)

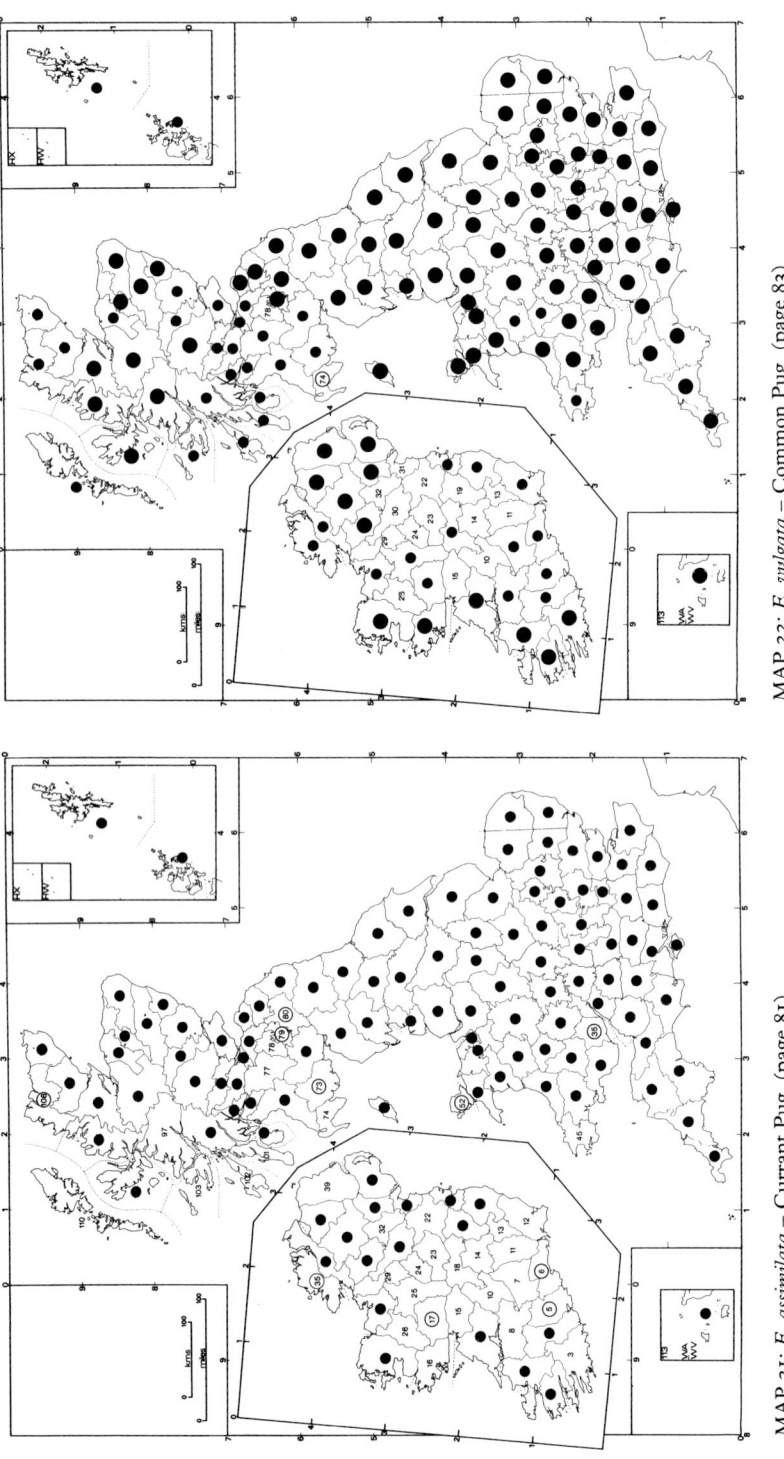

MAP 22: *E. vulgata* – Common Pug (page 83)

MAP 21: *E. assimilata* – Currant Pug (page 81)

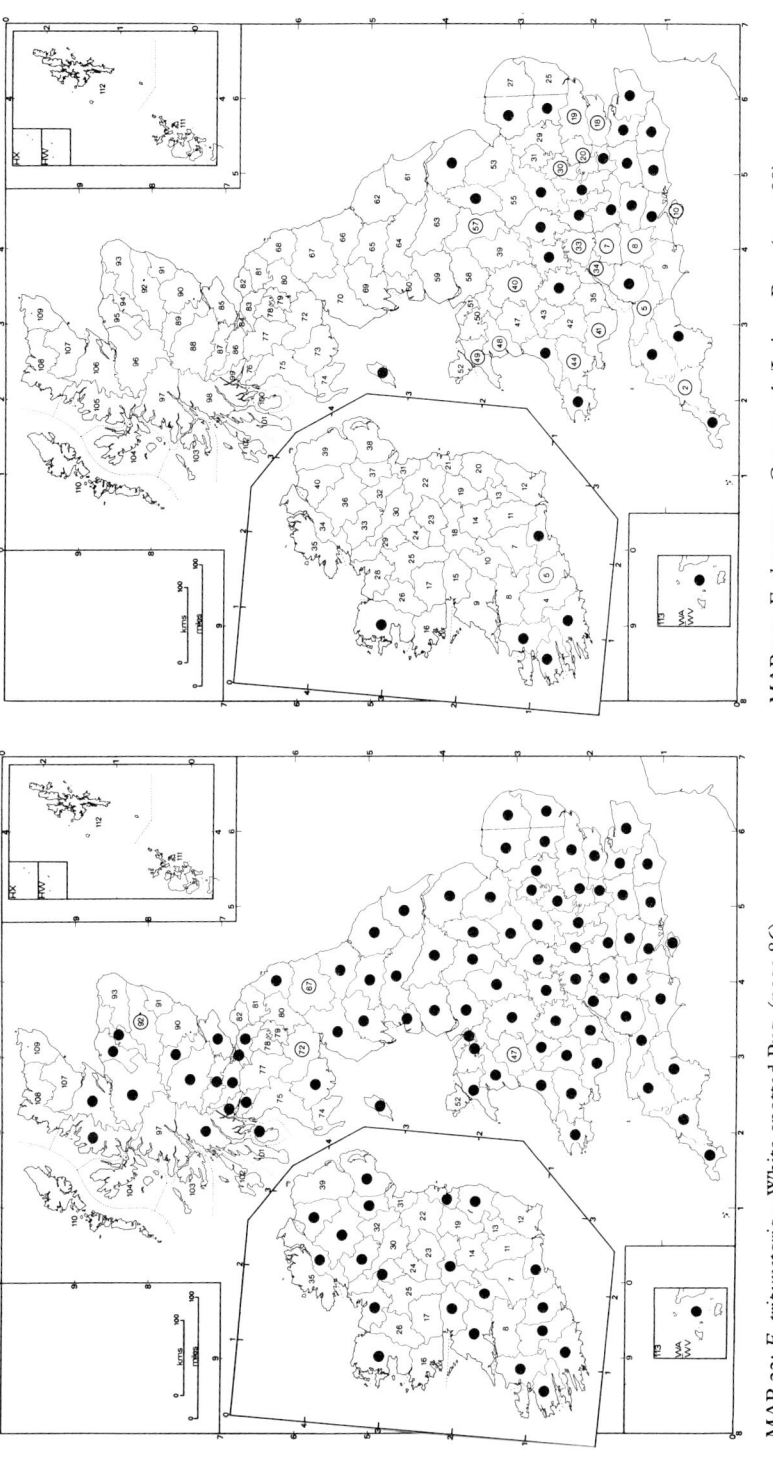

MAP 24: *E. denotata* – Campanula/Jasione Pug (page 88)

MAP 23: *E. tripunctaria* – White-spotted Pug (page 86)

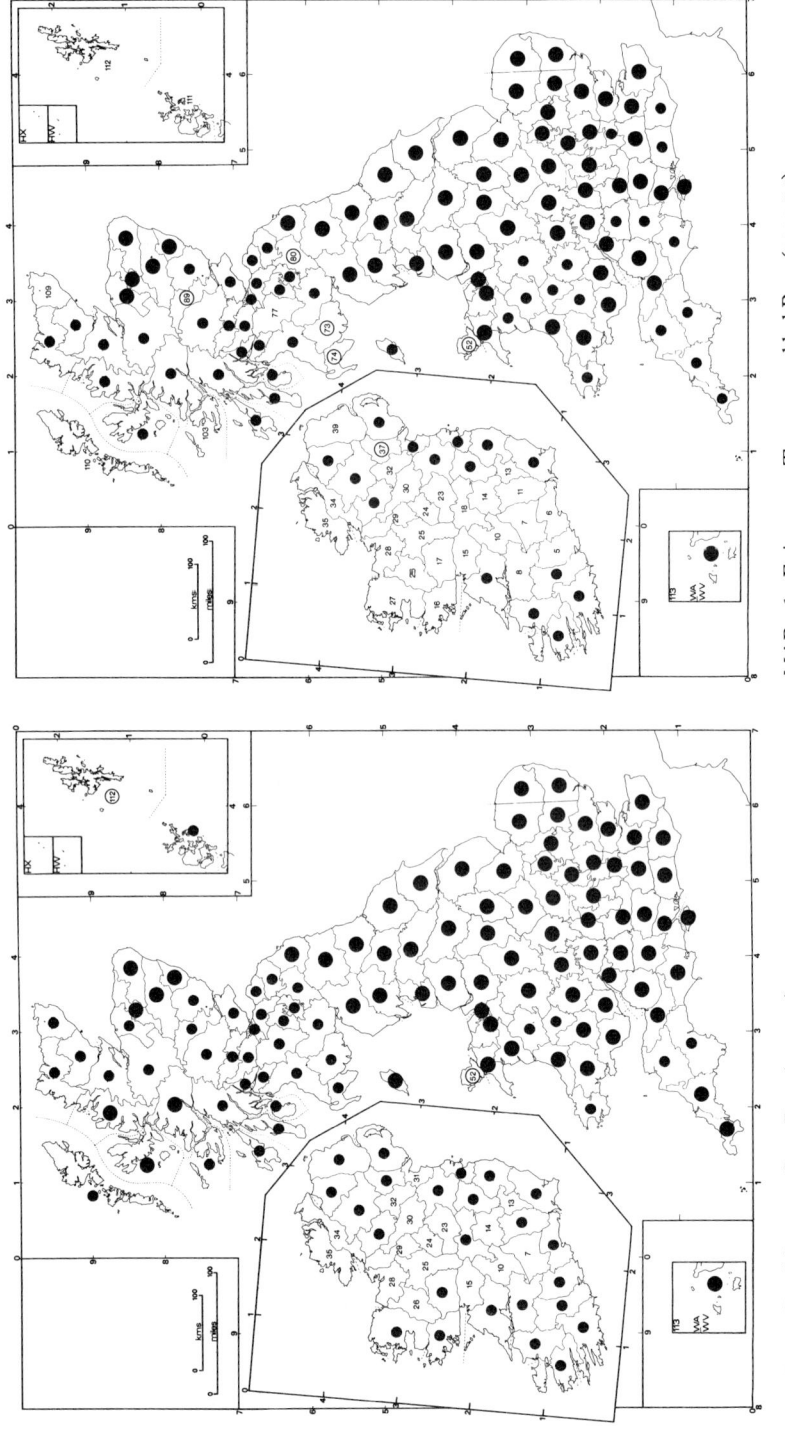

MAP 26: *E. icterata* – Tawny-speckled Pug (page 94)

MAP 25: *E. subfuscata* – Grey Pug (page 91)

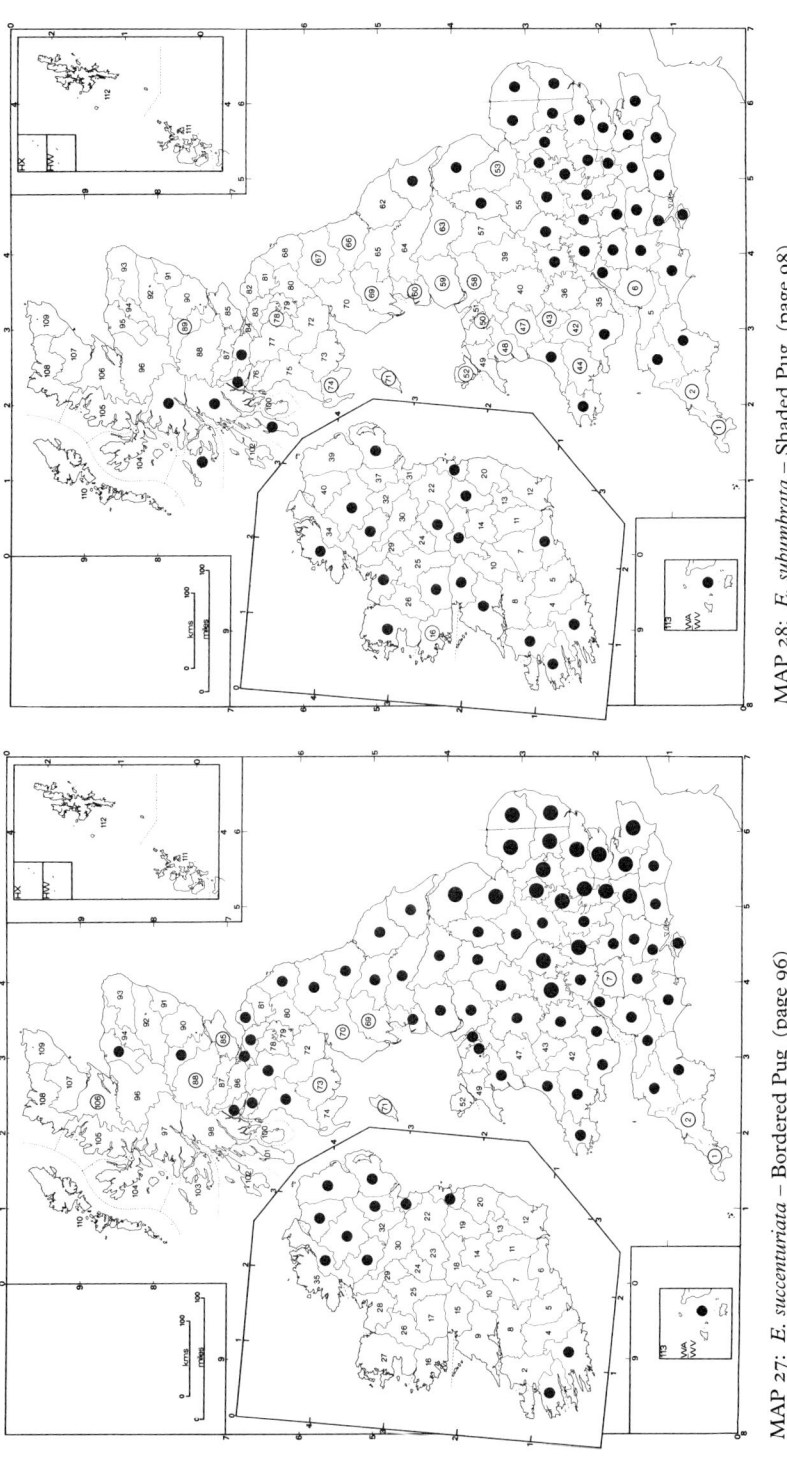

MAP 28: *E. subumbrata* – Shaded Pug (page 98)

MAP 27: *E. succenturiata* – Bordered Pug (page 96)

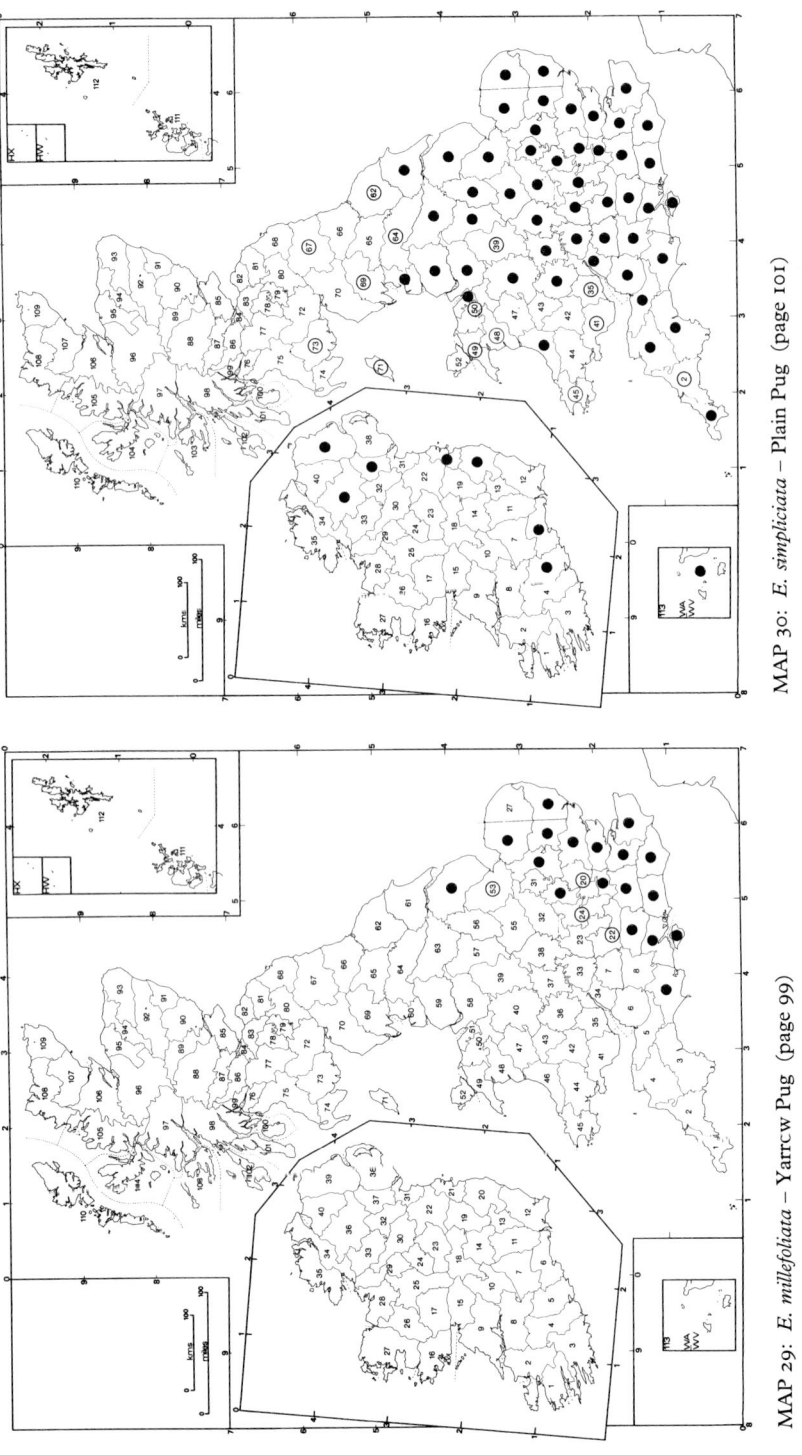

MAP 30: *E. simpliciata* – Plain Pug (page 101)

MAP 29: *E. millefoliata* – Yarrcw Pug (page 99)

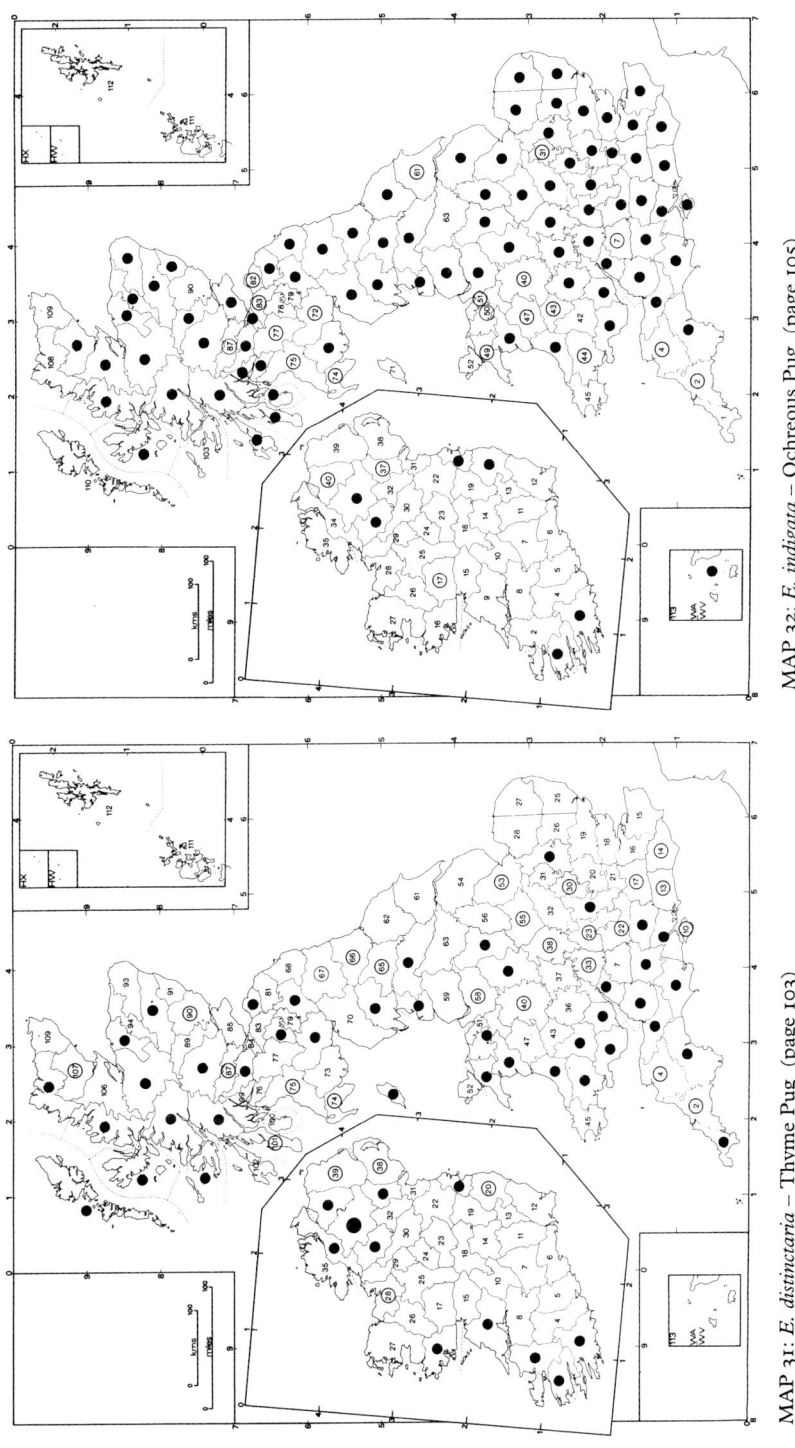

MAP 31: *E. distinctaria* – Thyme Pug (page 103)

MAP 32: *E. indigata* – Ochreous Pug (page 105)

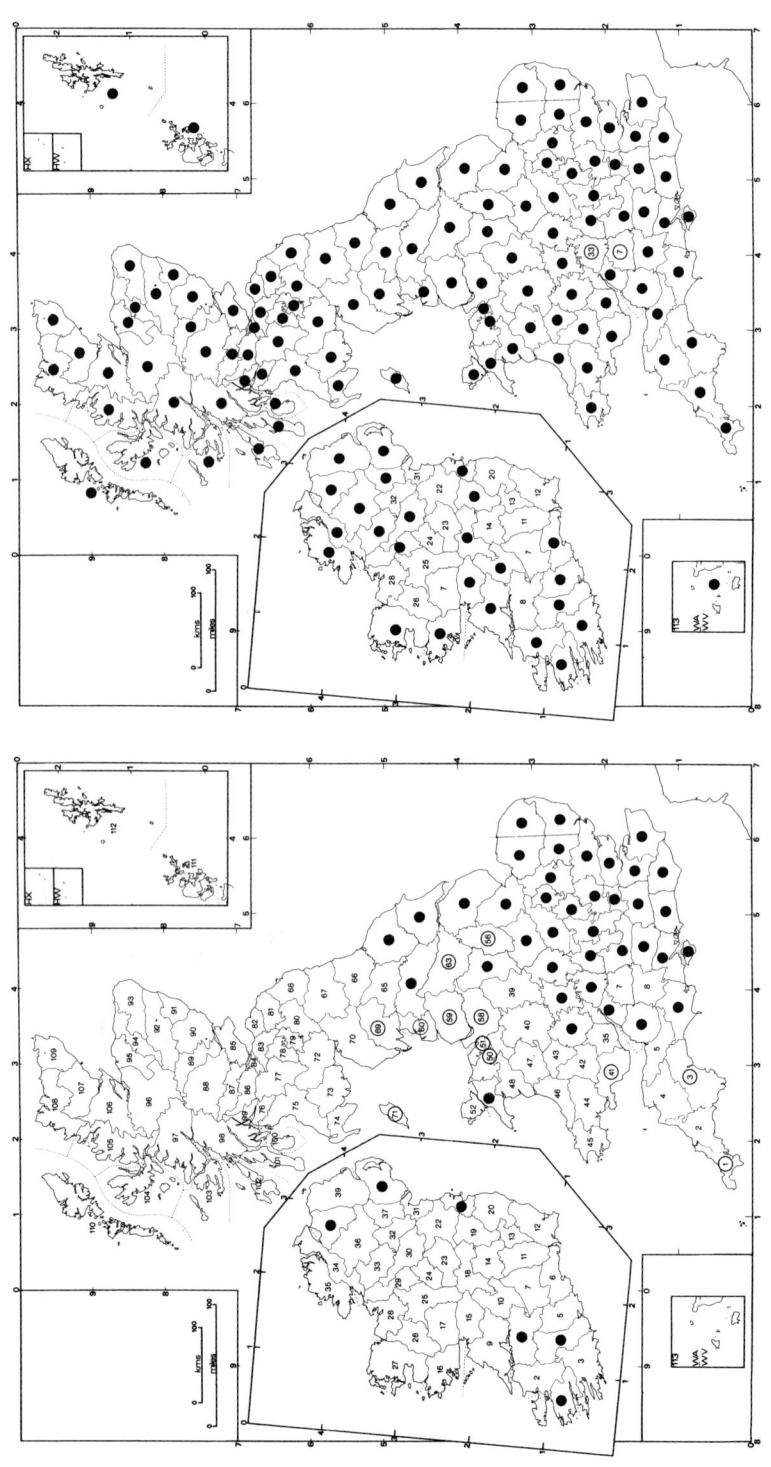

MAP 34: *E. nanata* – Narrow-winged Pug (page 108)

MAP 33: *E. pimpinellata* – Pimpinel Pug (page 106)

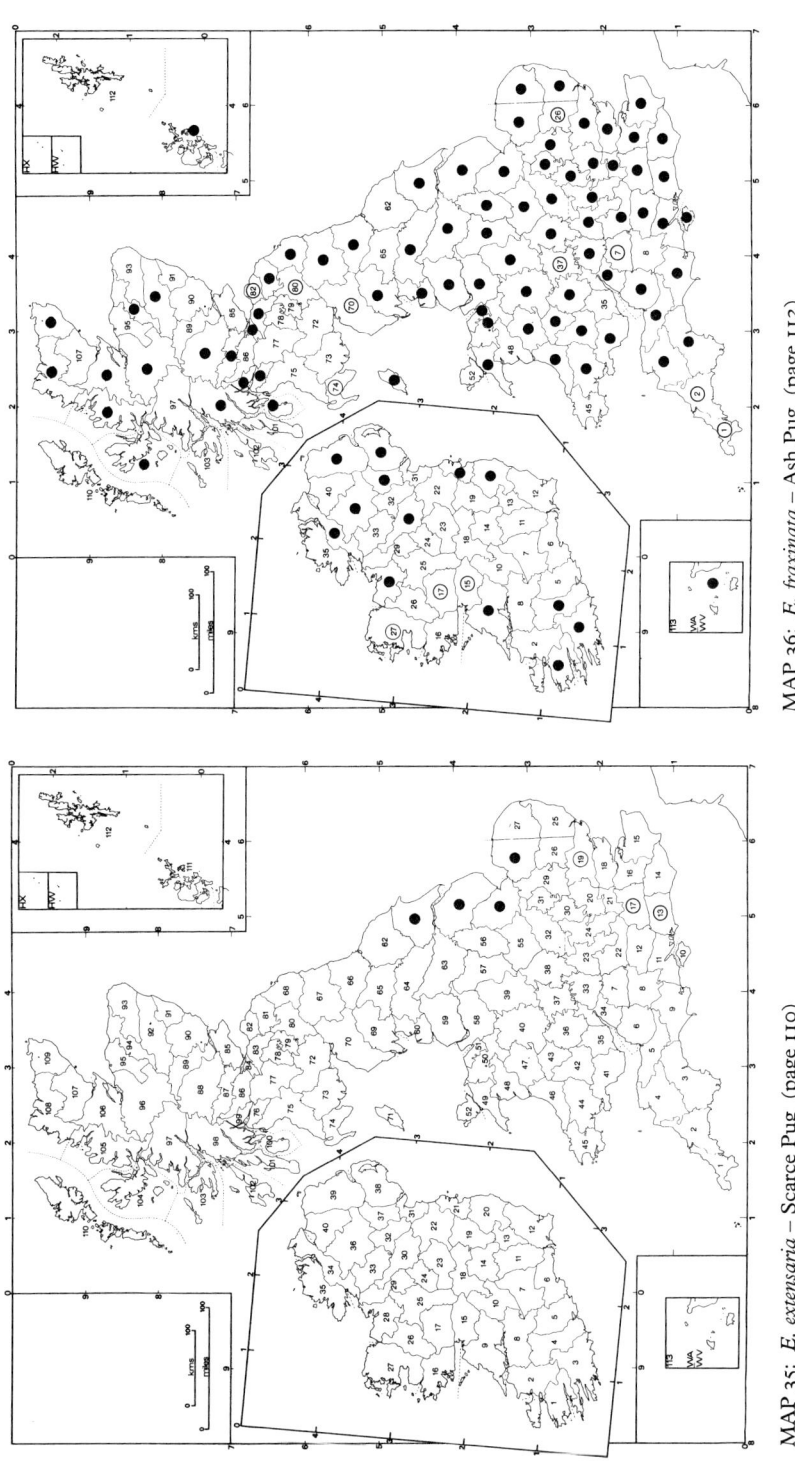

MAP 35: *E. extensaria* – Scarce Pug (page 110)

MAP 36: *E. fraxinata* – Ash Pug (page 112)

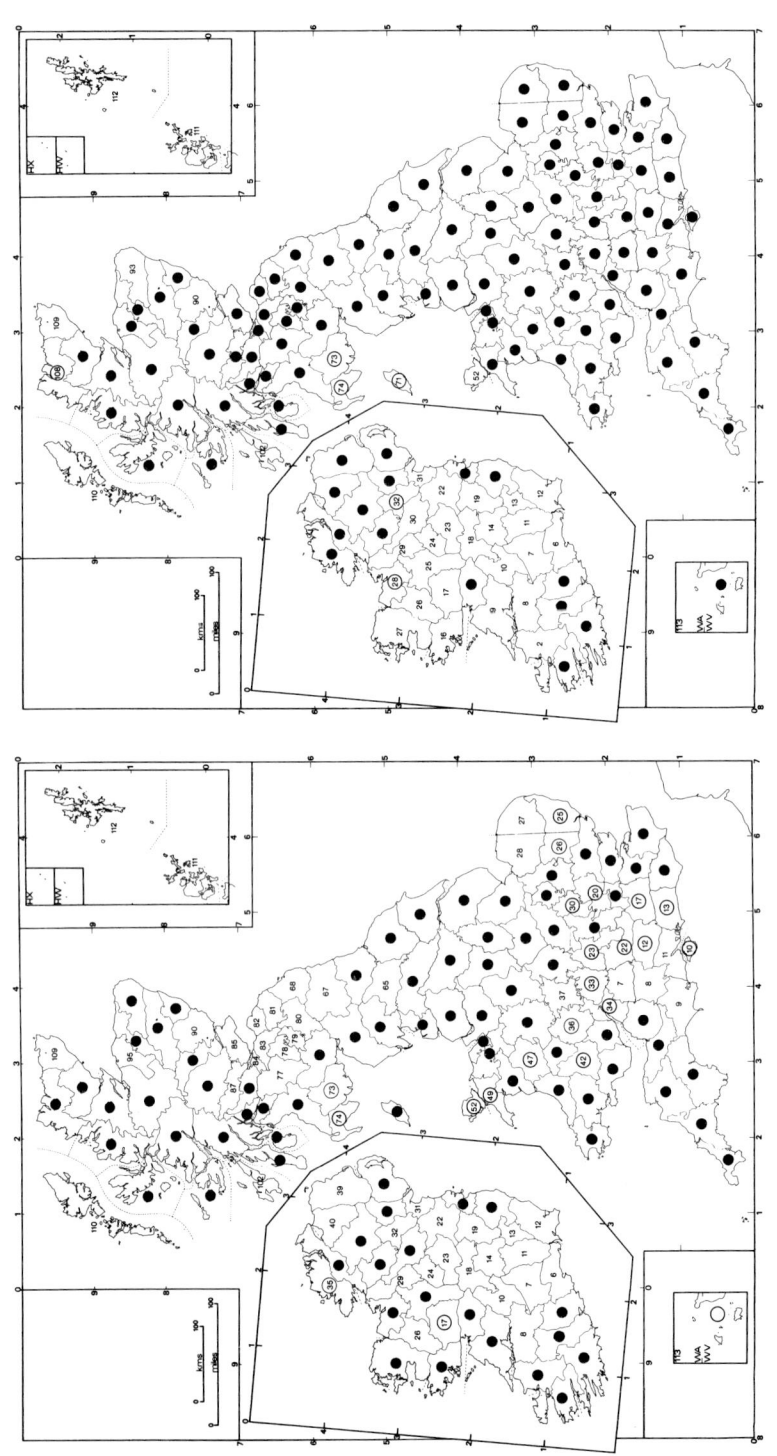

MAP 38: *E. abbreviata* – Brindled Pug (page 117)

MAP 37: *E. virgaureata* – Golden-rod Pug (page 114)

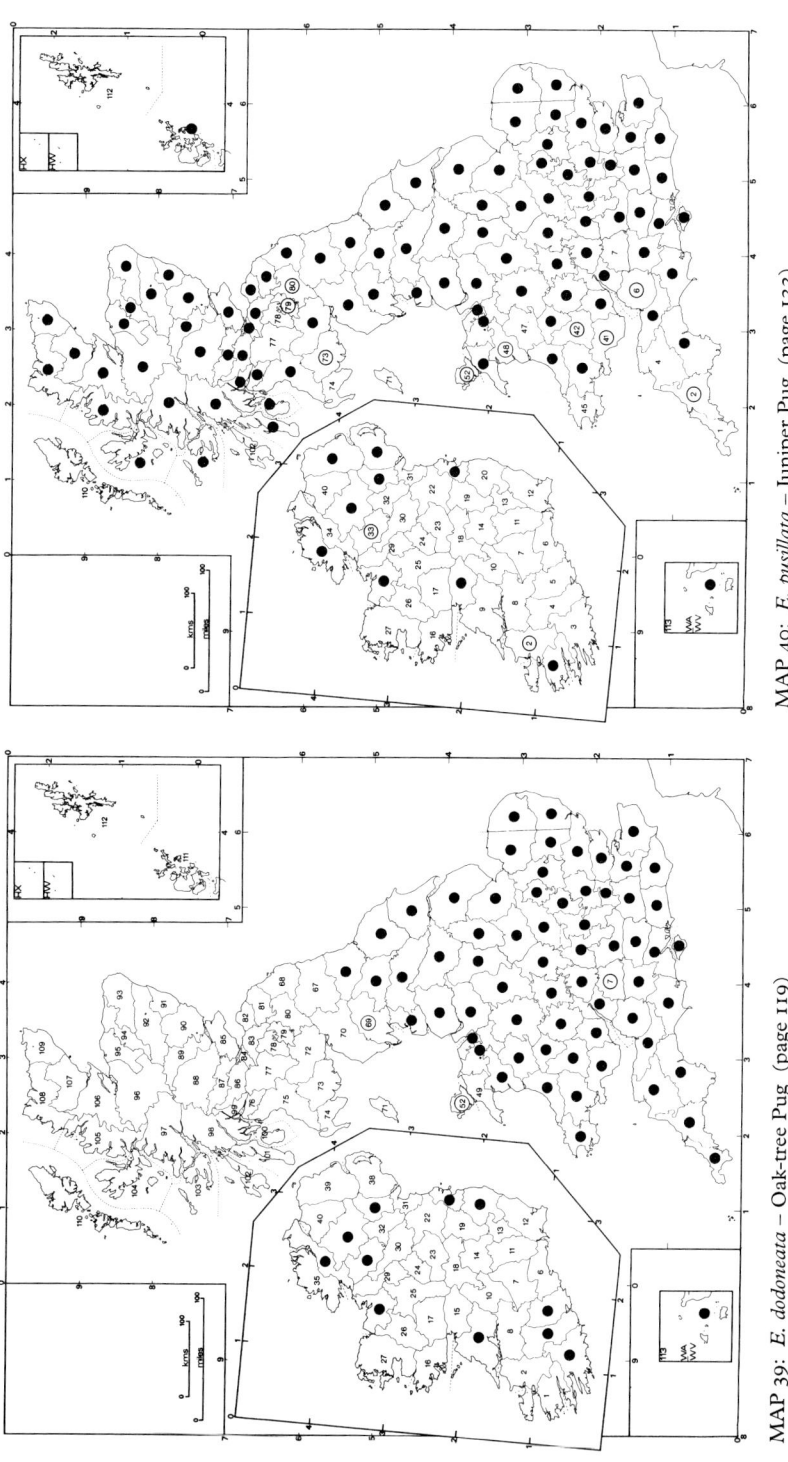

MAP 40: *E. pusillata* – Juniper Pug (page 122)

MAP 39: *E. dodoneata* – Oak-tree Pug (page 119)

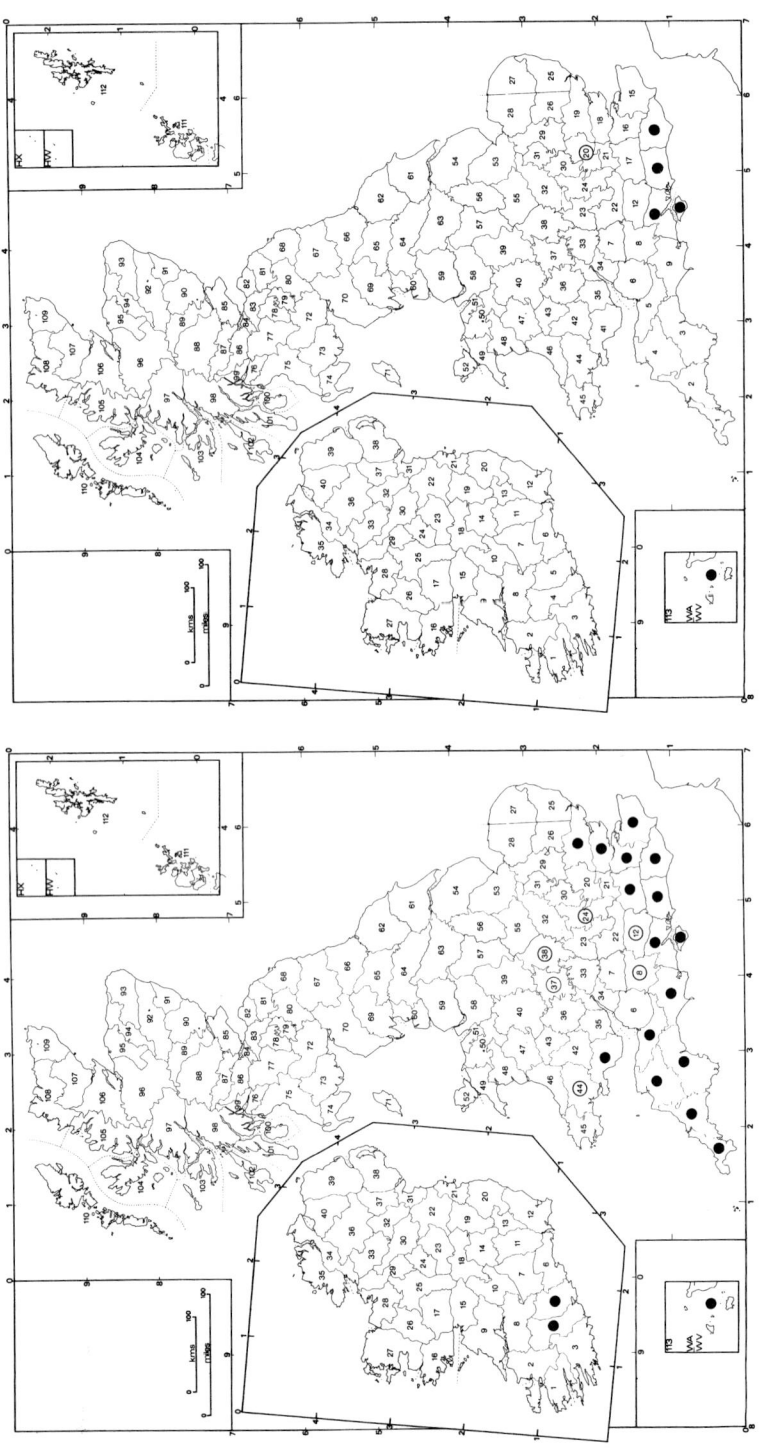

MAP 41: *E. phoeniceata* – Cypress Pug (page 124)

MAP 42: *E. ultimaria* – Channel Islands Pug (page 126)

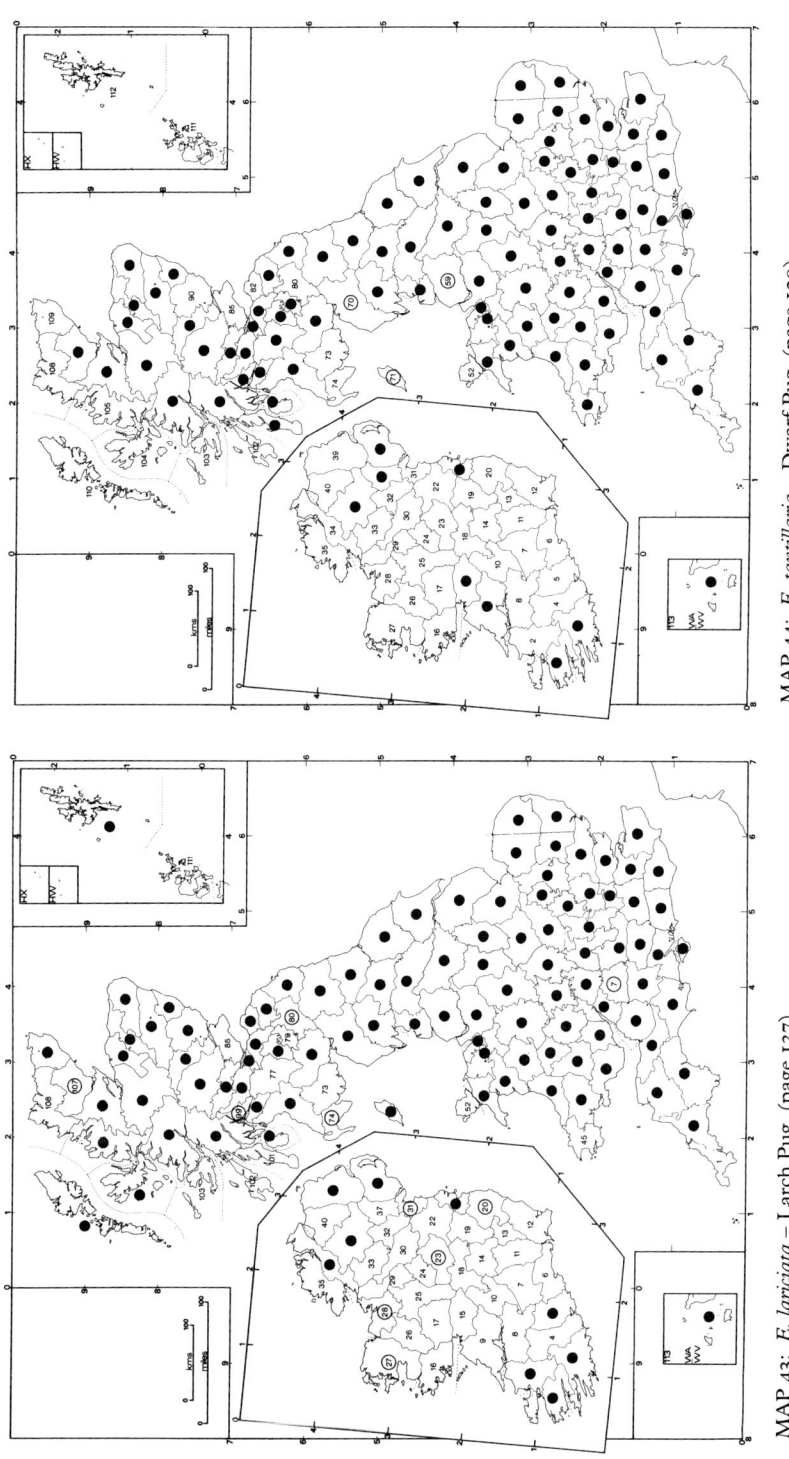

MAP 44: *E. tantillaria* – Dwarf Pug (page 129)

MAP 43: *E. lariciata* – Larch Pug (page 127)

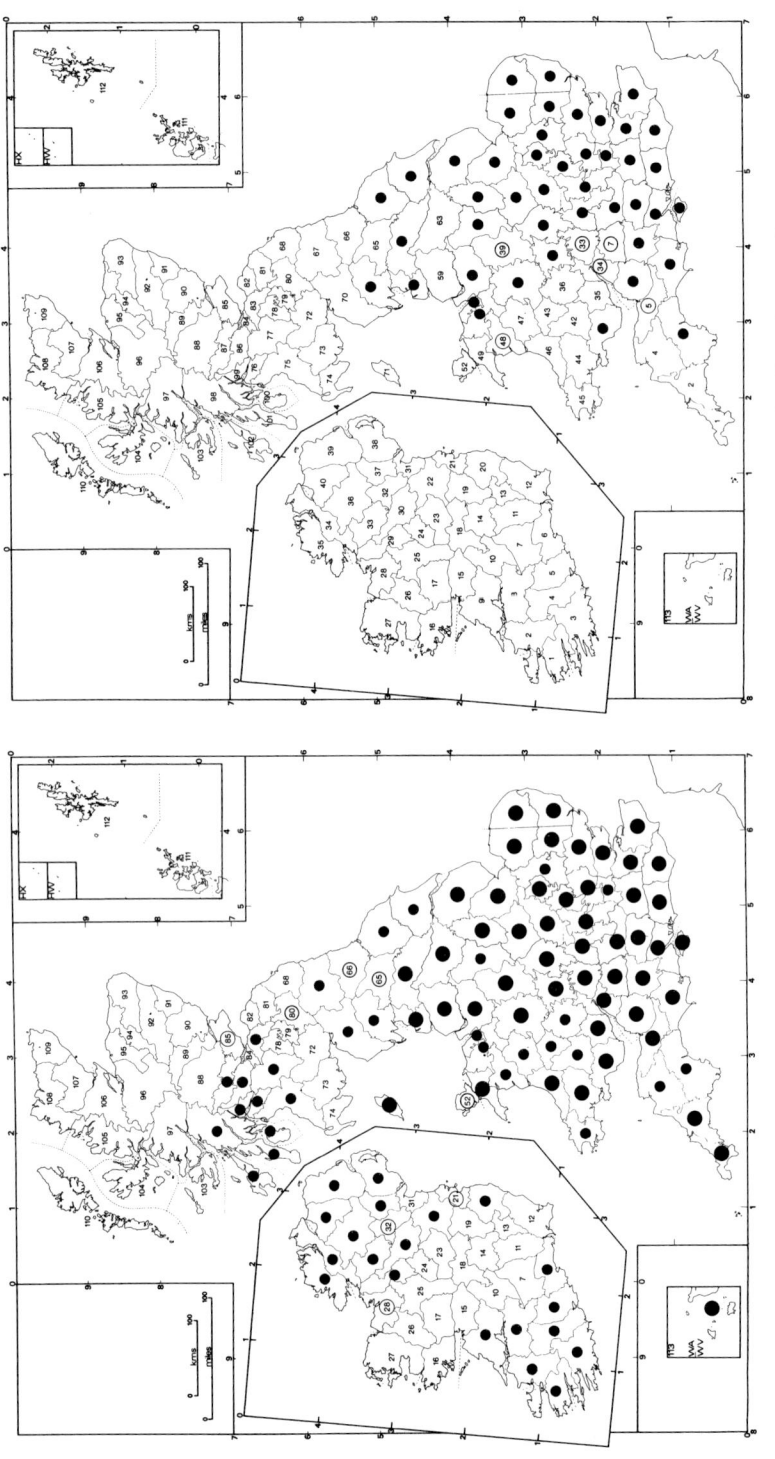

MAP 46: *Pasiphila chloerata* – Sloe Pug (page 132)

MAP 45: *Chloroclystis v-ata* – V-Pug (page 131)

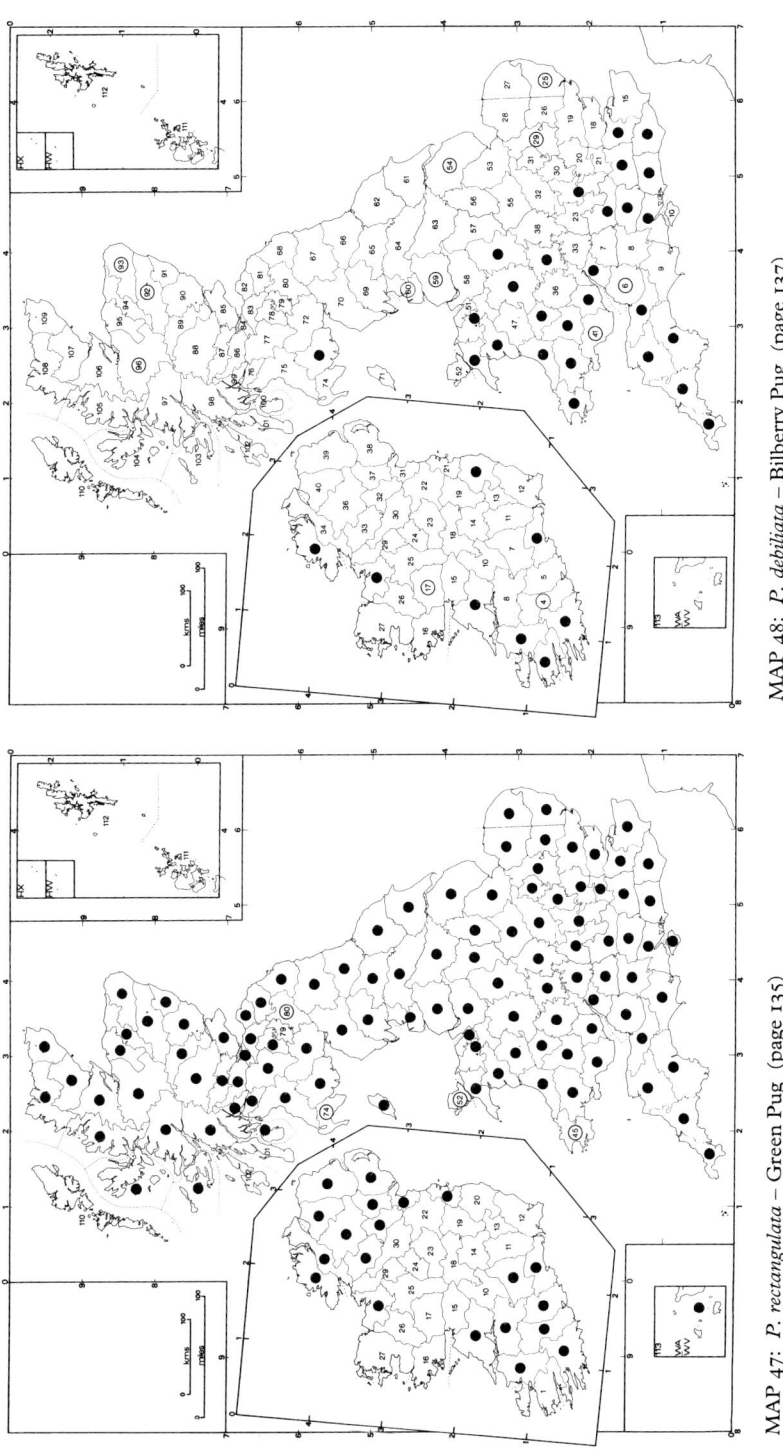

MAP 47: *P. rectangulata* – Green Pug (page 135)

MAP 48: *P. debiliata* – Bilberry Pug (page 137)

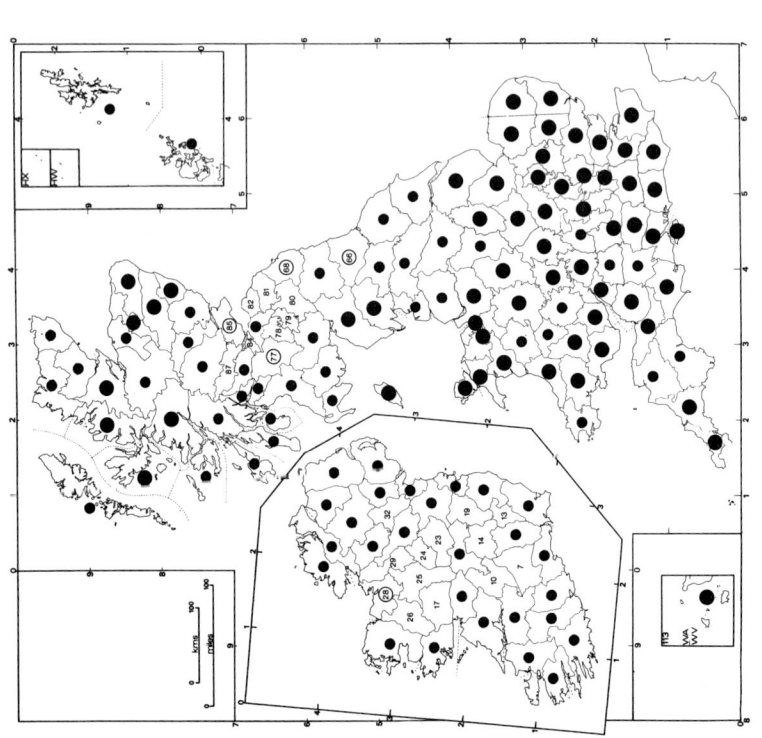

MAP 49: *Gymnoscelis rufifasciata* – Double-striped Pug (page 138)

Colour Plates

PLATE I

1. *Eupithecia tenuiata* (Hübner), Slender Pug, Maulden Woods, Beds., ex larva, sallow, em. vii.1985, A. M. Riley. (Page 33)
2. *E. tenuiata* (Hübner) f. *cinerea* Gregson, Slender Pug, Kinlochewe, Ross-shire, 1.viii.1988, (RIS trap 350), A. M. Riley. (Page 33)
3. *E. tenuiata* (Hübner) f. *johnsoni* Harrison, Slender Pug, Maulden Woods, Beds., ex larva, sallow, em. vii.1985, A.M. Riley. (Page 33)
4. *E. inturbata* (Hübner), Maple Pug, Farningham, Kent, em. vii.1977, B. Skinner. (Page 37)
5. *E. haworthiata* Doubleday, Haworth's Pug, Llanymynech, Powys, bred vi.1979, J. Reid. (Page 39)
6. *E. plumbeolata* (Haworth), Lead-coloured Pug, Pamber, Hants, 9.vi.1980, B. Skinner. (Page 41)
7. *E. abietaria* (Goeze), Cloaked Pug, Hamsterley Forest, Co. Durham, ex larva, *P. abies*, em. vi.1988, A. M. Riley. (Page 43)
8. *E. abietaria* (Goeze), Cloaked Pug (dark individual), Hamsterley Forest, Co. Durham, ex larva, *P. abies*, em. vi.1988, A. M. Riley. (Page 43)
9. *E. linariata* ([Denis & Schiffermüller]), Toadflax Pug, Eynsford, Kent, em. vi.1978, B. Skinner. (Page 45)
10. *E. pulchellata* Stephens, Foxglove Pug, Harpenden, Herts., 3.vii.1984, A. M. Riley. (Page 47)
11. *E. pulchellata* Stephens f. *hebudium* Sheldon, Foxglove Pug, [Rhydwyn], Anglesey, vi.1985, MV light, A. M. Riley. (Page 47)
12. *E. irriguata* (Hübner), Marbled Pug, Powerstock, Dorset, bred, iv.1988, J. Reid. (Page 49)
13. *E. exiguata* (Hübner), Mottled Pug, Harpenden, Herts., 5.vi.1986, A. M. Riley. (Page 50)
14. *E. exiguata* (Hübner) f. *muricolor* Prout, Mottled Pug, Moneymusk, Aberdeenshire, 20.vi.1997, A. M. Riley. (Page 51)
15. *E. insigniata* (Hübner), Pinion-spotted Pug, Royston, Herts., v.1982, J. Reid. (Page 52)
16. *E. valerianata* (Hübner), Valerian Pug, Stoke Ferry, Norfolk, bred vi.1980, J. Reid. (Page 53)
17. *E. pygmaeata* (Hübner), Marsh Pug, Clophill, Beds., ex larva, mouse-eared chickweed, em. v.1986, A. M. Riley. (Page 55)
18. *E. pygmaeata* (Hübner), Marsh Pug, S. Ronaldsay, Orkney Is., em. v.1978, B. Skinner. (Page 55)
19. *E. venosata venosata* (Fabricius), Netted Pug, Charterhouse, Som., em. v.1979, B. Skinner. (Page 57)
20. *E. venosata* (Fabricius) f. *hebridensis*, Parkinson Curtis, Netted Pug, Gruinart, Islay, 26.v.68, G. Prior. (Page 58)
21. *E. venosata* (Fabricius) subsp. *ochracea* Gregson, Netted Pug, Mainland, Orkney, 22.vi.1963, G. Prior & A. M. Riley. (Page 57)
22. *E. venosata* (Fabricius) subsp. *fumosae* Gregson, Netted Pug, Hillswick, Shetland, 14.vi.1963, G. Prior & A. M. Riley. (Page 58)
23. *E. venosata* (Fabricius), subsp. *plumbea* Huggins, Netted Pug, Slea Head, Co. Kerry, 22.v.1970, G. Prior & A. M. Riley. (Page 58)
24. *E. egenaria* Herrich-Schäffer, Fletcher's Pug, Tintern, Mon., em. v.1979, B. Skinner. (Page 59)
25. *E. centaureata* ([Denis & Schiffermüller]), Lime-speck Pug, Harpenden, Herts, ex ♂, 1990, A. M. Riley. (Page 63)
26. *E. trisignaria* Herrich-Schäffer, Triple-spotted Pug, Wicken, Cambs., bred, vii.1976, J. Reid. (Page 64)
27. *E. trisignaria* Herrich-Schäffer, Triple-spotted Pug, Wicken, Cambs., bred, vii.1976, J. Reid. (Page 64)
28. *E. trisignaria* Herrich-Schäffer f. *angelicata* Prout, Triple-spotted Pug, Wicken, Cambs., bred, vii.1976, J. Reid. (Page 64)
29. *E. intricata* Zetterstedt subsp. *arceuthata* (Freyer), Freyer's Pug, Harpenden, Herts., ex ♀, v.1991, A. M. Riley. (Page 67)
30. *E. intricata* Zetterstedt subsp. *millieraria* Wnukowsky, Edinburgh Pug, Aviemore, Inverness-shire, v.1970, B. Skinner. (Page 68)
31. *E. intricata* Zetterstedt subsp. *hibernica* Mere, Mere's Pug, Burren, Co. Clare, bred ex larva, em. v.1990, B. Skinner. (Page 68)
32. *E. absinthiata* (Clerck), Wormwood Pug, Houghton Regis, Beds., ex larva, ragwort, em. vi.1986, A. M. Riley. (Page 76)
33. *E. absinthiata* (Clerck) f. *goosensiata* Mabille, Ling Pug, Chobham, Surrey, em. vi.1979, B. Skinner. (Page 77)

Plate 1

PLATE 2

1. *Eupithecia satyrata* f. *satyrata* (Hübner), Satyr Pug, Maulden Woods, Beds., ab ovo, hawksbeard, em. v.1986, A. M. Riley. (Page 72)
2. *E. satyrata* f. *satyrata* (Hübner), Satyr Pug, Maulden Woods, Beds., ab ovo, hawksbeard, em. v.1986, A. M. Riley. (Page 72)
3. *E. satyrata* (Hübner) f. *callunaria* Doubleday, Satyr Pug, Arran, Buteshire, bred ex larva, em. v.1979, B. Skinner. (Page 72)
4. *E. satyrata* (Hübner) f. *curzoni* Gregson, Satyr Pug, Shetland, bred, em. vi.1976, J. Reid. (Page 72)
5. *E. cauchiata* (Duponchel), Doubleday's Pug, Usedom, Germany, ab ovo, 11.x.81, H.-J. Weigt. (Page 75)
6. *E. expallidata* Doubleday, Bleached Pug, Hamstreet, Kent, bred, 15.vii.80, J. Reid. (Page 79)
7. *E. assimilata* Doubleday, Currant Pug, Royston, Herts., bred, vi.1979, J. Reid. (Page 80)
8. *E. assimilata* Doubleday, Currant Pug (dark specimen), Royston, Herts., bred, vi.1979, J. Reid. (Page 80)
9. *E. vulgata* (Haworth), Common Pug, Ranmore, Surrey, em. vi.1978, B. Skinner. (Page 83)
10. *E. vulgata* (Haworth), Common Pug, Lymington, Hants, ex ♀, 28.v.1990, J. Riley. (Page 83)
11. *E. vulgata* (Haworth), Common Pug (small specimen), Yarmouth, I.o.W., ex ♀, 30.v.1990, A. M. Riley. (Page 83)
12. *E. vulgata* (Haworth), Common Pug (dark specimen), Houghton Regis, Beds., ex larva, ragwort, em. v.1986. (Page 83)
13. *E. vulgata* (Haworth) f. *atropicta* Dietze, Common Pug, Harpenden, Herts., ex larva, ragwort, em. v.1986, A. M. Riley. (Page 83)
14. *E. vulgata* (Haworth) f. *scotica* Cockayne, Common Pug, Chathill, Northumberland, 26.v.1988, RIS trap, A M. Riley. (Page 83)
15. *E. vulgata* (Haworth) f. *clarensis* Huggins, Common Pug, Burren, Co. Clare, 12.vi.1979, B. Skinner. (Page 83)
16. *E. tripunctaria* Herrich-Schäffer, White-spotted Pug, Santon Downham, Norfolk, 22.v.1983, J. Reid. (Page 83)
17. *E. tripunctaria* Herrich-Schäffer f. *angelicata* Barrett, White-spotted Pug, Cockayne Hatley, Beds., 8.v.1990, RIS trap, A. M. Riley. (Page 86)
18. *E. denotata denotata* Hübner, Campanula Pug, Surrey, bred, vi.90, J. Reid. (Page 88)
19. *E. denotata* (Hübner) subsp. *jasioneata* Crewe, Jasione Pug, Lizard, Cornwall, 1983, J. Reid. (Page 88)
20. *E. subfuscata* (Haworth), Grey Pug, Houghton Regis, Beds., ex larva, dandelion, em. v.1986, A. M. Riley. (Page 91)
21. *E. subfuscata* (Haworth) f. *obscurissima* Prout, Grey Pug, Harpenden, Herts.., 1, G1, 15.vi.1986, RES trap 22, A. M. Riley. (Page 91)
22. *E. millefoliata* Rössler, Yarrow Pug, Chichester, Sussex, bred, vii.1974, J. Reid. (Page 99)
23. *E. icterata* (Villers) f. *subfulvata* (Haw.), Tawny-speckled Pug, Lyme Regis, Dorset, bred, viii.1979, J. Reid. (Page 94)
24. *E. icterata* (Villers) f. *cognata* Stephens, Tawny-speckled Pug, Alexandria, Dunbartonshire, bred ex larva, vii.1985, J. Reid. (Page 94)
25. *E. icterata* (Villers) f. *grisescens* Lempke, Tawny-speckled Pug, Alexandria, Dunbartonshire, bred ex larva, vii.1985, J. Reid. (Page 94)
26. *E. succenturiata* (Linnaeus), Bordered Pug, Harpenden, Herts., 20.vi.1988, A. M. Riley. (Page 96)
27. *E. succenturiata* (Linnaeus) f. *obscurata* Lempke, Bordered Pug, Sharpenhoe, Beds., ex larva, yarrow, em. 25.vi.1985, A. M. Riley. (Page 96)
28. *E. subumbrata* [(Denis & Schiffermüller)], Shaded Pug, Harpenden, Herts, ex larva, ragwort, em. vi.1986, A. M. Riley. (Page 98)
29. *E. subumbrata* [(Denis & Schiffermüller)] f. *obscurata* Lempke, Shaded Pug, Maulden Woods, Beds., F2, hawkweed, em. i.1987, A. M. Riley. (Page 98)
30. *E. simpliciata* (Haworth), Plain Pug, I.o.W., bred, vii.1979, J. Reid. (Page 101)
31. *E. sinuosaria* Eversmann, Goosefoot Pug, Harpenden, Herts., 19–21.vi.1992, RIS Trap, Knott Wood, A. M. Riley. (Page 102)
32. *E. distinctaria* Herrich-Schäffer, subsp. *constrictata* Guenée, Thyme Pug, Aberystwyth, Cards., bred 22.vi.1980, J. Reid. (Page 103)

PLATE 2

PLATE 3

1. *Eupithecia indigata* (Hübner), Ochreous Pug, Mundford, Norfolk, vi.1978, J. Reid. (Page 105)
2. *E. pimpinellata* (Hübner), Pimpinel Pug, Royston, Herts., bred, viii.1977, J. Reid. (Page 106)
3. *E. pimpinellata* (Hübner), Pimpinel Pug, Royston, Herts., bred, viii.1979, J. Reid. (Page 106)
4. *E. pimpinellata* (Hübner), Pimpinel Pug (unicolorous form), Royston, Herts., bred, viii.1979, J. Reid. (Page 106)
5. *E. nanata* (Hübner) subsp. *angusta* Prout, Narrow-winged Pug, New Forest, v.1978, B. Skinner. (Page 108)
6. *E. nanata* (Hübner) subsp. *angusta* Prout, Narrow-winged Pug, Charmouth, Dorset, ex larva, v.1986, J. Reid. (Page 108)
7. *E. extensaria* (Freyer) subsp. *occidua* Prout, Scarce Pug, Brancaster, Norfolk, ex pupa, 24.xii.1984, A. M. Riley. (Page 110)
8. *E. fraxinata* Crewe, Ash Pug, Spurn Head, Yorks, ex ♀ F1, ex larva on ash, 1990, A. M. Riley. (Page 112)
9. *E. fraxinata* Crewe, Ash Pug, Spurn Head, Yorks, ex ♀ F1, ex larva on mugwort, 1990, A. M. Riley. (Page 112)
10. *E. fraxinata* Crewe f. *unicolor* Prout, Ash Pug, Spurn Head, Yorks, ex ♀ F1, ex larva on mugwort, 1990, A. M. Riley. (Page 112)
11. *E. virgaureata* Doubleday, Golden-rod Pug, Bradford, Derbys., em. vi.1978, B. Skinner. (Page 114)
12. *E. virgaureata* Doubleday f. *nigra* Lemke, Golden-rod Pug, Brampton Hall, Cumbria, ex ♀, viii.1992, A. M. Riley. (Page 114)
13. *E. abbreviata* Stephens, Brindled Pug, Alice Holt, Hants, iv.1975, B. Skinner. (Page 117)
14. *E. abbreviata* Stephens f. *nigra* Cockayne, Brindled Pug, Harpenden, Herts., 8.v.1985, A. M. Riley. (Page 117)
15. *E. dodoneata* Guenée, Oak-tree Pug, Whitstable, Kent, 19.iv.1993, B. Skinner. (Page 119)
16. *E. dodoneata* Guenée, Oak-tree Pug, Fowlmere, Cambs., 15.v.1985, ex larva, J. Reid. (Page 119)
17. *E. phoeniceata* (Rambur), Cypress Pug, Torquay, Devon, viii.1973, B. Skinner. (Page 124)
18. *E. ultimaria* Boisduval, Channel Islands Pug, Littlehampton, W. Sussex, bred ex larva, em. v.1997, B. Skinner. (Page 126)
19. *E. pusillata pusillata* [(Denis & Schiffermüller)], Juniper Pug, Teesdale, Co. Durham, em. vii.1979, B. Skinner. (Page 123)
20. *E. pusillata pusillata* [(Denis & Schiffermüller)] f. *nigrofasciata* Dietze, Juniper Pug, Harpenden, Herts., 8.viii.1984, A. M. Riley. (Page 123)
21. *E. pusillata pusillata* [(Denis & Schiffermüller)] f. *scotica* Dietze, Juniper Pug, Aviemore, Inverness-shire, em. 1974, B. Skinner. (Page 122)
22. *E. pusillata* (Denis & Schiffermüller)] subsp. *anglicata* Herrich-Schäffer, Kentish Tamarisk Pug, Dover, [Kent,] 1894. (Page 122)
23. *E. lariciata* (Freyer), Larch Pug, Kingsdown, Kent, v.1978, B. Skinner. (Page 127)
24. *E. lariciata* (Freyer) f. *nigra* Prout, Larch Pug, Kingsdown, Kent, 23.v.1978, B. Skinner. (Page 127)
25. *E. massiliata* Millière, Epping Pug, Epping, Essex, 2.iv.2002, B. Goodey. (Page 120)
26. *E. tantillaria* Boisduval, Dwarf Pug, Hamstreet, Kent, v.1982, B. Skinner. (Page 129)
27. *Chloroclystis v-ata* (Haworth), V-Pug, Sandwich, Kent, em. viii.1982, B. Skinner. (Page 131)
28. *C. v-ata* (Haworth), V-Pug, Swanage, Dorset, viii.1980, B. Skinner. (Page 131)
29. *Pasiphila chloerata* Mabille, Sloe Pug, Warden Hills, Beds., ex larva, blackthorn, em. vi.1985, A. M. Riley. (Page 132)
30. *P. debiliata* (Hübner), Bilberry Pug, Friday Street, Surrey, bred, vi.1981, J. Reid. (Page 137)
31. *P. rectangulata* (Linnaeus), Green Pug, Harpenden, Herts., ex larva, apple, em. vi.1985, A. M. Riley. (Page 135)
32. *P. rectangulata* (Linnaeus), Green Pug, Boldre, New Forest, em. vi.1978, B. Skinner. (Page 135)
33. *P. rectangulata* (Linnaeus) f. *anthrax* Dietze, Green Pug, Harpenden, Herts., ex larva, apple, em. vi.1987, A. M. Riley. (Page 135)
34. *P. rectangulata* (Linnaeus.) f. *nigrosericeata* Haworth, Green Pug, Harpenden, Herts., ex larva, apple, em. vi.1985, A. M. Riley. (Page 135)
35. *Gymnoscelis rufifasciata* (Haworth), Double-striped Pug, Harpenden, Herts., ex larva, ragwort, em. 19.iii.1986, A. M. Riley. (Page 138)
36. *G. rufifasciata* (Haworth), Double-striped Pug, Lakenheath, Suffolk, bred, 17.v.1984, J. Reid. (Page 138)

PLATE 3

PLATE 4

1. *Eupithecia venosata venosata* (Fabricius), Netted Pug, Charterhouse, Som., em. v.1979, B. Skinner. (Page 57)
2. *E. venosata* (Fabricius), subsp. *plumbea* Huggins, Netted Pug, Slea Head, Co. Kerry, 22.v.1970, G. Prior & A. M. Riley. (Page 58)
3. *E. venosata* (Fabricius) subsp. *ochracea* Gregson, Netted Pug, Mainland, Orkney, 22.vi.1963, G. Prior & A. M. Riley. (Page 57)
4. *E. venosata* (Fabricius) subsp. *fumosae* Gregson, Netted Pug, Hillswick, Shetland, 14.vi.1963, G. Prior & A. M. Riley. (Page 58)
5. *E. venosata* (Fabricius) subsp. *hebridensis*, Parkinson Curtis, Netted Pug, Gruinart, Islay, 26.v.1968, G. Prior. (Page 58)
6. *E. icterata* (Villers) f. *subfulvata* (Haworth), Tawny-speckled Pug, Lyme Regis, Dorset, bred, vii.1979, J. Reid. (Page 94)
7. *E. icterata* (Villers) f. *cognata* Stephens, Tawny-speckled Pug, Alexandria, Dunbartonshire, bred ex larva, vii.1985, J. Reid. (Page 94)
8. *E. icterata* (Villers) f. *grisescens* Lempke, Tawny-speckled Pug, Alexandria, Dunbartonshire, bred ex larva, vii.1985, J. Reid. (Page 94)
9. *E. exiguata* (Hübner), Mottled Pug, Harpenden, Herts., 5.vi.1986, A. M. Riley. (Page 50)
10. *E. exiguata* (Hübner) f. *muricolor* Prout, Mottled Pug, Moneymusk, Aberdeenshire, 20.vi.1997, A. M. Riley. (Page 51)
11. *E. succenturiata* (Linnaeus), Bordered Pug, Harpenden, Herts., 20.vi.1988, A. M. Riley. (Page 96)
12. *E. succenturiata* (Linnaeus) f. *obscurata* Lempke, Bordered Pug, Sharpenhoe, Beds., ex larva, yarrow, em. 25.vi.1985, A. M. Riley. (Page 96)
13. *E. phoeniceata* (Rambur), Cypress Pug, Torquay, Devon, viii.1973, B. Skinner. (Page 124)
14. *E. extensaria* (Freyer) subsp. *occidua* Prout, Scarce Pug, Brancaster, Norfolk, ex pupa, 24.xii.1984, A. M. Riley. (Page 110)
15. *Chloroclystis v-ata* (Haworth), V-Pug, Sandwich, Kent, em. viii.1982, B. Skinner. (Page 131)
16. *Gymnoscelis rufifasciata* (Haworth), Double-striped Pug, Harpenden, Herts., ex larva, ragwort, em. 19.iii.1986, A. M. Riley. (Page 138)
17. *G. rufifasciata* (Haworth), Double-striped Pug, Lakenheath, Suffolk, bred, 17.v.1984, J. Reid. (Page 138)
18. *E. sinuosaria* Eversmann, Goosefoot Pug, Harpenden, Herts., 19–21.vi.1992, RIS Trap, Knott Wood, A. M. Riley. (Page 102)
19. *E. nanata* (Hübner) subsp. *angusta* Prout, Narrow-winged Pug, New Forest, v.1978, B. Skinner. (Page 108)
20. *E. fraxinata* Crewe, Ash Pug, Spurn Head, Yorks, ex ♀ F1, ex larva on ash, 1990, A. M. Riley. (Page 112)
21. *E. fraxinata* Crewe, Ash Pug, Spurn Head, Yorks, ex ♀ F1, ex larva on mugwort, 1990, A. M. Riley. (Page 112)
22. *E. centaureata* ([Denis & Schiffermüller]), Lime-speck Pug, Harpenden, Herts., ex ♂, 1990, A. M. Riley. (Page 63)
23. *E. insigniata* (Hübner), Pinion-spotted Pug, Royston, Herts., v.1982, J. Reid. (Page 52)

PLATE 4

Figs 1–23: distinctively marked species, confusion unlikely

1. *Pasiphila rectangulata* (Linnaeus), Green Pug, Boldre, New Forest, em. vi.1978, B. Skinner. (Page 135)
2. *P. rectangulata* (Linnaeus), Green Pug, Harpenden, Herts., ex larva, apple, em. vi.1985, A. M. Riley. (Page 135)
3. *P. chloerata* Mabille, Sloe Pug, Warden Hills, Beds., ex larva, blackthorn, em. vi.1985, A. M. Riley. (Page 132)
4. *P. debiliata* (Hübner), Bilberry Pug, Friday Street, Surrey, bred, vi.1981, J. Reid. (Page 137)
5. *Chloroclystis v-ata* (Haworth), V-Pug, Swanage, Dorset, viii.1980 B. Skinner. (Page 131)
6. *Eupithecia pulchellata* Stephens, Foxglove Pug, Aviemore, Inverness-shire, em. 1977, B. Skinner. (Page 47)
7. *E. pulchellata* Stephens, Foxglove Pug, Harpenden, Herts., 3.vii.1984, A. M. Riley. (Page 47)
8. *E. pulchellata* Stephens f. *hebudium* Sheldon, Foxglove Pug, [Rhydwyn,] Anglesey, vi.1985, MV light, A. M. Riley. (Page 47)
9. *E. linariata* ([Denis & Schiffermüller]), Toadflax Pug, Eynsford, Kent, em.vi.1978, B. Skinner. (Page 45)
10. *E. irriguata* (Hübner), Marbled Pug, Powerstock, Dorset, bred, iv.1988, J. Reid. (Page 49)
11. *E. tantillaria* Boisduval, Dwarf Pug, Hamstreet, Kent, v.1982, B. Skinner. (Page 129)
12. *E. vulgata* (Haworth) f. *clarensis* Huggins, Common Pug, Burren, Co. Clare, 12.vi.1979, B. Skinner. (Page 83)
13. *E. subumbrata* [(Denis & Schiffermüller)], Shaded Pug, Harpenden, Herts., ex larva, ragwort, em. vi 1986, A. M. Riley. (Page 98)
14. *E. pusillata pusillata* [(Denis & Schiffermüller)] f. *scotica* Dietze, Juniper Pug, Aviemore, Inverness-shire, em. 1974, B. Skinner. (Page 122)
15. *E. pusillata* (Denis & Schiffermüller)] subsp. *anglicata* Herrich-Schäffer, Kentish Tamarisk Pug, Dover, [Kent,] 1894. (Page 122)
16. *E. abietaria* (Goeze), Cloaked Pug (dark individual), Hamsterley Forest, Co. Durham, ex larva, *P. abies*, em. vi.1988, A. M. Riley. (Page 43)
17. *E. abietaria* (Goeze), Cloaked Pug, Hamsterley Forest, Co. Durham, ex larva, *P. abies*, em. vi.1988, A. M. Riley. (Page 43)

PLATE 5

Figs 1–5: green or with greenish tinge

Figs 6–9: conspicuous broad median fasciae

Figs 10–15: white or pale grey with darker borders

Figs 16,17: large with prominent forewing discal spot

1. *Eupithecia trisignaria* Herrich-Schäffer f. *angelicata* Prout, Triple-spotted Pug, Wicken, Cambs., bred, vii.76, J. Reid. (Page 64)
2. *E. subfuscata* (Haworth) f. *obscurissima* Prout, Grey Pug, Harpenden, Herts., 15.vi.1986, RES trap 22, A. M. Riley. (Page 91)
3. *E. virgaureata* Doubleday f. *nigra* Lemke, Golden-rod Pug, Brampton Hall, Cumbria, ex ♀, viii.1992, A. M. Riley. (Page 114)
4. *E. pusillata pusillata* [(Denis & Schiffermüller)] f. *nigrofasciata* Dietze, Juniper Pug, Harpenden, Herts., 8.viii.1984, A. M. Riley. (Page 123)
5. *E. vulgata* (Haworth) f. *atropicta* Dietze, Common Pug, Harpenden, Herts., ex larva, ragwort, em. v.1986, A. M. Riley. (Page 83)
6. *E. fraxinata* Crewe f. *unicolor* Prout, Ash Pug, Spurn Head, Yorks., ex ♀ F1, ex larva on mugwort, 1990, A. M. Riley. (Page 112)
7. *E. tripunctaria* Herrich-Schäffer f. *angelicata* Barrett, White-spotted Pug, Cockayne Hatley, Beds., 8.v.1990, RIS trap, A. M. Riley. (Page 86)
8. *E. lariciata* (Freyer) f. *nigra* Prout, Larch Pug, Kingsdown, Kent, 23.v.1978, B. Skinner. (Page 127)
9. *E. abbreviata* Stephens f. *nigra* Cockayne, Brindled Pug, Harpenden, Herts., 8.v.1985, A. M. Riley. (Page 117)
10. *Pasiphila rectangulata* (Linnaeus) f. *anthrax* Dietze, Green Pug, Harpenden, Herts., ex larva, apple, em. vi.1987, A. M. Riley. (Page 135)
11. *P. rectangulata* (Linnaeus) f. *nigrosericeata* Haworth, Green Pug, Harpenden, Herts., ex larva, apple, em. vi.1985, A. M. Riley. (Page 135)
12. *E. abbreviata* Stephens, Brindled Pug, Alice Holt, Hants, iv.1975, B. Skinner. (Page 117)
13. *E. dodoneata* Guenée, Oak-tree Pug, Whitstable, Kent, 19.iv.1993, B. Skinner. (Page 119)
14. *E. satyrata* f. *satyrata* (Hübner), Satyr Pug, Maulden Woods, Beds., ab ovo, hawksbeard, em. v.1986, A. M. Riley. (Page 72)
15. *E. cauchiata* (Duponchel), Doubleday's Pug, Usedom, Germany, ab ovo, 11.x.81, H.-J. Weigt. (Page 75)
16. *E. nanata* (Hübner) subsp. *angusta* Prout, Narrow-winged Pug, Charmouth, Dorset, ex larva, v.1986, J. Reid. (Page 108)
17. *E. satyrata* (Hübner) f. *curzoni* Gregson, Satyr Pug, Shetland, bred, em. vi.1976, J. Reid. (Page 72)
18. *E. intricata* Zetterstedt subsp. *hibernica* Mere, Mere's Pug, Burren, Co. Clare, bred ex larva, em. v.1990, B. Skinner. (Page 68)

PLATE 6

Figs 1–11: melanic forms: black or extremely dark grey or brown with few markings

Figs 12,13: distinct arrow markings on forewing postmedian line

Figs 14,15: pale greyish brown with striae absent or inconspicuous

Figs 16–18: conspicuous light and dark banding on forewings

PLATE 7

1. *Eupithecia distinctaria* Herrich-Schäffer, subsp. *constrictata* Guenée, Thyme Pug, Aberystwyth, Cards., bred 22.vi.1980, J. Reid. (Page 103)
2. *E. ultimaria* Boisduval, Channel Islands Pug, Littlehampton, W. Sussex, bred ex larva, em. v.1997, B. Skinner. (Page 126)
3. *E. indigata* (Hübner), Ochreous Pug, Mundford, Norfolk, vi.1978, J. Reid. (Page 105)
4. *E. dodoneata* Guenée, Oak-tree Pug, Fowlmere, Cambs., 15.v.1985, ex larva, J. Reid. (Page 119)
5. *E. tenuiata* (Hübner) f. *johnsoni* Harrison, Slender Pug, Maulden Woods, Beds., ex larva, sallow, em. vii.1985, A.M. Riley. (Page 33)
6. *E. tenuiata* (Hübner), Slender Pug, Maulden Woods, Beds., ex larva, sallow, em. vii.1985, A. M. Riley. (Page 33)
7. *E. tenuiata* (Hübner) f. *cinerea* Gregson, Slender Pug, Kinlochewe, Ross-shire, 1.viii.1988, (RIS trap 350), A. M. Riley. (Page 33)
8. *E. valerianata* (Hübner), Valerian Pug, Stoke Ferry, Norfolk, bred vi.1980, J. Reid. (Page 53)
9. *E. vulgata* (Haworth), Common Pug (small specimen), Yarmouth, I.o.W., ex ♀, 30.v.1990, A. M. Riley. (Page 83)
10. *E. inturbata* (Hübner), Maple Pug, Farningham, Kent, em. iii.1977, B. Skinner. (Page 37)
11. *E. haworthiata* Doubleday, Haworth's Pug, Llanymynech, Powys, bred vi.1979, J. Reid. (Page 39)
12. *E. plumbeolata* (Haworth), Lead-coloured Pug, Pamber, Hants, 9.vi.1980, B. Skinner. (Page 41)
13. *E. pygmaeata* (Hübner), Marsh Pug, Clophill, Beds., ex larva, mouse-eared chickweed, em. v.1986, A. M. Riley. (Page 55)
14. *E. pygmaeata* (Hübner), Marsh Pug, S. Ronaldsay, Orkney Is., v.1978, B. Skinner. (Page 55)
15. *E. absinthiata* (Clerck) f. *goosensiata* Mabille, Ling Pug, Chobham, Surrey, em. vi.1979, B. Skinner. (Page 77)
16. *E. absinthiata* (Clerck), Wormwood Pug, Houghton Regis, Beds., ex larva, ragwort, em. vi.1986, A. M. Riley. (Page 76)
17. *E. trisignaria* Herrich-Schäffer, Triple-spotted Pug, Wicken, Cambs., bred, vii.76, J. Reid. (Page 64)
18. *E. assimilata* Doubleday, Currant Pug, Royston, Herts., bred, vi.1979, J. Reid. (Page 80)
19. *E. expallidata* Doubleday, Bleached Pug, Hamstreet, Kent, bred, 15.vii.80, J. Reid. (Page 79)
20. *E. assimilata* Doubleday, Currant Pug (dark specimen), Royston, Herts., bred, vi.1979, J. Reid. (Page 80)
21. *E. denotata denotata* Hübner, Campanula Pug, bred, vi.90, J. Reid. (Page 88)
22. *E. vulgata* (Haworth), Common Pug, Ranmore, Surrey, em. vi.1978, B. Skinner. (Page 83)
23. *E. vulgata* (Haworth) f. *scotica* Cockayne, Common Pug, Chathill, Northumberland, 26.v.1988, RIS trap, A M. Riley. (Page 83)
24. *E. pimpinellata* (Hübner), Pimpinel Pug, Royston, Herts., bred, viii.1979, J. Reid. (Page 106)

PLATE 7

Figs 1–14: small size; less than 18 mm wingspan

Figs 15–24: plain brown with few, if any striae; forewing discal spot obvious

PLATE 8

1. *Eupithecia denotata* (Hübner) subsp. *jasioneata* Crewe, Jasione Pug, Lizard, Cornwall, 1983, J. Reid. (Page 88)
2. *E. virgaureata* Doubleday, Golden-rod Pug, Bradford, Derbys., em. vi.1978, B. Skinner. (Page 114)
3. *E. subfuscata* (Haworth), Grey Pug, Houghton Regis, Beds., ex larva, dandelion, em. v.1986, A. M. Riley. (Page 91)
4. *E. lariciata* (Freyer), Larch Pug, Kingsdown, Kent, v.1978, B. Skinner. (Page 127)
5. *E. tripunctaria* Herrich-Schäffer, White-spotted Pug, Santon Downham, Norfolk, 22.vi.1983, J. Reid. (Page 86)
6. *E. intricata* Zetterstedt subsp. *arceuthata* (Freyer), Freyer's Pug, Harpenden, Herts., ex ♀, v.1991, A. M. Riley. (Page 67)
7. *E. pimpinellata* (Hübner), Pimpinel Pug (unicolorous form), Royston, Herts., bred, viii.1979, J. Reid. (Page 106)
8. *E. egenaria* Herrich-Schäffer, Fletcher's Pug, Tintern, Mon., em. 1979, B. Skinner. (Page 59)
9. *E. satyrata* f. *satyrata* (Hübner), Satyr Pug, Maulden Woods, Beds. ab ovo, hawksbeard, em. v.1986, A. M. Riley. (Page 72)
10. *E. vulgata* (Haworth), Common Pug (dark specimen), Pamber, Hants, 9.vi.1980, B. Skinner. (Page 83)
11. *E. vulgata* (Haworth), Common Pug (worn specimen), Houghton Regis, Beds., ex larva, ragwort, em. v.1986, A. M. Riley. (Page 83)
12. *E. trisignaria* Herrich-Schäffer, Triple-spotted Pug, Wicken, Cambs., bred, vii.76, J. Reid. (Page 64)
13. *E. subumbrata* [(Denis & Schiffermüller)] f. *obscurata* Lempke, Shaded Pug, Maulden Woods, Beds., F2, hawkweed, em. i.1987, A. M. Riley. (Page 98)
14. *E. dodoneata* Guenée, Oak Tree Pug, (worn specimen), Maulden Woods, Beds., 18.v.1985, A. M. Riley. (Page 119)
15. *E. pusillata pusillata* [(Denis & Schiffermüller)], Juniper Pug, Teesdale, Co. Durham, em. viii.1979, B. Skinner. (Page 123)
16. *E. intricata* Zetterstedt subsp. *millieraria* Wnukowsky, Edinburgh Pug, Aviemore, Inverness-shire, v.1970, B. Skinner. (Page 68)
17. *E. vulgata* (Haworth), Common Pug, Lymington, Hants, ex ♀, 28.v.1990, J. Reid. (Page 83)
18. *E. satyrata* (Hübner) f. *callunaria* Doubleday, Satyr Pug, Arran, Buteshire, bred ex larva, em. v.1979, B. Skinner. (Page 72)
19. *E. millefoliata* Rössler, Yarrow Pug, Chichester, Sussex, bred, vii.1974, J. Reid. (Page 99)
20. *E. pimpinellata* (Hübner), Pimpinel Pug, Royston, Herts., bred, viii.1979, J. Reid. (Page 106)
21. *E. simpliciata* (Haworth), Plain Pug, I.o.W., bred, vii.1979, J. Reid. (Page 101)

PLATE 8

Figs 1–15: grey or greyish brown with obvious striae; forewing discal spot obvious

Figs 16–21: brown with obvious striae

PLATE A

1. *Eupithecia tenuiata* (Hübner), Slender Pug, Ashtead, Surrey, May 1991. (Page 33)
2. *E. inturbata* (Hübner), Maple Pug, Chessington, Surrey, July 1996. (Page 37)
3. *E. haworthiata* Doubleday, Haworth's Pug, Oxshott Heath, Surrey, July 1998. (Page 39)
4. *E. plumbeolata* (Haworth), Lead-coloured Pug, Blean, Kent, June 1983. (Page 41)
5. *E. abietaria* (Goeze), Cloaked Pug, Kyloe, Northumberland, ex larva, em. May 1985. (Page 43)
6. *E. linariata* ([Denis & Schiffermüller]), Toadflax Pug, Lakenheath, Suffolk, June 1982. (Page 45)
7. *E. pulchellata* Stephens, Foxglove Pug, Surbiton, Surrey, September 1981. (Page 47)
8. *E. pulchellata* Stephens, Foxglove Pug, Rehaghy Wood, nr Aughnacloy, Co. Tyrone, May 1999. (Page 47)
9. *E. irriguata* (Hübner), Marbled Pug, Lyndhurst, Hampshire, ex larva, em. May 1987. (Page 49)
10. *E. exiguata* (Hübner), Mottled Pug, Aughinlig, nr Moy, Co. Armagh, June 1999. (Page 50)
11. *E. insigniata* (Hübner), Pinion-spotted Pug, Stawford, Norfolk, May 1986. (Page 52)
12. *E. valerianata* (Hübner), Valerian Pug, Westbere, Kent, May 1984. (Page 53)
13. *E. pygmaeata* (Hübner), Marsh Pug, Hockwold, Norfolk, May 1983. (Page 55)
14. *E. venosata* (Fabricius), Netted Pug, Ardnamurchan, Argyll, ex larva, em. June 1985. (Page 57)
15. *E. venosata* (Fabricius), Netted Pug, Doolin, Co. Clare, May 1996. (Page 57)

Photographs:
Figs 1–7, 9, 11–15 © Jim Porter;
Figs 8, 10 © Robert Thompson

PLATE A

PLATE B

1. *Eupithecia egenaria* Herrich-Schäffer, Fletcher's or Pauper Pug, Tintern, Monmouthshire, May 1983. (Page 59)
2. *E. centaureata* ([Denis & Schiffermüller]), Lime-speck Pug, Murlough NNR, Dundrum, Co. Down, July 1999. (Page 62)
3. *E. trisignaria* Herrich-Schäffer, Triple-spotted Pug, Dartry Lodge, Ballycullen, Co. Armagh, July 1999. (Page 64)
4. *E. intricata* (Zetterstedt) subsp. *arceuthata* (Freyer), Freyer's Pug, Surbiton, Surrey, June 1986. (Page 67)
5. *E. satyrata* (Hübner), Satyr Pug, Ranmore, nr Dorking, Surrey, June 1989. (Page 72)
6. *E. absinthiata* (Clerck), Wormwood Pug, Aughinlig, nr Moy, Co. Armagh, July 1999. (Page 76)
7. *E. absinthiata* (Clerck), f. *goossensiata* Mabille, Ling Pug, Beeley, Derbyshire, July 1993. (Page 76)
8. *E. expallidata* Doubleday, Bleached Pug, Fernhurst, West Sussex, July 1986. (Page 79)
9. *E. assimilata* Doubleday, Currant Pug, Aughinlig, nr Moy, Co. Armagh, May 2001. (Page 81)
10. *E. vulgata* Haworth, Common Pug, Surbiton, Surrey, May 1984. (Page 83)
11. *E. vulgata* Haworth, Common Pug, Helen's Bay, Co. Down, May 1999. (Page 83)
12. *E. tripunctaria* Herrich-Schäffer, White-spotted Pug, Annagarriff Wood NNR, Peatlands, Co. Armagh, June 2001. (Page 86)
13. *E. denotata denotata* (Hübner), Campanula Pug, Bury St Edmunds, Suffolk, May 1989. (Page 88)
14. *E. denotata jasioneata* Crewe, Jasione Pug, Sennen, Cornwall, ex larva, em. June 1985. (Page 89)
15. *E. subfuscata* (Haworth), Grey Pug, Rehaghy Wood, nr Aughnacloy, Co. Tyrone, June 1999. (Page 91)

Photographs:

Figs 1, 4, 5, 7, 8, 10, 13, 14 © Jim Porter;
Figs 2, 3, 6, 9, 11, 12, 15 © Robert Thompson

PLATE B

PLATE C

1. *Eupithecia icterata* (Villers), Tawny-speckled Pug, Surbiton, Surrey, July 1981. (Page 94)
2. *E. succenturiata* (Linnaeus), Bordered Pug, Black Mountain, Dundonald, Co. Down, July 1994. (Page 96)
3. *E. subumbrata* ([Denis & Schiffermüller]), Shaded Pug, Sandwich, Kent, June 1984. (Page 98)
4. *E. millefoliata* (Rössler), Yarrow Pug, Murston, Sittingbourne, Kent, June 1983. (Page 99)
5. *E. simpliciata* (Haworth), Plain Pug, Chessington, Surrey, August 1998. (Page 101)
6. *E. sinuosaria* (Eversmann), Goosefoot Pug, Alpen, Südtirol, Italy, June 2000. (Page 102)
7. *E. distinctaria* subsp. *constrictata* Guenée, Thyme Pug, Dovedale, Derbyshire, ex larva, em. May 1985. (Page 103)
8. *E. indigata* (Hübner), Ochreous Pug, Chobham, Surrey, April 1982. (Page 105)
9. *E. pimpinellata* (Hübner), Pimpinel Pug, Killard Point NNR, Strangford, Co. Down, July 1999. (Page 106)
10. *E. nanata* (Hübner) subsp. *angusta* Prout, Narrow-winged Pug, Annagarritt Wood NNR, Peatlands, Co. Armagh, June 1999. (Page 108)
11. *E. nanata* (Hübner) subsp. *angusta* Prout, Narrow-winged Pug, Chessington, Surrey, August 1998. (Page 108)
12. *E. extensaria* (Freyer) subsp. *occidua* Prout, Scarce Pug, nr Hunstanton, Norfolk, ex larva, em. April 1993. (Page 110)
13. *E. fraxinata* Crewe, Ash Pug, Camber, East Sussex, ex larva, em. June 1985. (Page 112)
14. *E. fraxinata* Crewe, Ash Pug, Aughinlig, nr Moy, Co. Armagh, July 1999. (Page 112)
15. *E. virgaureata* Doubleday, Golden-rod Pug, Karlsruhe, Baden-Württemberg, Germany, April, 1992. (Page 114)

Photographs:

Figs 1, 3–5, 7, 8, 11, 13 © Jim Porter;
Figs 2, 9, 10, 14 © Robert Thompson;
Figs 6, 15 © Ulrich Ratzel;
Fig 12 © Paul Waring

PLATE C

PLATE D

1. *Eupithecia abbreviata* Stephens, Brindled Pug, Rostrevor Oakwood NNR, Co. Down, March 2000. (Page 117)
2. *E. dodoneata* Guenée, Oak-tree Pug, Oxford Island NNR, Lurgan, Co. Armagh, April 2002. (Page 119)
3. *E. pusillata* ([Denis & Schiffermüller]), Juniper Pug, Ranmore, nr Dorking, Surrey, May 1987. (Page 122)
4. *E. phoeniceata* (Rambur), Cypress Pug, Aldwick Bay, Bognor, West Sussex, September 1987. (Page 124)
5. *E. ultimaria* Boisduval, Channel Islands Pug, Brighton, East Sussex, August 1996. (Page 126)
6. *E. lariciata* (Freyer), Larch Pug, Breen Oakwood NNR, Co. Antrim, August 2001. (Page 127)
7. *E. tantillaria* Boisduval, Dwarf Pug, Gomshall, Surrey, May 1999. (Page 129)
8. *Chloroclystis v-ata* (Haworth), V-Pug, The Argory (National Trust), nr Moy, Co. Armagh, April 1999. (Page 131)
9. *C. v-ata* (Haworth), V-Pug, Ashtead, Surrey, June 1987. (Page 131)
10. *Pasiphila chloerata* Mabille, Sloe Pug, Effingham, Surrey, May 1982. (Page 132)
11. *P. rectangulata* (Linnaeus), Green Pug, Karlsruhe-Hochstetten, Baden-Württemberg, Germany, ex larva, June 1991. (Page 135)
12. *P. rectangulata* (Linnaeus), Green Pug (melanic form), Surbiton, Surrey, June 1986. (Page 135)
13. *P. debiliata* (Hübner), Bilberry Pug, Abinger, Surrey, June 1986. (Page 137)
14. *Gymnoscelis rufifasciata* (Haworth), Double-striped Pug, Chessington, Surrey, May 1997. (Page 138)
15. *G. rufifasciata* (Haworth), Double-striped Pug, Ballynasollus, Co. Tyrone, July 1999. (Page 138)

Photographs:
Figs 1, 2, 6, 8, 15 © Robert Thompson;
Figs 3–5, 7, 9, 10, 12–14 © Jim Porter;
Fig 11 © Ulrich Ratzel

PLATE D

Appendix I

Glossary

aberration (abbr. **ab**.) – see introductory chapter on infraspecific terminology.

aedeagus – the tubular portion of the male genitalia containing the vesica and cornuti.

allopatric – occupying separate geographic areas.

anal – hindmost.

angulate – forming an angle >.

antemedian – basad of the middle (of the wing).

anterior – towards the head.

apex – the tip of the wing; junction of the costa and the termen.

apical – adjacent to, or in the area of, the apex.

auctorum (abbr. **auctt**.) – of authors other than the original nomenclator.

basad – closer to the base (of the wing).

basal – at the base (of the wing); adjacent to the thorax.

biangulate – forming two angles.

biciliate – having two rows of cilia.

bivoltine – two generations per year.

bursa copulatrix – the bulbous anterior portion of the female genitalia.

cell – ovoid area of the wings enclosed by nervures. Costal and from base to middle.

chevron – arrow-shaped marking.

chitin – hard material forming exoskeleton, prothoracic and anal plates of some larvae and features of the genitalia such as the cornuti and signa.

ciliate – bearing cilia; having a hairy appearance.

cilium (pl. **cilia**) – fine hair-like structure or specialized hair-like wing scale.

clasper – one of a pair of structures of the male genitalia used for clasping the female during copulation; one of the hindmost pair of legs of a larva.

cline – a continuous spatial gradation of characters from one morphological form to another.

concolorous – of the same colour.

conspecific – of the same species.

cornutus (pl. **cornuti**) – spine-like structures attached to the male vesica within the aedeagus.

costa – the leading edge of the forewing; the front margin.

costad – closer to the costa.

costal – at, or adjacent to, the costa.

cremaster – series of hook-like structures on the anal segment of a pupa.

cross-lines – transverse linear markings on the wings.

cryptic – protective colouring or appearance making concealment easier.

curved – smoothly arched. Not forming an angle.

dentate – toothed.

diapause – state of suspended development during the winter months.

discal – in the outer portion of the cell.

distad – away from the base; closer to the termen.

distal – furthest from the base.

dorsum – the trailing edge of the wing; the hind margin.

ductus bursae – the tube leading from the bursa copulatrix.

ecdysis – the moulting process during which the outer skin is shed.

eversible – of a structure which everts, e.g. the aedeagus.

fascia (pl. **fasciae**) – a transverse band.

form (abbr. **f**.) – see introductory chapter on infraspecific terminology.

frass – larval excretions.
fuscous – dusky grey.
geniculate – abruptly bent forming an obtuse angle.
gynandromorph – part male and part female.
imago – the adult insect; the reproductive stage of metamorphosis.
infraspecific – within the species taxon.
instar – the stage between ecdyses or moults.
interneural – between the wing veins.
irrorate – finely dusted (with scales); covered with minute spots.
larva (pl. **larvae**) – caterpillar.
linear – in the form of a line.
median – transversely across the middle.
melanochroic – general darkening of colours. Not to be confused with 'melanic' in which black pigment dominates.
mesothorax – the second thoracic segment.
metathorax – the third (posterior) thoracic segment.
morphology – the study of the form or appearance of an animal and/or its constituent parts.
mosaic – exhibiting male and female or polymorphic markings.
multivoltine – several generations per year.
nervure – wing vein.
neural – referring to the wing veins.
obsolete – absent, or almost so. Usually refers to markings on the wings or larval tegument.
overwinter – to pass the winter at a certain stage of metamorphosis.
oviposit – to lay eggs.
ovum (pl. **ova**) – egg.
patagium – the area between head and thorax.
polymorphic – having two or more forms within a given population.
posterior – towards the tail.
postmedian – distad of the middle (of the wing).

prepupa – the inactive larval stage immediately prior to pupation.
prolegs – the abdominal legs of a larva.
prothorax – the first (anterior) thoracic segment.
pupa (pl. **pupae**) – chrysalis.
race – see introductory chapter on infraspecific terminology.
R.E.S. – Rothamsted Experimental Station.
R.I.S. – Rothamsted Insect Survey.
sclerotized – composed of a strong dark layer of chitin.
sensu – in the opinion of; according to.
seta (pl. **setae**) – a stiff, superficially hair-like bristle.
setose – bearing setae.
signum – patch of thorn-like processes inside the bursa copulatrix.
sinuate – wavy.
spatial – pertaining to geographical area.
spiracle – the opening of the respiratory tract.
spiracular – adjacent or pertaining to the spiracles.
stria (pl. **striae**) – a fine line.
strigulate – covered with short fine transverse lines.
subbasal – adjacent to the base (of the wing).
subcostal – adjacent to the costa (of the wing).
subdorsal – adjacent to the dorsum (of the wing).
subspecies (abbrev. **subsp.**) – see introductory chapter on infraspecific terminology.
subterminal – adjacent to the termen (of the wing).
suffused – shaded.
sympatric – occupying the same geographical area.
synonym – alternative name given to the same taxon.
taxon (pl. **taxa**) – a unit of classification (species, subspecies, form etc.).
temporal – pertaining to time.
termen – the outer edge of the wing; the edge furthest from the thorax.

GLOSSARY

thoracic – pertaining to the thorax.
tornal – in the area of the tornus.
tornus – the angle between the dorsum and the termen.
truncate – ending abruptly.
tubercle – small wart-like structure.
type – the actual specimen described by the author of a taxon.
unicolorous – of the same colour or shade.
univoltine – one generation per year.

valva (pl. **valvae**) – one of the pair of clasping organs of the male genitalia; clasper.
variety – See introductory chapter on infraspecific terminology.
venation – the arrangements of wing veins.
vesica – the eversible part of the male genitalia inside the aedeagus.
wingspan – twice the distance from the centre of the thorax to the forewing apex.

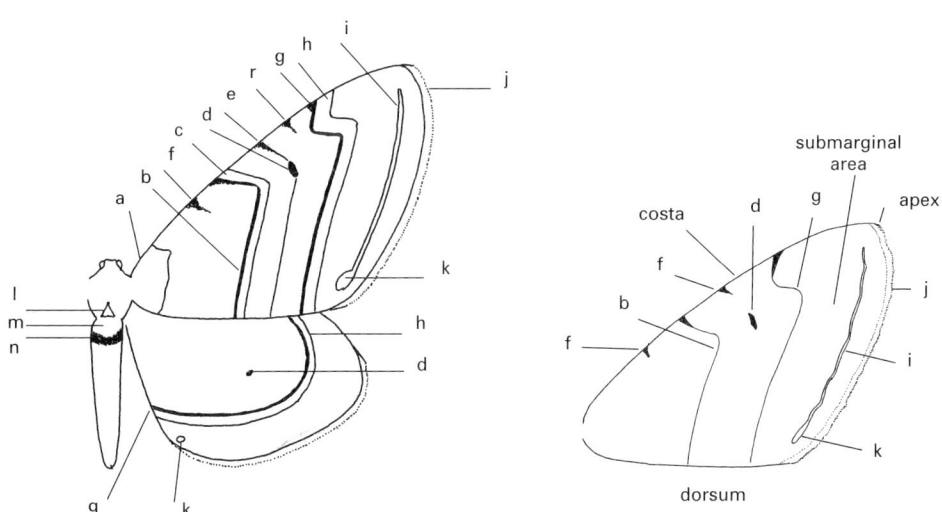

Text figures 34, 35 Key markings and wing areas of a typical eupitheciine moth
(**a**) basal patch (**b**) antemedian line (**c**) antemedian fascia (**d**) discal spot
(**e**) median line (**f**) costal spots (**g**) postmedian line (**h**) postmedian fascia
(**i**) subterminal line (**j**) fringe (**k**) tornal spot (**l**) metathoracic spot (**m**) abdominal band
(**n**) abdominal sub-basal band

223

Appendix II

Table of phenology

O = ovum, L = larva, P = pupa, I = imago. Characters in brackets signify assumed or suspected presence, or partial broods and emergences.

	J	F	M	A	M	J	J	A	S	O	N	D
E. tenuiata	O	OL	OL	L	P	PIO	IO	IO	O	O	O	O
E. inturbata	O	O	O	O	OLP	P	PIO	IO	O	O	O	O
E. haworthiata	P	P	P	P	P	PIO	IOL	LP	P	P	P	P
E. plumbeolata	P	P	P	P	PIO	IOL	L	LP	P	P	P	P
E. abietaria	P	P	P	P	P	IO	IOL	(I)L	LP	P	P	P
E. linariata	P	P	P	P	PI	IO	IO	IOL	(I)LP	LP	P	P
E. pulchellata	P	P	P	P(I)	PIO	IO	(I)OL	(I)LP	P	P	P	P
E. irriguata	P	P	P	PIO	IO	IL	LP	P	P	P	P	P
E. exiguata	P	P	P	P	PIO	IO	L	L	LP	LP	P	P
E. insigniata	P	P	P	P	IO	IOL	LP	P	P	P	P	P
E. valerianata	P	P	P	P	P	IO	IOL	L	(L)P	P	P	P
E. pygmaeata	P	P	P	P	PI	IOL	LP	P(I)	P	P	P	P
E. venosata	P	P	P	P	PIO	IOL	IL	LP	P	P	P	P
E. egenaria	P	P	P	P	IO	IOL	L	P	P	P	P	P
E. centaureata	P	P	P	P	PI	IOL	IOL	IOL	IOLP	ILP	P	P
E. trisignaria	P	P	P	P	P	PI	IO	IOL	LP	LP	P	P
E. intricata	P	P	P	P	PIO	IO	IOL	L	LP	P	P	P
E. satyrata	P	P	P	P	PIO	IO	IOL	L	LP	P	P	P
E. absinthiata	P	P	P	P	PI	PIO	IO	IOL	L	LP	P	P
E. assimilata	P	P	P	P	PI	IOL	OLP(I)	PI	IOL	LP(I)	P	P
E. expallidata	P	P	P	P	P	PI	IO	IO	L	LP	P	P
E. vulgata	P	P	P	P	IO	IOL	ILP	(I)P	P	P	P	P
E. tripunctaria	P	P	P	P	PIO	IOL	LP	PIO	IOL	LP	P	P
E. denotata	P	P	P	P	P	PI	IO	L	LP	LP	P	P
E. subfuscata	P	P	P	P	PIO	IO	IOL	L(I)	L	LP	P	P

TABLE OF PHENOLOGY

	J	F	M	A	M	J	J	A	S	O	N	D
E. icterata	P	P	P	P	P	PI	PI	IO	OL	LP	P	P
E. succenturiata	P	P	P	P	P	P	PI	IO	OL	LP	P	P
E. subumbrata	P	P	P	P	PI	PI	IOL	L	LP	P	P	P
E. millefoliata	P	P	P	P	PI	PIO	IO	OL	L	LP	LP	P
E. simpliciata	P	P	P	P	P	PIO	IO	IOL	L	P	P	P
E. distinctaria	P	P	P	P	P	PIO	IO	L	LP	P	P	P
E. indigata	P	P	P	PI	IO	IOL	L	L	LP	P	P	P
E. pimpinellata	P	P	P	P	P(I)	PIO	IO	OL	L	LP	P	P
E. nanata	P	P	P	PIO	IO	O	L(I)	(IO)LP	LPI	LPI	P	P
E. extensaria	P	P	P	P	P(I)	PIO	IOL	L	LP	P	P	P
E. fraxinata	P	P	P	P	PI	IOL	LPI	PIO	LP	P	P	P
E. virgaureata	P	P	P	P	IO	IOL	LPI	PIO	OL	LP	P	P
E. abbreviata	P	P	P	IO	IO	L	LP	P	P	P	P	P
E. dodoneata	P	P	P	PI	PIO	IOL	L	LP	P	P	P	P
E. pusillata	O	O	OL	OL	L	LPI	PIO	IO	IO	O	O	O
E. phoeniceata	L(P)	LP	LP	P	P	P	P(I)	PIO	IOL	IOL	L	L(P)
E. lariciata	P	P	P	P	I	IOL	L(I)	LP(I)	P(I)	P	P	P
E. tantillaria	P	P	P	PI	IO	IO	(IO)L	LP	P	P	P	P
C. v ata (south)	P	P	P	PIO	IOL	IOLP	IOLP	IOL	LP	LP	P	P
C. v-ata (north)	P	P	P	P	P	IPO	IOL	L	PL	P	P	P
P. chloerata	O	O	OL	LP	LPI	IO	(I)O	O	O	O	O	O
P. rectangulata	O	O	O	OL	LP	PIO	IO	IO	O	O	O	O
P. debiliata	O	O	O	OL	LP	PIO	IO	IO	O	O	O	O
G. rufifasciata	P	P	PIO	IO	IOL	LPI	PIO	IO	IOL	LP(I)	P	P

Appendix III

Index of larval foodplants with associated eupitheciine species

With the exception of a few species or cultivars which do not occur naturally in the British Isles, the botanical nomenclature conforms to that of Stace (1991). The nomenclature of alien species follows that of Clement & Foster (1994).

Records without literature references have been confirmed by the present authors.

    * pers.comm.   # in captivity   † considered doubtful

Abies alba (European silver-fir)   *E. abietaria* (Styles, 1961; Winter, 1983)
A. grandis (giant fir)   *E. tantillaria* (Winter, 1990; M. C. Townsend*)
A. procera (noble fir)   *E. abietaria* (Rutherford, 1988; Winter, 1990)
Acer campestre (field maple)   *E. inturbata*
     E. exiguata (Allan, 1949)
A. pseudoplatanus (sycamore)   *E. exiguata* (Skinner, 1984; Owen, 1991)
Achillea millefolium (yarrow)   *E. absinthiata*
     E. icterata
     E. millefoliata
     E. subfuscata
     E. succenturiata
     E. centaureata (Newman, 1869)
     E. vulgata (Chalmers-Hunt, 1968–71; C. W. Plant*)
     E. pimpinellata (Wilson, 1880)†
     E. virgaureata (Tutt, 1906c; Prout, 1907b)†
Alnus glutinosa (alder)   *E. exiguata* (Newman, 1896)
Angelica sylvestris (wild angelica)   *E. trisignaria*
     E. tripunctaria (Newman, 1869; C. W. Plant*)
     E. subfuscata (Carrington, 1871; Wilson, 1880)
     E. centaureata (Carrington, 1871; C. W. Plant*)
     C. v-ata (Newman, 1869)
     E. pimpinellata (Carrington, 1871; Tutt, 1906c)†
     E. virgaureata (Cue, 1963)†
Anthriscus cerefolium (garden chervil)   *E. tripunctaria*# (Allan, 1949)
A. sylvestris (cow parsley)   *E. tripunctaria* (Tutt, 1906c)
     E. virgaureata (Crewe, 1862k; Prout, 1907b; South, 1961; Cue, 1963)
     G. rufifasciata (Allan, 1949); (# Newman, 1869)
Antirrhinum majus (snapdragon)   *E. linariata* (Chalmers-Hunt, 1968–71)
Arctostaphylos sp. (bearberry)   *E. satyrata* (Allan, 1949)

INDEX OF LARVAL FOODPLANTS

Artemisia abrotanum (southernwood)   *E. absinthiata* (C. W. Plant*)
                                        E. icterata (C. W. Plant*)
                                        E. succenturiata (C. W. Plant*)
                                        E. extensaria#
A. absinthium (wormwood)   *E. absinthiata* (Allan, 1949)
                                        E. fraxinata#? (Allan, 1949)
                                        E. icterata (Gardner, 1907)
                                        E. succenturiata (Chalmers-Hunt, 1968–71;
                                          C. W. Plant*)
A. campestris (field wormwood)   *E. fraxinata*#? (Allan, 1949)
A. vulgaris (mugwort)   *E. centaureata* (C. W. Plant*)
                                        E. absinthiata
                                        E. subfuscata
                                        E. succenturiata
                                        E. icterata (Prout, 1896)
                                        C. v-ata (Allan, 1949)
                                        E. fraxinata (Gregson, 1874a; Prout, 1904)
                                          Confirmed#
Aster amellus (aster, 'King George') *E. centaureata* (C. W. Plant*)
                                        E. absinthiata (C. W. Plant*)
Aster novi-belgii (hairy Michaelmas-daisy) *E. absinthiata*
                                        E. centaureata (C. W. Plant*)
                                        E. expallidata# (Crewe, 1860f)
A. tripolium (sea aster)   *E. absinthiata* (Allan, 1949; C. W. Plant*)
                                        E. centaureata (Chalmers-Hunt, 1968–71;
                                          C. W. Plant*)
                                        G. rufifasciata (C. W. Plant*)
Atriplex laciniata (frosted orache)   *E. simpliciata*
A. patula (common orache)   *E. simpliciata* (Tutt, 1906d; Allan, 1949)
Ballota nigra (black horehound)   *E. absinthiata* (Allan, 1949)
Bellis perennis (daisy)   *G. rufifasciata* (Wilson, 1880)
Berberis vulgaris (barberry)   *E. exiguata* (Newman, 1869)
                                        E. subfuscata (G. M. Haggett*)
Betula sp. (birch)   *E. irriguata* (Scorer, 1913)
Buddleja davidii (butterfly-bush)   *E. exiguata* (Owen, 1991)
                                        G. rufifasciata (Hammond, 1952)
Bupleurum falcatum (sickle-leaved hare's-ear) *E. pimpinellata* (Tutt, 1906c)
Corylus avellana (hazel)   *E. subfuscata* (Wilson, 1880)
Calluna vulgaris (heather)   *E. absinthiata*
                                        E. nanata
                                        G. rufifasciata
                                        E. centaureata (Wilson, 1880)
                                        E. satyrata (Gregson, 1884b; Allan, 1949)
                                        E. subfuscata (Wilson, 1880; Tugwell, 1892)
                                        E. virgaureata (Tutt, 1906c; Prout, 1907b)
Campanula sp. (bellflower)   *E. subumbrata* (Scorer, 1913)
C. glomerata (clustered bellflower)   *E. denotata*
                                        E. absinthiata (Allan, 1949)
                                        E. centaureata (Newman, 1869)

227

C. *latifolia* (giant bellflower)   *E. denotata* (Crewe, 1870)
C. *medium* (Canterbury-bells)   *E. denotata* (Crewe, 1870)
C. *patula* (spreading bellflower)   *E. denotata* (Crewe, 1870)
C. *persicifolia* (peach-leaved bellflower)   *E. denotata* (Crewe, 1870)
C. *rapunculoides* (creeping bellflower)   *E. denotata* (Crewe, 1870)
C. *rapunculus* (rampion bellflower)   *E. denotata* (Crewe, 1870)
C. *rotundiflora* (harebell)   *E. centaureata* (Wilson, 1880)
    E. denotata (Wilson, 1880)
C. *trachelium* (nettle-leaved bellflower)   *E. denotata* (Crewe, 1870)
Castanea sativa (sweet chestnut)   *C. v-ata* (South, 1961)
Centaurea nigra (common knapweed)   *E. satyrata*
    E. absinthiata (Allan, 1949)
    E. centaureata (Allan, 1949)
    E. subumbrata (Newman, 1869)
Centaurium erythraea (common centaury)   *E. centaureata* (Wilson, 1880)
    E. satyrata (Wilson, 1880)
    E. subfuscata (Wilson, 1880)
Centranthus ruber (red valerian)   *E. centaureata*[#]
    E. satyrata[#]
    E. valerianata[#] (Hammond, 1952)
Cerastium arvense (field mouse-ear)   *E. pygmaeata*
C. *tomentosum* (snow-in-summer)   *E. pygmaeata* (Scorer, 1913)
Chamaecyparis lawsoniana (Lawson's cypress)   *E. intricata*
    E. phoeniceata[#]
    E. tantillaria (Hatcher, 1989; Winter, 1990)
Chamaemelum nobile (chamomile)   *E. succenturiata* (Allan, 1949)
Chenopodium album (fat-hen)   *E. simpliciata*
C. *bonus-henricus* (good-King-Henry)   *E. simpliciata*
C. *vulvaria* (stinking goosefoot)   *E. simpliciata* (Allan, 1949)
Chrysanthemum frutescens (Marguerite)   *G. rufifasciata* (Chalmers-Hunt, 1968–71)
Cicuta virosa (cowbane)   *E. absinthiata* (Allan, 1949)
    E. centaureata (Allan, 1949)
    E. tripunctaria (Tutt, 1906c)
Clematis vitalba (traveller's-joy)   *E. haworthiata*
    C. v-ata
    G. rufifasciata
    E. absinthiata (Allan, 1949)
    E. centaureata (Allan, 1949)
    E. plumbeolata (Wilson, 1880)
Convolvulus arvensis (field bindweed)   *G. rufifasciata* (Allan, 1949)
Cornus sanguinea (dogwood)   *E. exiguata* (Tutt, 1906d)
Crataegus monogyna (hawthorn)   *E. abbreviata*
    E. dodoneata
    E. exiguata
    E. insigniata
    E. subfuscata
    E. vulgata
    G. rufifasciata
    E. virgaureata (Prout, 1907b; Haggett, 1968a)

INDEX OF LARVAL FOODPLANTS

Crepis sp. (hawk's-beard)
C. capillaris (smooth hawk's-beard)
C. vesicaria (beaked hawk's-beard)
×*Cupressocyparis leylandii* (Leyland cypress)

Cupressus/Chamaecyparis sp. (cypress)

Cupressus macrocarpa (Monterey cypress)

Cytisus scoparius (broom)

Dahlia pinnata (dahlia)

Daucus carota carota (wild carrot)
Dendranthema sp. (florist's chrysanthemum)

Descurainia sophia (flixweed)
Digitalis purpurea (foxglove)
Epilobium sp. (willowherb)
Erica sp. (heath)

E. cinerea (bell heather)
E. tetralix (cross-leaved heath)

Eupatorium cannabinum (hemp-agrimony)

Euphrasia sp. (eyebright)

Fagus sylvatica (beech)
Filipendula ulmaria (meadowsweet)
Foeniculum vulgare (fennel)
Fraxinus excelsior (ash)

Galeopsis sp. (hemp-nettle)
G. tetrahit (common hemp-nettle)
Galium mollugo (hedge bedstraw)

G. verum (lady's bedstraw)

C. v-ata (Prout, 1908; Allan, 1949; C. W. Plant*)
P. rectangulata (de Worms, 1978)
E. satyrata (Wilson, 1880)
E. fraxinata (Prout, 1904)
E. subumbrata (Newman, 1869)
E. intricata
E. phoeniceata#
E. intricata
E. pusillata (Skinner, 1984)
E. indigata# (Crewe, 1863a)
E. intricata
E. phoeniceata
E. subfuscata (G. M. Haggett*)
G. rufifasciata (Scorer, 1913; Allan, 1949)
E. centaureata (Chalmers-Hunt, 1968–71)
E. virgaureata (Allan, 1949)
E. subumbrata (Carrington, 1885b)
E. satyrata (Allan, 1949)
E. subumbrata (Tutt, 1906b)
E. succenturiata (Prout, 1896; Allan, 1949)
E. icterata# (Prout, 1896)
E. subumbrata (Skinner, 1984)
E. pulchellata
E. subfuscata (Wilson, 1880)
E. centaureata (Wilson, 1880)
E. satyrata (Wilson, 1880)
E. subfuscata (Wilson, 1880)
E. absinthiata (Allan, 1949)
E. absinthiata (Allan, 1949)
E. nanata (Chalmers-Hunt, 1968–71)
E. centaureata
C. v-ata
G. rufifasciata
E. absinthiata (Newman, 1869)
E. subumbrata (Tutt, 1906b)
E. plumbeolata# (Allan, 1949)
E. irriguata (Tutt, 1906b 'more rarely')
E. satyrata (Duddington & Johnson, 1983)
E. tripunctaria (Hammond, 1952)
E. fraxinata
E. exiguata (Newman, 1869; Owen, 1991; C. W. Plant*)
E. satyrata (Allan, 1949)
E. subfuscata (Chalmers-Hunt, 1968–71)
E. satyrata (Crewe, 1860i)
E. subumbrata (Newman, 1869)
E. satyrata (G. M. Haggett*)
E. subumbrata (G. M. Haggett*)

Gentianella sp. (gentian)
G. amarella (autumn gentian)

Gentianella campestris (field gentian)

Gladiolus sp. (gladiolus)
Globularia vulgaris (globularia)
Helianthemum nummularium (common rock-rose)
Heracleum sphondylium (hogweed)

Hippophae rhamnoides (sea-buckthorn)
Humulus lupulus (hop)

Hypericum sp. (St John's-wort)

H. perforatum (perforate St John's-wort)
Iberis sempervirens (perennial candytuft)
Ilex aquifolium (holly)
Jasione montana (sheep's-bit)

Juniperus communis (juniper)

J. phoenicea (Phoenician juniper)
J. sabina (common savin)
Knautia arvensis (field scabious)

Larix decidua (larch)

L. ×*eurolepis* (hybrid larch)
L. kaempferi (Japanese larch)
Laserpitium latifolium (broad-leaved sermountain)
Lathyrus pratensis (meadow vetchling)
Leontodon hispidus (rough hawkbit)

Leucanthemum vulgare (ox-eye daisy)

E. virgaureata (Prout, 1907b)
E. satyrata (Crewe, 1860i)
E. subumbrata (Newman, 1869)
E. satyrata (Crewe, 1860i)
E. subumbrata (Newman, 1869)
E. centaureata (South, 1961)
E. subumbrata (Tutt, 1906b)
E. satyrata (Allan, 1949)

E. trisignaria
E. tripunctaria
E. vulgata (Chalmers-Hunt, 1968–71; C. W. Plant*)
E. virgaureata (Prout, 1907b)†
E. fraxinata
E. assimilata
E. exiguata (G. M. Haggett*)
E. satyrata (Allan, 1949)
E. subfuscata (Wilson, 1880)
E. subumbrata (Tutt, 1906b; Agassiz *et al.*, 1981)
C. v-ata (Allan, 1949)
E. exiguata (Owen, 1991)
G. rufifasciata
E. denotata
E. subumbrata (Tutt, 1906b)
E. intricata
E. pusillata
E. indigata# (Crewe, 1863a)
E. phoeniceata (South, 1961)
E. intricata (Hammond, 1952)
E. satyrata
E. subumbrata
E. absinthiata (Gregson, 1874b)
E. centaureata (Wilson, 1880; Chalmers-Hunt, 1968–71; C. W. Plant*)
E. lariciata
E. tantillaria (Duddington & Johnson, 1983)
E. indigata# (Tutt, 1906c; Allan, 1949)
E. abietaria (Wilson, 1880)
E. nanata (Browne, 1968)
E. lariciata (G. M. Haggett*)
E. lariciata (G. M. Haggett*)
E. tripunctaria (Tutt, 1906c)

E. subfuscata (Wilson, 1880)
E. satyrata
E. subumbrata
E. satyrata (Hodgkinson, 1885)

Ligustrum vulgare (wild privet)
Limonium vulgare (common sea-lavender)
Linaria vulgaris (common toadflax)

Lychnis flos-cuculi (ragged-Robin)
Lysimachia vulgaris (yellow loosestrife)

Lythrum salicaria (purple-loosestrife)
Malus domestica (apple)
M. sylvestris (crab apple)

Melampyrum arvense (field cow-wheat)
M. pratense (common cow-wheat)

Neottia nidus-avis (bird's-nest orchid)
Nothofagus sp. (southern beech)
Odontites vernus (red bartsia)
Ononis repens (common restharrow)

Origanum vulgare (wild marjoram)

Paeonia officinalis (peony)
Pastinaca sativa var. *sylvestris* (wild parsnip)

Pedicularis palustris (marsh lousewort)
Pelargonium sp. (geranium)
Petroselinum crispum (garden parsley)
Peucedanum officinale (hog's fennel)
Picea abies (Norway spruce)

P. sitchensis (Sitka spruce)

Pilosella aurantiaca (fox-and-cubs)
Pimpinella major (greater burnet-saxifrage)

E. exiguata (Wilson, 1880)
E. centaureata (J. Steeden*)
E. linariata
E. virgaureata (Wilson, 1880)
E. subfuscata (Wilson, 1880)
E. absinthiata (Wilson, 1880)
C. v-ata (Allan, 1949)
E. virgaureata (Wilson, 1880)
C. v-ata (Allan, 1949)
P. rectangulata
E. insigniata
P. rectangulata
C. v-ata
E. plumbeolata (Allan, 1949)
E. plumbeolata
E. virgaureata (Cue, 1963)
E. abbreviata (Curtis, 1934b)
E. abbreviata
E. plumbeolata# (Allan, 1949)
E. centaureata (Wilson, 1880)
E. satyrata (Wilson, 1880)
E. subfuscata (Wilson, 1880)
E. absinthiata (Allan, 1949)
E. satyrata (Crewe, 1860i)
E. subumbrata (Newman, 1869)
G. rufifasciata (Wilson, 1880)
E. distinctaria# (Allan, 1949)
E. exiguata (Owen, 1991)
E. tripunctaria
E. trisignaria
E. centaureata (Wilson, 1880)
E. subfuscata (Wilson, 1880)
E. subumbrata (Chalmers-Hunt, 1968–71)
E. satyrata (Palmer, 1975)
E. centaureata (Wilson, 1880)
G. rufifasciata (Wilson, 1880)
E. centaureata (Chalmers-Hunt, 1968–71)
E. abietaria
E. tantillaria
E. lariciata (Crewe, 1865a; Newman, 1869; Scorer, 1913)
E. nanata (Browne, 1968)†
E. abietaria (Barbour, 1985)
E. satyrata (Winter, 1974; 1983)
E. tantillaria (Styles, 1961; Winter, 1983)
G. rufifasciata (Winter, 1974; 1983)
E. subumbrata
E. absinthiata (Allan, 1949)
E. centaureata (Newman, 1869)

P. major (cont'd)

P. saxifraga (burnet-saxifrage)

Pinus nigra (Austrian pine)
P. contorta (lodgepole pine)

P. sylvestris (Scots pine)

Plantago major (greater plantain)

Polygonum aviculare (knotgrass)

Populus nigra (black poplar)
Potentilla fruticosa (shrubby cinquefoil)
P. reptans (creeping cinquefoil)
Prunella vulgaris (selfheal)

Prunus avium (wild cherry)
P. spinosa (blackthorn)

Pseudotsuga menziesii (Douglas fir)

Pteridium aquilinum (bracken)
Pyrus communis (pear)

P. pyraster (wild pear)
Quercus sp. (oak)

Q. coccifera
Q. ilex (evergreen oak)

Q. petraea (sessile oak)
Q. robur (pedunculate oak)

E. pimpinellata (Allan, 1949)
E. trisignaria (Firmin et al., 1975)
E. absinthiata
E. pimpinellata
E. centaureata (Newman, 1869; C. W. Plant*)
E. subumbrata (Tutt, 1906b)
E. trisignaria (Allan, 1949)
E. tantillaria (Hammond & Smith, 1957)
E. indigata (Winter, 1990)
E. satyrata (Winter, 1974; 1983)
G. rufifasciata (Winter, 1974; 1983)
E. indigata
E. tantillaria (Hammond, 1952)
E. abietaria# (M. C. Townsend*)
E. nanata (Browne, 1968)
E. absinthiata (Allan, 1949)
E. centaureata (Allan, 1949)
E. centaureata (Chalmers-Hunt, 1968–71)
E. satyrata (Allan, 1949)
E. exiguata (Owen, 1991)
E. exiguata (Owen, 1991)
G. rufifasciata (Crewe, 1862c)
E. satyrata (Crewe, 1860i)
E. subumbrata (Newman, 1869)
P. rectangulata
E. exiguata
E. subfuscata
P. chloerata
P. rectangulata
E. insigniata (Tutt, 1906b; Scorer, 1913)
E. irriguata (Wilson, 1880)
E. virgaureata (Prout, 1907)
E. tantillaria
E. intricata (Hammond, 1952)
E. nanata (Browne, 1968)†
E. subfuscata (Tutt, 1906b)
P. rectangulata
C. v-ata
P. rectangulata (C. W. Plant*)
E. vulgata (Wilson, 1880)
E. insigniata (Wilson, 1880)
E. virgaureata (Allan, 1949)
E. massiliata (Mironov, 2003)
E. dodoneata (Chalmers-Hunt, 1968–71)
E. massiliata (Mironov, 2003)
E. abbreviata
E. abbreviata
E. dodoneata

Q. *suber* (cork oak)
Rhinanthus minor (yellow-rattle)

Ribes nigrum (black currant)

R. rubrum (red currant)
R. sanguineum (flowering currant)
R. uva-crispa (gooseberry)
Rosa cultivar (rose)

Rubus fruticosus (bramble)

R. idaeus (raspberry)

Salix sp. (sallow)

S. aurita (eared willow)
S. caprea (goat willow)

S. cinerea (grey willow)
Sambucus nigra (elder)

Saxifraga granulata (meadow saxifrage)

Scabiosa sp. (scabious)

S. atropurpurea (sweet scabious)
S. columbaria (small scabious)

Sedum sp. (stonecrop)
S. telephium (orpine)
Senecio erucifolius (hoary ragwort)

S. jacobaea (common ragwort)

E. *exiguata*
E. *irriguata*
E. *massiliata* (Mironov, 2003)
E. *plumbeolata* (Allan, 1949)
E. *satyrata* (Allan, 1949)
E. *assimilata*
E. *exiguata* (Newman, 1869)
E. *assimilata* (Allan, 1949)
E. *vulgata* (Owen, 1991)
E. *exiguata* (Owen, 1991)
E. *exiguata* (Owen, 1991)
G. *rufifasciata*
E. *subfuscata* (Wilson, 1880)
C. *v-ata* (Chalmers-Hunt, 1970)
G. *rufifasciata* (G. M. Haggett*)
E. *tripunctaria*#
E. *vulgata*#
E. *vulgata* (Tutt, 1906d; Allan, 1949)
E. *insigniata* (Tutt, 1906b)
E. *satyrata* (Scorer, 1913)
E. *virgaureata*# (Prout, 1907b)
E. *tenuiata* (Allan, 1949)
E. *exiguata*
E. *tenuiata*
E. *vulgata* (Scorer, 1913)
E. *tenuiata* (Tutt, 1906d)
E. *tripunctaria*
C. *v-ata* (Skinner, 1984)
E. *absinthiata* (Allan, 1949)
E. *centaureata* (Allan, 1949)
E. *fraxinata* (Robson, 1902; Scorer, 1913; South, 1961)
G. *rufifasciata* (Crewe, 1862c)
E. *absinthiata* (Allan, 1949)
E. *centaureata* (Newman, 1969; Wilson, 1880)
E. *subumbrata* (Evans & Evans, 1973)
E. *subfuscata* (Tutt, 1906c)
E. *vulgata* (Allan, 1949)
E. *absinthiata* (Newman, 1869)
E. *centaureata* (Newman, 1869)
E. *absinthiata*
E. *centaureata*
E. *satyrata*
E. *subfuscata*
E. *subumbrata*
E. *tripunctaria*
E. *virgaureata*
E. *vulgata*
C. *v-ata*

S. jacobaea (cont'd)   E. expallidata#
                                            E. icterata (Sheldon, 1896; Allan, 1949; 'in Scotland')
S. squalidus (Oxford ragwort)   E. centaureata (C. W. Plant*)
                                            E. absinthiata (C. W. Plant*)
S. vulgaris (groundsel)   E. absinthiata (Wilson, 1880)
                                            E. centaureata (Allan, 1949)
Seriphidium maritimum (sea wormwood)   E. extensaria
                                            E. succenturiata (Tutt, 1906b; Allan, 1949)
Silaum silaus (pepper-saxifrage)   E. centaureata (Newman, 1869)
Silene acaulis (moss campion)   E. venosata (Allan, 1949)
S. dioica (red campion)   E. subfuscata (Wilson, 1880)
                                            E. venosata (Newman, 1869)
                                            E. virgaureata (Prout, 1907b)
S. gallica (small-flowered catchfly)   E. venosata (Allan, 1949)
S. latifolia (white campion)   E. venosata (Evans & Evans, 1973)
                                            E. subfuscata (Wilson, 1880)
S. nutans (Nottingham catchfly)   E. venosata (Allan, 1949)
S. uniflora (sea campion)   E. venosata
S. vulgaris (bladder campion)   E. venosata
                                            E. vulgata (Allan, 1949)
Solidago canadensis (Canadian goldenrod)   E. absinthiata
                                            E. centaureata
                                            E. satyrata
                                            E. vulgata (Evans & Evans, 1973)
                                            E. expallidata#
                                            E. virgaureata#
                                            G. rufifasicata (Crewe, 1862c)
S. virgaurea (goldenrod)   E. absinthiata
                                            E. centaureata
                                            E. expallidata (Crewe, 1860g)
                                            E. satyrata
                                            E. subfuscata
                                            E. virgaureata
                                            E. subumbrata (Tutt, 1906b)
                                            E. tripunctaria (Agassiz et al., 1981)
                                            E. vulgata (Chalmers-Hunt, 1968–71)
                                            C. v-ata (Newman, 1869)
                                            E. pimpinellata (Wilson, 1880)
Sorbus aucuparia (rowan)   E. exiguata (Agassiz et al., 1981; Owen, 1991)
                                            G. rufifasciata (Wilson, 1880; Tutt, 1906d)
Spiraea ×vanhouttei (spiraea)   E. exiguata (Owen, 1991)
Stellaria holostea (greater stitchwort)   E. pygmaeata (Allan, 1949)
Succisa pratensis (devil's-bit scabious)   E. satyrata
                                            E. absinthiata (Wilson, 1880)
                                            E. centaureata (Wilson, 1880)
                                            E. subfuscata (Wilson, 1880)
                                            G. rufifasciata (Newton & Meredith, 1984)
Symphoricarpos albus (snowberry)   E. exiguata (Tutt, 1906d)

Syringa vulgaris (lilac)	*E. fraxinata*# (G. M. Haggett*)
Tagetes erecta (African marigold)	*E. centaureata* (Harris, 1766)
Tamarix gallica (tamarisk)	*E. fraxinata*
	G. rufifasciata
	E. intricata (South, 1961)
Tanacetum parthenium (feverfew)	*E. icterata* (West, 1994)
T. vulgare (tansy)	*E. absinthiata*
	E. icterata (Prout, 1896; Freer, 1896)
	E. succenturiata (Prout, 1896; Allan, 1949)
Taraxacum officinale (dandelion)	*E. centaureata*#
	E. subfuscata#
	E. vulgata#
Thuja orientalis (Chinese thuja)	*E. intricata* (Hammond, 1952)
	E. pusillata (Skinner, 1984)
T. plicata (western red-cedar)	*E. tantillaria* (Hatcher, 1989; Winter, 1990)
Thymus vulgaris (thyme)	*E. distinctaria*
	C. v-ata (Allan, 1949)
Tilia × vulgaris (lime)	*E. egenaria*#
T. cordata (small-leaved lime)	*E. egenaria*
T. platyphyllos (large-leaved lime)	*E. egenaria*
Tsuga heterophylla (western hemlock-spruce)	*E. tantillaria* (Hatcher, 1989; Winter, 1990)
Ulex europaeus (gorse)	*G. rufifasciata*
Urtica sp. (nettle)	*E. subfuscata* (Wilson, 1880)
Vaccinium myrtillus (bilberry)	*P. debiliata*
	E. vulgata (Scorer, 1913; Allan, 1949)
	E. nanata (Browne, 1968)
Valeriana dioica (marsh valerian)	*E. valerianata* (Allan, 1949; Duddington & Johnson, 1983)
V. officinalis (common valerian)	*E. valerianata*
	E. satyrata (Wilson, 1880; Symes, 1958)
	C. v-ata (Wilson, 1880; Sutton & Beaumont, 1989)
Verbascum sp. (mullein)	*E. satyrata* (Allan, 1949)
V. lychnitis (white mullein)	*E. icterata* (Wilson, 1880)
Viburnum lantana (wayfaring tree)	*C. v-ata* (G. M. Haggett*)
V. tinus (laurustinus)	*E. fraxinata*#
Weigela florida (weigelia)	*E. exiguata* (Owen, 1991)

References and Bibliography

Although all available county lists have been consulted, space does not permit their inclusion here unless they contain important information such as foodplant lists or significant historical records.

AGASSIZ, D. J. L. & other members of the BENHS, 1981. *An identification guide to the British pugs.* [Good colour photographs of all the species; good brief text; dichotomous key and genitalia drawings less useful.]

ALLAN, P. B. M., 1949. *Larval foodplants.* Watkins and Doncaster, London.

ANDERSON, J., 1890. Eupithecia togata, E. venosata and Emmelesia albulata, two years in the pupa. *Entomologist* **23**: 260. [Also notes capture of *E. abietaria* in Shetland and comments on variation in this species.]

ANON, 1878. Varieties of Lepidoptera at the national entomological exhibition. *Entomologist* **11**: 168–170. [First record (with illustration) of *E. tripunctaria* f. *angelicata*.]

ATMORE, E. A., 1891. Eupithecia pygmaeata probably double-brooded. *Entomologist's Record & Journal of Variation* **2**: 258. [Account of finding adults in August.]

BAKER, G., 1875. Eupithecia minutata larvae feeding on *Achillea millefolium*. *Entomologist* **8**: 109. [Notes on *E. absinthiata* f. *goossensiata* larvae feeding on yarrow.]

BARBOUR, D. A., 1976. Macrolepidoptera of Banffshire. *Entomologist's Record & Journal of Variation* **88**: 1–11.

———, 1985. Two records of *E. abietaria* Goeze. *Entomologist's Record & Journal of Variation* **97**: 146. [Records of larvae feeding on sitka spruce in Inverness and Norway spruce in Northumberland.]

BARRETT, C. G., 1877. Description of Eupithecia albipunctata var. *angelicata*. *Entomologist's monthly Magazine* **13**: 278.

———, 1887. Eupithecia extensaria Freyer in Norfolk. *Entomologist's monthly Magazine* **24**: 114.

———, 1889. Note on Eupithecia extensaria. *Entomologist's monthly Magazine* **25**: 258. [Detailed notes on larval forms and crypsis; description of pupa.]

———, 1904. *The Lepidoptera of the British Islands* **9**. Lovell Reeve, London. [Colour illustrations of all the species known at that time. Useful text.]

BASTELBERGER, M. J., 1907. Tephroclystia pyreneata and T. tenuiata. *Iris* **20**: 263. [Original descriptions of *E. tenuiata* ab. *niveipicata* and *E. pulchellata* ab. *reducta*.]

———, 1908. Weiterer neue Geometridenaus meiner Sammlung. *Internationale Entomologische Zeitschrift* **2**: 98. [Original descriptions of *E. innotata* ab. *rotundata* and *E. abbreviata* f. *hirschkei*.]

BAYNES, E. S. A., 1964. *A revised catalogue of Irish Macrolepidoptera.* Classey, Hampton. [A comprehensive account of all pugs known from Ireland to date of publication.]

———, 1970. *Supplement to A revised catalogue of Irish Macrolepidoptera.* Classey, Hampton.

BEIRNE, B. P., 1947. The origin and history of the British Macro-Lepidoptera. *Transactions of the Royal Entomological Society of London.* **98**(7): 273–372.

BENHS, 1989. Annual Exhibition, 1988. *British Journal of Entomology & Natural History* **2**: pl. 3. [Colour illustration of *G. rufifasciata* ab. *mediopallens*.]

———, 1990. Annual Exhibition, 1989. *British Journal of Entomology & Natural History* **3**: pl. 4. [Illustration of *E. phoeniceata* f. *mediofasciata* nov.]

BERGMAN, A., 1955. *Die Gross-Schmetterlinge Mitteldeutschlands* **5**. Urania, Leipzig. [Account of 55 spp., some British; photographs of adults and habitats. Text in German.]

BIRD, G. W., 1875. Eupithecia knautiata.

Entomologist **8**: 87–88. [Remarks on Gregson's (1874b; 1875a & b) claims on the specific status of *E. knautiata* (=*E. absinthiata*).]

BLESZYNSKI, S., 1965. *Klucze do oznaczania owadów Polski* [Keys for the identification of Polish insects] **27**: Lepidoptera, Part 46b; Geometridae: *Eupithecia*. pp.193–305. [Excellent line drawings of adults; genitalia drawings less clear. Text in Polish.]

BOISDUVAL, J. A., 1840. *Genera et Index Methodicus Europaeorum Lepidopterorum*. Paris. [Original descriptions of *E. ultimaria* and *E. tantillaria*; description of *E. nanata* ab. *pauxillaria* Rambur.]

BORKHAUSEN, M. B., 1794. *Naturgeschichte der Europäischen Schmetterlinge* **5**. Frankfurt. [Original description of *C. rectangulata* ab. *cydoneata*.]

BRADLEY, J. D., 1998. *Checklist of Lepidoptera recorded from the British Isles*. D. J. & M. J. Bradley, Fordingbridge & Newent. [Revised edition, 2000.]

—— & FLETCHER, D. S., 1987. *Indexed List of British Butterflies and Moths*. Kedleston Press, Kent.

BRIGGS, J., 1979. The Cloaked Pug in South Westmorland (VC 69). *Entomologist's Record & Journal of Variation* **91**: 220. [Discovery of *E. abietaria* at Witherslack with editorial note listing all *E. abietaria* records since 1945.]

BROWNE, F. G., 1968. *Pests and Diseases of Forest Plantation Trees*. Clarendon Press, Oxford. [List of supposed and apparently unsubstantiated larval foodplants of *E. nanata*.]

BUCKLER, W., 1899. *Larvae of the British Butterflies and Moths* **8**. Ray Society, London. [Good colour illustrations of most pug species.]

BURROWS, C. R. N., 1915. Notes of the taxonomic value of the genital armature in Lepidoptera. *Entomologist's Record & Journal of Variation* **27**: 40–43. [Discussion on methods of genitalia representation in taxonomic works with reference to Petersen, 1909.]

CAMPBELL, C., 1862. *Eupithecia debiliata*. *Zoologist* **20**: 8209. [First suggestion of bilberry as larval foodplant of *P. debiliata*.]

CARRINGTON, J. T., 1881. Description of Plate, Figs 6 & 7. *Entomologist* **14**: 303.

[Colour plate illustrating *E. ultimaria* Dup. (=*E. pusillata* subsp. *anglicata*).]

——, 1882. Re-occurrence of *Eupithecia extensaria*. *Entomologist* **15**: 67.

——, 1885a. Collecting the genus *Eupithecia*. *Entomologist* **18**: 108–112. [Notes on collecting and breeding the pugs with account of the discovery of *E. valerianata* larvae.]

——, 1885b. Collecting the genus *Eupithecia*. *Entomologist* **18**: 139–146. [Notes on collecting and breeding the pugs.]

——, 1871. Notes on the genus *Eupithecia*. *Entomologist's monthly Magazine* **7**: 213. [List of larvae found on Angelica.]

CATTERMOLE, P. A., 1986. Delayed emergence in *Eupithecia*. *Entomologist's Record & Journal of Variation* **98**: 230. [Note on *E. haworthiata* pupa lying over two winters.]

CHALMERS-HUNT, J. M., 1968–71. *The Butterflies and Moths of Kent* **3**. Chalmers-Hunt, London. [County account including comprehensive foodplants list.]

——, 1970. The butterflies and moths of the Isle of Man. *Transactions of the Society for British Entomology* **19**(1). [County account including comprehensive foodplants list.]

——, 1980. The cloaked pug (*Eupithecia abietaria* Goeze): further records. *Entomologist's Record & Journal of Variation* **92**: 25. [Lists four post-war records additional to those listed under Briggs, 1979.]

CHANT, J & BENTLEY, –, 1833. Entomological tour. *Entomological Magazine* **1**: 184. [First published British record of *Pasiphila debiliata* (described as *E. nigropunctata*).]

CHATELAIN, R. G., 1977. Foodplant of the Juniper Pug (*E. pusillata* D.& S.). *Entomologist's Record & Journal of Variation* **89**: 347. [Capture of *E. pusillata* where the foodplant is absent.]

CLEMENTS, E. J. & FOSTER, M. C., 1994. *Alien Plants of the British Isles*. Botanical Society of the British Isles, London.

CLERCK, C., 1759. *Icones Insectorum variorum cum nominibus eorum trivialibus locisque*. Stockholm. [Original description of *E. absinthiata*.]

COCKAYNE, E. A., 1951a. *Aspitates gilvaria*

237

Fabr. ssp. *burrenensis* ssp. nov. *Entomologist's Gazette* **2**: 100–101 + Pl. 2. [Includes a description of *E. goossensiata* (=*absinthiata*) ab. *obscura* Cockayne]

———, 1951b. Two new subspecies of British Lepidoptera. *Entomologist* **84**: 154. [Original description of *E. vulgata* f. *scotica*.]

———, 1952a. A search for larvae of *Eupithecia actaeata* Walderdorff and *Eupithecia immundata* Zeller. *Entomologist's Record & Journal of Variation* **64**: 11–12. [Account of unsuccessful search for these species in Yorkshire.]

———, 1952b. A search for larvae of *Eupithecia actaeata* Walderdorff and *Eupithecia immundata* Zeller. A postscript. *Entomologist's Record & Journal of Variation* **64**: 104–105. [List of further possible localities.]

———, 1953. Aberrations of British Geometridae. *Entomologist's Record & Journal of Variation* **65**: 167–168 + Pl. XII. [Original descriptions of *E. linariata* ab. *punctata*; *E. pulchellata* ab. *guttata*; *E. satyrata* abs., *trilineata* and *nigra*; *E. icterata* ab. *goodsoni*; *E. extensaria* f. *albescens*; *E. abbreviata* f. *nigra*; *E. lariciata* ab. *virgata*; *P. rectangulata* ab. *effusa*; *P. debiliata* f. *albescens*.]

———, 1954. Aberrations of British Lepidoptera. *Entomologist's Record & Journal of Variation* **66**: 66 + Pl. 2. [Original description of *E. venosata* ab. *basingrata*.]

COWIN, W. S., 1946. *Eupithecia alliaria* (Lep.: Hydriomenidae): a moth new to the British list. *Entomologist's monthly Magazine*, **82**: 91. [Reputed discovery of this species in the Isle of Man, later proved to be an erroneous identification.]

CREWE, H. HARPUR, 1854. List of *Eupithecia* etc. reared from larvae. *Zoologist* **12**: 4370. [Short account of rearing *E. satyrata*, *E. fraxinata*, *E. subumbrata*, *E. absinthiata*, *E. tenuiata* (almost certainly erroneous as stated to have been reared on thyme) and *G. rufifasciata*.]

———, 1859a. Description of the larva of *Eupithecia assimilata*. *Zoologist* **17**: 6579.

———, 1859b. Description of the larva of *Eupithecia haworthiata*. *Zoologist* **17**: 6609–6610.

———, 1859c. Description of the larva of *Eupithecia innotata*. *Zoologist* **17**: 6610.

———, 1859d. Description of the larva of *Eupithecia pimpinellata*. *Zoologist* **17**: 6694–6695. [=*E. virgaureata* Dbl.]

———, 1859e. Description of the larva of *Eupithecia vulgata*. *Zoologist* **17**: 6695.

———, 1859f. Description of the larva of *Eupithecia absinthiata*. *Zoologist* **17**: 6734–6735.

———, 1859g. Description of the larva of *Eupithecia denotata*. *Zoologist* **17**: 6735. [=*E. pimpinellata*.]

———, 1859h. Mr Gregson's criticism on the description of the larva of *Eupithecia assimilata*. *Zoologist* **17**: 6753.

———1859i. Description of the larva of *Eupithecia subnotata*. *Zoologist* **17**: 6769–6770. [=*E. simpliciata*.]

———, 1859j. Additional remarks on the larva of *Eupithecia innotata*. *Zoologist* **17**: 6770.

———, 1859k. Description of the larva of *Eupithecia centaureata*. *Zoologist* **17**: 6770.

———, 1859l. Description of the larva of *Eupithecia sobrinata*. *Zoologist* **17**: 6789. [=*E. pusillata*.]

———, 1859m. Description of the larva of *Eupithecia exiguata*. *Zoologist* **17**: 6789.

———, 1859n. Additional remarks on the larva of *Eupithecia assimilata*. *Zoologist* **17**: 6790.

———, 1859o. Description of the dorsally-blotched larvae of *Eupithecia assimilata*. *Zoologist* **17**: 6790.

———, 1860a. Description of the larva of *E. linariata*. *Zoologist* **18**: 6817.

———, 1860b. Description of the larva of *Eupithecia subfulvata*. *Zoologist* **18**: 6817. [=*E. icterata*.]

———, 1860c. Description of the larva of *Eupithecia tenuiata*. *Zoologist* **18**: 6868.

———, 1860d. Description of the larva of *Eupithecia nanata*. *Zoologist* **18**: 6868–6869.

———, 1860e. Description of the larva of *Eupithecia castigata*. *Zoologist* **18**: 6904. [=*E. subfuscata*.]

———, 1860f. Description of the larva of *Eupithecia minutata*. *Zoologist* **18**: 6904. [=*E. absinthiata* f. *goossensiata*.]

———, 1860g. *Eupithecia expallidata* bred from goldenrod. *Zoologist* **18**: 7005.

———, 1860h. Description of the larva of *Eupithecia rectangulata*. *Zoologist* **18**: 7107. [=*P. rectangulata*.]

———, 1860i. Description of the larva of *Eupithecia expallidata*. *Zoologist* **18**: 7107–7108.
———, 1860j. Description of a variety of the larva of *Eupithecia assimilata*. *Zoologist* **18**: 7107.
———, 1860k. Description of the larva of *Eupithecia satyrata*. *Zoologist* **18**: 7215.
———, 1860l. Description of the larva of *Eupithecia helveticata*. *Zoologist* **18**: 7215–7216. [=*E. intricata* subsp. *millieraria*.]
———, 1860m. Description of the larva of *Eupithecia subumbrata*. *Zoologist* **18**: 7216.
———, 1860n. Occurrence of *Eupithecia helveticata* in Buckinghamshire. *Zoologist* **18**: 7251. [First record of *E. intricata* subsp. *arceuthata*, though described herein as *E. helveticata* (=subsp. *millieraria*).]
———, 1860o. Description of the larva of *Eupithecia abbreviata*. *Zoologist* **18**: 7251.
———, 1860p. Note on the pupation of *Eupithecia tenuiata*. *Zoologist* **18**: 7251–7252.
———, 1860q. Observations: *Eupithecia innotata*. *Entomologist's Weekly Intelligencer* **8**: 20–21. [Account of rearing *E. fraxinata* on *Viburnum tinus*.]
———, 1861a. Description of the larva of *Eupithecia pumilata*. *Zoologist* **19**: 7323. [=*Gymnoscelis rufifasciata*.]
———, 1861b. Oviposition of *Eupithecia sobrinata*. *Zoologist* **19**: 7408. [Account of achieving captive oviposition in *E. pusillata*.]
———, 1861c. Description of the larva of *Eupithecia tripunctaria*. *Zoologist* **19**: 7567–7568. [First record of *E. tripunctaria* in Britain with descriptions of two larval forms.]
———, 1861d. Description of the larva of *Eupithecia trisignata*. *Zoologist* **19**: 7568–7569. [=*E. trisignaria*.]
———, 1861e. Description of the larva of *Eupithecia pusillata*. *Zoologist* **19**: 7762. [=*E. tantillaria*.]
———, 1861f. Description of the larva of *Eupithecia distinctata*. *Zoologist* **19**: 7762. [=*E. distinctaria*.]
———, 1861g. Doublebroodedness of *Eupithecia assimilata*. *Zoologist* **19**: 7762. [In captivity.]
———, 1861h. Occurrence of the larvae of *Eupithecia trisignata* and *E. tripunctata* in Buckinghamshire. *Zoologist* **19**: 7762. [Account of finding *E. trisignaria* and *E. tripunctaria* on *Angelica sylvestris* seeds.]
———, 1861i. Description of the larva of *Eupithecia subfulvata*. *Zoologist* **19**: 7796. [=*E. icterata*.]
———, 1861j. Description of the larva of *Eupithecia succenturiata*. *Zoologist* **19**: 7796–7799. [Three colour forms described.]
———, 1861k. Notes on *Eupithecia* larvae. *Entomologist's Annual* **7**: 126–146. [First larval descriptions of 25 pug species, previously published in the *Zoologist* but with some amendments. *E. icterata* was also described but was further amended in Crewe (1862a).]
———, 1862a. Notes on *Eupithecia* larvae. *Entomologist's Annual* **8**: 38–49. [First full larval descriptions of *E. tripunctaria*, *E. trisignaria*, *E. dodoneata*, *E. tantillaria*, *E. distinctaria*, *E. icterata* and *E. succenturiata*. Some had been amended since publication in the *Zoologist*.]
———, 1862b. Discussion on various *Eupithecia*. *Entomologist's Annual* **8**: 44–49. [Brief discussion on status of *E. succenturiata*, *E. subumbrata*, *E. plumbeolata*, *E. haworthiata*, *E. intricata*, *E. satyrata*, *E. tripunctaria*, *E. trisignaria*, *E. fraxinata*, *E. pimpinellata*, *E. virgaureata*, *E. pulchellata*, *E. indigata*, *E. abbreviata*, *E. insigniata*, *E. irriguata*, *E. expallidata*, *E. absinthiata* (+ f. *goossensiata*), *E. pygmaeata*, *E. tantillaria*, *E. assimilata*, *E. vulgata*, *P. debiliata* and *C. v-ata*.]
———, 1862c. Foodplants of *Eupithecia pumilata*. *Zoologist* **20**: 7971–7972. [=*G. rufifasciata*.]
———, 1862d. *Eupithecia arceuthata* Frey. and *Eupithecia helveticata* Bdv.: are they distinct? *Zoologist* **20**: 8052. [Discussion on the status of *E. intricata* subspp. *arceuthata* and *millieraria*.]
———, 1862e. Description of the larva of *Eupithecia viminata*. *Zoologist* **20**: 8174. [=*E. valerianata*.]
———, 1862f. Description of the larva of *Eupithecia indigata*. *Zoologist* **20**: 8174.
———, 1862g. Larva of *Eupithecia debiliata*. *Zoologist* **20**: 8209. [Discussion about possible larval foodplants of *P. debiliata*.]
———, 1862h. Double-broodedness of *Eupithecia virgaureata*. *Zoologist* **20**:

8209. [First record of bivoltinism in *E. virgaureata*.]
——, 1862i. Larva of *Eupithecia pulchellata*. *Zoologist* **20**: 8209. [Public request for livestock of this species as the larva was unknown at this time. The author suggested that Willowherb might be the natural foodplant.]
——, 1862j. Is *Eupithecia tripunctata* double-brooded? *Zoologist* **20**: 8209–8210. [Recorded capture of second brood *E. tripunctaria*.]
——, 1862k. Double-broodedness of *Eupithecia virgaureata*. *Weekly Entomologist* **1**: 18–19. [Rearing two broods in captivity.]
——, 1862l. Larvae of *Eupithecia pulchellata*. *Weekly Entomologist* **1**: 19. [As Crewe 1862i.]
——, 1862m. Larva of *Eupithecia debiliata*. *Weekly Entomologist* **1**:19. [Discussion about possible larval foodplants of *P. debiliata*.]
——, 1862n. Is *Eupithecia tripunctata* double-brooded? *Weekly Entomologist* **1**: 19–20. [As Crewe, 1862j.]
——, 1862o. Occurrence of *Eupithecia fraxinata* (Crewe) in Great Britain. *Weekly Entomologist* **1**: 134–136. [Supposed first record of *E. fraxinata*.]
——, 1862p. Occurrence of *Eupithecia arceuthata* Frey. in Great Britain. *Weekly Entomologist* **1–2**: 179–180. [First official record of *E. intricata* subsp. *arceuthata* with first description of larva comparative with subsp. *millieraria*.]
——, 1863a. Notes on some of the genus *Eupithecia*. *Entomologist's Annual* **9**: 116–128. [First larval descriptions of *E. valerianata*, *E. indigata*; larval and adult descriptions of *E. intricata* ssp. *arceuthata*; discussion of the specific status of *E. fraxinata/innotata*; account of the discovery of *E. intricata* subsp. *arceuthata*.]
——, 1863b. Description of the larva of *Eupithecia arceuthata* of Freyer. *Zoologist* **21**: 8313. [Description of the larva of *E. intricata* subsp. *arceuthata*.]
——, 1863c. Occurrence of *E. arceuthata* in Britain. *Zoologist* **21**: 8342. [Repeat of Crewe (1862p).]
——, 1863d. Description of the larva of *Eupithecia helveticata* Bdv. *Zoologist* **21**: 8343. [Larval description of *E. intricata* subsp. *millieraria*.]

——, 1863e. Description of the imago and larva of *Eupithecia fraxinata*, an English species new to Science, and also of those of *E. innotata*, with which it has hitherto been confounded. *Zoologist* **21**: 8405–8407.
——, 1863f. Description of the larva of *Eupithecia debiliata*. *Zoologist* **21**: 8648. [=*P. debiliata*.]
——, 1864a. Notes on the larva, pupa and foodplant of *Eupithecia pulchellata*. *Entomologist's monthly Magazine* **1**: 95.
——, 1864b. Description of the larva and pupa of *Eupithecia lariciata*. *Entomologist's monthly Magazine* **1**: 141–142.
——, 1864c. Occurrence of *Eupithecia campanulata* H.-S. in Bucks: description of the larva and pupa. *Entomologist's monthly Magazine* **1**: 142–143. [First record of *E. denotata* subsp. *denotata*.]
——, 1865a. Notes on Eupitheciae. *Entomologist's Annual* **11**: 117–127. [Larval descriptions of *E. pulchellata*, *E. lariciata*, *E. denotata denotata* subsp. and *P. debiliata*; descriptions of several forms of *E. fraxinata* larva; references to larval descriptions in the *Zoologist*.]
——, 1865b. Notes on the larva and foodplant of *Eupithecia plumbeolata*. *Entomologist's monthly Magazine* **2**: 90.
——, 1868. Description of the larvae of *Eupithecia consigniata*. *Entomologist's monthly Magazine* **5**: 72. [=*E. insigniata*.]
——, 1870. Foodplants of *Eupithecia campanulata*. *Entomologist's monthly Magazine* **7**: 143. [Discussion of the foodplants of *E. denotata* subsp. *denotata* listing 10 known *Campanula* host spp.]
——, 1871. Description of the larva of *Eupithecia irriguata*. *Entomologist* **5**: 348.
——, 1872a. Description of the larva of *Eupithecia subciliata*. *Entomologist's monthly Magazine* **9**: 16–17. [Original description of the larva of *E. inturbata*.]
——, 1872b. Description of the larva of *Eupithecia pygmaeata*. *Entomologist* **6**: 166.
——, 1872c. Description of the larva of *Eupithecia pygmaeata*. *Zoologist* (Second Series) **7**: 3193. [First larval description of *E. pygmaeata* (simultaneous with Crewe, 1872b).]
——, 1872d. Description of the larva of *Eupithecia togata*. *Zoologist* (Second Series) **7**: 3276. [=*E. abietaria*.]

———, 1872e. Description of the larva of *E. togata*. *Entomologist's monthly Magazine* **9**: 114–115. [=*E. abietaria*.]
———, 1874a. *Eupithecia knautiata* of Gregson =*E. minutata* of Hübner. *Entomologist* **7**: 290–291. [Remarks on Gregson's (1874b; 1875a & b) claims on the specific status of *E. knautiata* (=*E. absinthiata*).]
———, 1874b. Foodplant of *Eupithecia innotata*. *Entomologist* **7**: 291. [Discussion of Gregson's (1874a) claim that *E. innotata* larvae feed on *Artemisia campestris*.]
———, 1875. Description of the larva of *Eupithecia togata*. *Entomologist* **8**: 297–298. [=*E. abietaria*.]
———, 1881. *Eupithecia jasioneata* Crewe: a species new to science. *Entomologist* **14**: 198–199 + Pl. 1, Figs 4 & 5. [Original description of *E. denotata* subsp. *jasioneata*.]
CUE, P., 1961. The discovery of the larva of *Eupithecia innotata* in Britain on Sea Buckthorn with notes on its habits, etc. *Entomologist's Record & Journal of Variation* **73**: 210–211.
———, 1963. *Eupithecia virgaureata*. *Entomologist's Record & Journal of Variation* **75**: 228–229. [List of foodplants.]
CULOT, J., 1919. *Les Noctuelles et Géomètres d'Europe* **4**. Geneva. [Good diagnostic notes; excellent paintings of adults. Text in French.]
———, 1987. *Les Noctuelles et Géomètres d'Europe* **4**. Apollo Books, Svendborg. [As above but reprinted. The plates remain excellent.]
CURTIS, J., 1824–39. *British Entomology; or illustrations and descriptions of the genera of Insects found in Great Britain and Ireland*. London.
———, 1829. *A Guide to the Arrangement of British Insects*. Westley, London. [List of British Lepidoptera including names of 51 pugs, 28 of which have since been synonymized.]
CURTIS, W. PARKINSON, 1934a. *Eupithecia distinctaria* H.-S. (Lep.). *Journal of the Society for British Entomology* **1**(6): 166–168. [Discussion on the specific status of *E. distinctaria* H.-S., *E. constrictata* Guen. and *E. sextiata* Mill.]
———, 1934b. *Eupithecia abbreviata* Steph. (Lep.) feeding on *Neottia nidus-avis* Linn. *Journal of the Society for British Entomology* **1**(7): 173–174.
———, 1934c. *Eupithecia distinctaria* H.-S. (Lep.). *Journal of the Society for British Entomology* **1**(8): 226–228. [Descriptions and figures of male abdominal plate.]
———, 1944a. *Eupithecia venosata* Fabr. race *hebridensis* n.f. (Lep.). *Journal of the Society for British Entomology* **2**(5) 169. [Original description of *E. venosata* f. *hebridensis*.]
———, 1944b. *Eupithecia tripunctaria* H.-S. (=*albipunctata* Haw.) (Lep.). *Journal of the Society for British Entomology* **2**(5): 169–170.
DADD, E. M., 1906a. A puzzling group of Eupitheciids. *Entomologist's Record & Journal of Variation* **18**: 259–260. [Discussion on *E. fraxinata* complex.]
———, 1906b. Another puzzling group of Eupitheciids. *Entomologist's Record & Journal of Variation* **18**: 261. [Discussion of the specific status of *E. icterata* and *E. succenturiata*.]
DANNEHL, F. 1925. Neue Formen und Lokalrassen. *Entomologische Zeitschrift* **39**: 16. [Original descriptions of *E. plumbeolata* ab. *flaveolata* and *E. satyrata* ab. *contrastata*.]
———, 1927. Beiträge zur Lepidopteren-Fauna Sudtirols. *Entomologische Zeitschrift* **41**: 272–284, 309–319. [Original descriptions of *E. plumbeolata* abs. *plumbalbeolata*, *flaveolata*, *explicata* and *lividata*; *E. tenuiata* abs. *coaequata* and *fuscosparsata*, *E. pusillata* ab. *achromata* and *C. rectangulata* ab. *brunneata*. Also a further description of *E. satyrata* ab. *contrastata*.]
DE WORMS, C. G. M., 1951. New additions to the British Macrolepidoptera during the past half century. *Entomologist's Gazette* **2**: 153–168. [Short account of the discovery of *E. millefoliata* in Britain.]
———, 1952. *Eupithecia millefoliata* Rössler: an early record. *Entomologist's Record & Journal of Variation* **64**: 53. [This record pre-dates by six years what was considered by de Worms (1951) to be the first British capture.]
———, 1953. Abundance of larvae of *E. millefoliata* Rössler. *Entomologist's Record & Journal of Variation* **65**: 328. [Note on how to find larvae of *millefoliata*.]

———, 1958. *Eupithecia alliaria* Stdgr. (Lep., Geometridae) as a British Species: a correction. *Entomologist's monthly Magazine* **94**: 67. [Re-examination of *E. alliaria*, proving it to be *E. pygmaeata*.]

———, 1963. Recent additions to the British macrolepidoptera. *Entomologist's Gazette* **14**: 101-119. [Review of first captures of *E. egenaria* and *E. phoeniceata*.]

———, 1978. Further recent additions to the British Macrolepidoptera: (6) *Chloroclystis chloerata* Mab. *Entomologist's Gazette* **29**: 23. [Review of discovery with a note on old specimens in the de Worms collection which had been incorrectly identified as *C.* (=*Pasiphila*) *rectangulata*.]

——— & MESSENGER, J. L., 1960. *Eupithecia phoeniciata* Rambur (Lep.: Geometridae): a pug new to Great Britain. *Entomologist* **93**: 93-94 + 2 pl.

[DENIS, M. & SCHIFFERMÜLLER, I.], 1775. *Ankündigung eines systematischen Werkes von den Schmetterlingen der Wiener Gegend*. Vienna. [Original descriptions of *E. linariata*, *E. centaureata*, *E. subumbrata* and *E. pusillata*.]

DERENNE, F., 1932. *Chloroclystis rectangulata* L. ab. *ochrea*, nov. ab. *Lambillionea* **32**: 156. [=*P. rectangulata*.]

DIETZE, C., 1870. Description of the larva of *Eupithecia irriguata* Hb. *Entomologist's monthly Magazine* **8**: 14-15.

———, 1871a. On the habits of *E. subciliata*. *Entomologist's monthly Magazine* **8**: 290. [Reputed discovery of the larva of *E. inturbata* feeding on oak.]

———, 1871b. Verzeichniss der in der Umgegend von Frankfurt a. M. und Wiesbaden gefundenen *Eupithecia*-Arten. *Stettiner Entomologische Zeitung* **32**: 209-211. [Original description of *E. innotata* ab. *suspectata* (=*E. fraxinata* ?).]

DIETZE, K., 1910. *Die Biologie der Eupithecien* I. Berlin. [Excellent colour illustrations of larvae of most British species; adults less useful. Text published separately (Dietze, 1913).]

———, 1913. *Die Biologie der Eupithecien* II. Berlin. [Excellent text (in German) covering the biology of most British species. Concluding from Dietze, 1910.]

DIOSZEGHY, L., 1929/30. *Die Lepidopterenfauna des Retyezat-Gebirges*. *Verhandlungen und Mitteilungen des Siebenbürg Vereins für Naturwissenschaften zu Hermennstadt* **78-80**: 188-289 + Pl. 1. [Original description of *E. pusillata* ab. *rittichi*.]

DJAKANOV, A., 1929. Entomologische Ergebnisse der schwedischen Kamtchatka-Expedition 1920-1922. *Arkiv för Zoologi* **21**: 1-24. [Original description of *E. succenturiata* ab. *malaisei*.]

DOBRÉE, N. F., 1875. *Eupithecia extensaria*. *Entomologist* **8**: 133. [Discussion of possible origin.]

DONOVAN, C., 1936. *A catalogue of the Macrolepidoptera of Ireland*. Cheltenham. [A comprehensive account of all pugs known from Ireland to date of its publication.]

———, N.D. Supplement to above. Privately published.

DONOVAN, E., 1799. *The natural history of British Insects* **8**. Rivington, London. [First British record of *E. venosata*.]

DOUBLEDAY, H., 1850a. *Synonymic List of British Lepidoptera ... excepting the family Tineidae*. Van Voorst, London. [First literature record of *E. tenuiata*; 35 names listed, of which seven are now synonyms.]

———, 1850b. Descriptions of Lepidopterous insects of the genera *Hypenodes*, *Eupithecia* and *Spilonota* recently discovered in Britain. *Zoologist* **8**: App. cv-cvi. [Original descriptions of first British *E. pygmaeata* (as *E. palustraria*) and *E. satyrata* f. *callunaria* (as *E. callunaria*).]

———, 1856. A list of the British *Eupithecia* with notes on some of the species. *Zoologist* **14**: 5139. [Original descriptions of *E. haworthiata*, *E. assimilata* and *E. expallidata*; First British record of *E. haworthiata*, *E. inturbata*, *E. dodoneata*, *E. assimilata*, *E. tantillaria*, *E. expallidata* and *E. distinctaria*.]

———, 1857. *Eupithecia helveticaria* in Britain. *Zoologist* **15**: 5436-5437. [First record of *E. intricata* subsp. *millieraria*.]

———, 1858a. Occurrence of *Eupithecia pernotata* in England. *Zoologist* **16**: 5963. [Short account of the capture of *E. cauchiata*.]

———, 1858b. *Eupithecia viminata* Doubleday. *Zoologist* **16**: 6103. [First British record of *E. valerianata*.]

———, 1861. Notes on new or little-known Eupithecias. *Zoologist* **19**: 7566-7567.

[Original description of *E. virgaureata*; first British record of *E. trisignaria* with other notes on *E. absinthiata*, *E. icterata*, *E. pimpinellata*, *E. tripunctaria*, *E. distinctaria*, *E. vulgata*, *E. pusillata* subsp. *anglicata*, *E. haworthiata*, *E. pulchellata*, *E. expallidata* and *E. abbreviata*.]
——, 1864. *Eupithecia lariciata* Freyer. *Entomologist's monthly Magazine* **1**: 94. [Review of discovery in Britain.]
——, 1875. *Eupithecia extensaria* taken in Yorkshire. *Entomologist* **8**: 108–109. [Account of first British record with editorial by E. Newman discussing European status and describing the adult.]
DOUGLAS, J. W., 1851. Entomological localities. *Zoologist* **9**: 3246–3248. [First British record of *E. indigata*.]
DRAUDT, M., 1905. Zur Kenntnis der Eupithecien-Eier. *Iris* **18**(1): 280–320. [Description of the ova of most British species with some photographs. Text in German.]
DUDDINGTON, J. & JOHNSON, R., 1983. *The Butterflies and Moths of Lincolnshire and South Humberside*. Lincolnshire Naturalists' Union, Lincoln. [County account with useful list of foodplants.]
DUPONCHEL, P. A., 1831. *Histoire Naturelle de Lépidoptères ou Papillons de France*. Paris. [Original description of *E. cauchiata*.]
EMMET, A. M., 1981, *Eupithecia egenaria* H.-S. – an ancient relic? *Entomologist's Record & Journal of Variation* **93**: 177. [Discussion of the possible origins of *E. egenaria* and its foodplant *Tilia cordata*.]
EVANS, L. K. & EVANS, K. G. W., 1973. *A Survey of the Macro-lepidoptera of Croydon and N.E. Surrey*. Evans & Evans, Croydon.
FABRICIUS, J. C., 1787. *Mantissa Insectorum ... Tom. II*. Hafnia [Copenhagen]. [Original description of *E. venosata* subsp. *venosata*.]
FARREN, W., 1892. Notes on the dates of appearance of *E. pygmaeata*. *Entomologist's Record & Journal of Variation* **3**: 19–20. [Discussion on the possibility of *E. pygmaeata* having a protracted flight period.]
FIBIJER, M., 1979. *Eupithecia absinthiata* Cl./*goossensiata* Mab. én eller to arter? *Lepidoptera* (N.S.) **3**: 203–206. [Brief discussion of specific status of *E. absinthiata* f. *goossensiata*. Text in Danish. Some photographs but not comparative.]
——, 1980. *Eupithecia innotata* Hufn. og *Eupithecia innotata* ssp. *fraxinata* Crewe. *Lepidoptera* (N.S.) **3**: 256. [Discussion on *E. innotata/fraxinata*. Text in Danish.]
FIRMIN, J., PYMAN, G. A. & others, 1975. *A Guide to the Butterflies and Larger Moths of Essex*. Essex Naturalists' Trust, Colchester.
FISCHER, C., 1947a. Comment acquérir la connaissance générale d'un groupe de papillons. *Bulletin de la Société Entomologique de Mulhouse* **1947**: 36–39. [An attempt to sort pugs into groups of superficially similar species: explanation of groupings plus Group One – large central band edged with white. Text in French. No explanatory drawings.]
——, 1947b. Comment acquérir la connaissance générale d'un groupe de papillons. *Bulletin de la Société Entomologique de Mulhouse* **1947**: 49–51. [As Fischer (1947a): Group Two – white with dark blotches at base, costa and/or termen of forewing; Group Three – postmedian line situated in place of submarginal line.]
——, 1947c. Comment acquérir la connaissance générale d'un groupe de papillons. *Bulletin de la Société Entomologique de Mulhouse* **1947**: 55–57. [As Fischer (1947a): Group Four – moderately straight median line elbowed at costa, antemedian line strongly elbowed near elongated discal spot and diverging towards median line at inner edge and strong marks on veins at postmedian lines; Group Five – median line elbowed above discal spot and antemedian line angulated and distant from discal spot and median line.]
——, 1947d. Comment acquérir la connaissance générale d'un groupe de papillons. *Bulletin de la Société Entomologique de Mulhouse* **1947**: 68–71. [As Fischer (1947a): Group Six – postmedian and submarginal lines very close, former sometimes denated and closer to median line at dorsum than at costa; broad, ill-defined median band.]
——, 1947e. Comment acquérir la connaissance générale d'un groupe de papillons. *Bulletin de la Société Entomologique*

de Mulhouse **1947**: 76–79. [As Fischer (1947a): Group Seven – fasciae fine, well defined parallel to postmedian line and perpendicular to dorsum; median line elbowed at costa and postmedian line closer to submarginal than median line.]

———, 1947f. Comment acquérir la connaissance générale d'un groupe de papillons. *Bulletin de la Société Entomologique de Mulhouse* **1947**: 84–87. [As Fischer (1947a): Group Eight – postmedian line equal distance from submarginal and median lines or closer to latter and clear; median fascia comprising fine lines elbowed around discal spot and antemedian fascia gently curved from costa to dorsum.]

———, 1947g. Comment acquérir la connaissance générale d'un groupe de papillons. *Bulletin de la Société Entomologique de Mulhouse* **1947**: 92–95. [As Fischer (1947a): Group Nine – overall lack of markings, small hindwings; all lines elbowed at radial vein, slightly curved at costa and obliquely angled at dorsum and broad pale postmedian line which is equidistant between median and submarginal lines.]

———, 1947h. Comment acquérir la connaissance générale d'un groupe de papillons. *Bulletin de la Société Entomologique de Mulhouse* 1947: 107–108. [As Fischer (1947a): Group Ten – postmedian line rounded at costa, not angled and distant from submarginal line.]

FLETCHER, D. S., 1963. Some additions and corrections to the 1961 edition of South's *Moths of the British Isles*. Coridon, Series B. No. 4: 1–7. [Review of the discoveries of *E. egenaria* and *E. phoeniceata*.]

FOLTIN, H., 1938. Einige neue Formen vol Faltern aus Oberösterreich: (*Eupithecia linariata*). *Zeitschrift des Österreichen Entomologen-Vereines* **22**: 126. [Original descriptions of *E. linariata* ab. *reducta* and ab. *flavofasciata*.]

FREER, R., 1892. Seasonal variation of larvae. *Entomologist's Record & Journal of Variation* **3**: 279. [Short notes on colour forms of the larvae of *E. absinthiata*, *E. assimilata* and *E. lariciata*.]

FREER, F., 1896. *Eupithecia subfulvata* and *E. succenturiata*. *Entomologist's Record & Journal of Variation* **7**: 254–255. [Discussion of the specific status of *E. icterata* and *E. succenturiata*.]

FREYER, C. F., 1836. *Neuere Beiträge zur Schmetterlingskunde mit Abbildungen nach der Natur*. II. Augsburg [Description of *E. fraxinata* 'race' *tamarisciata*.]

———, 1842. *Neuere Bieträge zur Schmetterlingskunde* ... IV. Augsburg. [Original descriptions of *E. intricata* subsp. *arceuthata* and *E. lariciata*.]

———, 1844. *Neuere Beiträge zur Schmetterlingskunde* ... V. Augsburg. [Original description of *E. extensaria*.]

FUCHS, A., 1900. Alte und neue Gross–schmetterlinge der Europäischen Fauna. *Jahrbuch des Nassauischen Vereins Naturkunde* **55**: 67–80. [Original description of *E. denotata* ab. *solidaginis*.]

FUCHS, F., 1901. *Eupithecia satyrata*. *Entomologische Zeitung* **62**: 379. [Original description of *E. satyrata* ab. *strandi*.]

———, 1904. Neue Schmetterlingsformen. *Societas Entomologica Stuttgart* **19**: 18. [Original descriptions of *E. denotata* ab. *ochraceata* and *E. indigata* ab. *tristrigata*.]

GARDNER, J., 1907. *Eupithecia succenturiata* and *E. subfulvata*. *Entomologist's Record & Journal of Variation* **19**: 23–24. [Discussion of the specific status of *E. succenturiata* and *E. icterata*.]

GOATER, B., 1974. *Butterflies and Moths of Hampshire and the Isle of Wight*. Classey, Faringdon. (1992. *Supplement: additions and corrections*. UK Nature Conservation No. 7. JNCC, Peterborough.)

GOEZE, J. E. A., 1781. *Abhandlungen zür Geschichte der Insekten*. Nuremberg. [Original description of *E. abietaria*.]

GOODEY, B., 2003. *Eupithecia massiliata* Dardoin & Millière (Lep.: Geometridae) – a pug moth new to the British fauna in Epping Forest. *Entomologist's Record & Journal of Variation* **115**(4): 167–170.

GOODSON, A. L. [unpublished]. Aberrational forms of the British Lepidoptera. Natural History Museum, London.

GREENE, J., 1854. List of Lepidoptera captured near Aylesbury, Buckinghamshire. *Zoologist* **12**: 4184–4188. [First British record of *E. satyrata* f. *satyrata* (referred to as *E. fagicolaria*).]

———, 1859. The larva of *Eupithecia assimilata*. *Zoologist* **17**: 6791.

GREGSON, C. S., 1859a. Note on the larva of *Eupithecia assimilata*. *Zoologist* **17**: 6695.

———, 1859b. Reply to Mr Crewe's note on the larva of *Eupithecia assimilata*. *Zoologist* **17**: 6790–6791.
———, 1874a. Notes on *Eupithecia innotata*. *Entomologist* **7**: 68–69. [Reputed discovery of larvae of *E. innotata* on Mugwort.]
———, 1874b. Description of an *Eupithecia* new to science, together with notes on its life-history. *Entomologist* **7**: 255–257. [Description of *E. knautiata* sp. nov. (=*E. absinthiata*).]
———, 1875a. Note on *Eupithecia knautiata* mihi. *Entomologist* **8**: 38–41. [Reply to critical letters regarding specific status of *E. knautiata* (=*E. absinthiata*).]
———, 1875b. *Eupithecia knautiata*. *Entomologist* **8**: 199. [Description of rearing experiments of larvae of *E. knautiata* (=*E. absinthiata*).]
———, 1876. An attempt to arrange the British Eupithecidae by their larval characteristics. *Entomologist* **9**: 8–10. [Groups the species according to the shape of the larvae, as follows: i – shortish, broad from head to anus; ii – short-attenuate; iii – medium and stout; iv – long, cylindrical, generally tapering to the head; v – slender.]
———, 1884a. Description of an *Eupithecia* new to Science – *Eupithecia curzoni* mihi. *Entomologist* **17**: 230–231. [Description of *E. satyrata* f. *curzoni*.]
———, 1884b. Description of an *Eupithecia* new to Science. *Eupithecia curzoni*. *Young Naturalist* **5**: 230–231. [Description of *E. satyrata* f. *curzoni*. Published simultaneously with Gregson, 1884a.]
———, 1885. *Eupithecia curzoni*. *Entomologist* **18**: 52–53. [Further note on *E. satyrata* f. *curzoni*. Published the year after his description of the same moth (Gregson, 1884a).]
———, 1886. Notes on the Orkney (Hoy) Lepidoptera. *Young Naturalist* **7**: 128. [First description of *E. venosata* subsp. *ochracae*.]
———, 1887. Notes on Eupitheciae: *E. irriguata* and *E. venosata* var. *fumosae* and var. *bandanae*. *Young Naturalist* **8**: 110–111. [First description of *E. venosata* subsp. *fumosae* and ab. *bandanae*.]
———, 1888. Description of *E. tenuiata* ab. *cinerae*. *Young Naturalist* **9**: 104.
GUENÉE, A., 1857. *Histoire Naturelle des Insectes. Species général des Lépidoptères*. Tom. X, Uranides et Phalénites, Part II. Paris. [Original descriptions of *E. distinctaria* subsp. *constrictata* and *E. dodoneata*.]
HAGGETT, G. M., 1951. *Eupithecia intricata* Zett. f. *arceuthata* Freyer (Lepidoptera: Geometridae). *Entomologist* **84**: 58–60. [Review of the distribution and biology of *E. intricata* subsp. *arceuthata*.]
———, 1955. *Eupithecia millefoliata*; *E. extensaria*; *E. intricata arceuthata*. *Proceedings & Transactions of the South London entomological and natural History Society*. **195**: 159–163. [Account of the biology of these three species.]
———, 1963. *Eupithecia innotata* Hufn. and *E. fraxinata* Crewe (Lep.: Geometridae). *Entomologist's Gazette* **14**: 13–23. [Discussion of the specific status in Britain of *E. innotata* and *E. fraxinata*.]
———, 1968a. *Eupithecia virgaureata* Dbl. & *E. satyrata* Hb. *Proceedings of the British entomological and natural History Society* **1**(2): 99–102. [Descriptions, taxonomy and biology.]
———, 1968b. *Eupithecia phoeniceata* Ramb. *Proceedings of the British entomological and natural History Society* **1**(2): 102–104. [Account of biology and description of larva.]
———, 1968c. *Eupithecia intricata* Zett. sspp. *hibernica* and *arceuthata*. *Proceedings of the British entomological and natural History Society* **1**(2): 104–105. [Account of biology with larval descriptions.]
———, 1968d. *Eupithecia egenaria* H.-S.: Fletcher's Pug. *Proceedings of the British entomological and natural History Society* **1**(2): 105–108. [Account of the biology with first description of larva.]
———, 1980a. *Eupithecia pusillata* D.& S. away from natural juniper. *Entomologist's Record & Journal of Variation* **92**: 144. [Short note on finding *E. pusillata* where juniper does not occur.]
———, 1980b. *Chloroclystis chloerata* Mab. the sloe pug. *Proceedings of the British entomological and natural History Society* **13** (3/4): 105. [Detailed description of the larva and preferred habitat of *Pasiphila chloerata*.]
———, 1980c. The occurrence and foodplant in Britain of *Eupithecia insigniata* (Hb.) (Lepidoptera, Geometridae). *En-*

*tomologist's Gaz*ette **31**: 221–226. [Detailed account of the early stages of *E. insigniata*.]

——, 1981a. Larvae of the British Lepidoptera not figured by Buckler. British Entomological and Natural History Society, London. [Colour illustrations, with accompanying text, of *E. egenaria*, *E. intricata*, *E. millefoliata*, *E. extensaria* subsp. *occidua*, *E. virgaureata*, *E. phoeniciata*, and *C. chloerata* and a note on *E. satyrata*.]

——, 1981b. *Eupithecia egenaria* H.-S.: a recent arrival? *Entomologist's Record & Journal of Variation* **93**: 236. [Discussion of the possible British origins of *E. egenaria* with information on the Breckland colonies.]

——, 1989a. *Chloroclystis rectangulata* L. (Lep.: Geometridae) ab. *pilcheri* ab. nov. *Entomologist's Record & Journal of Variation* **101**: 65. [Aberration of *P. rectangulata*.]

——, 1989b. Double-brooded *Eupithecia tripunctaria* H.-S. (Lep.: Geom.). *Entomologist's Record & Journal of Variation* **101**: 184–185. [Short review of voltinism.]

——, 1992. The larval food of *Eupithecia* species *indigata* Hb., *dodoneata* Guen. and *abbreviata* Steph. (Lep.: Geometridae). *Entomologist's Record & Journal of Variation* **104**: 39–42. [Observations on the biology of these three species.]

—— & CHALMERS-HUNT, J. M., 1961. *Eupithecia innotata* Hfn. on Sea Buckthorn. *Entomologist's Record & Journal of Variation* **73**: 211–212. [Includes an account of rearing two broods of *E. fraxinata*.]

—— & MERE, R., 1964. A note on the biology of *Eupithecia egenaria* H.-S. (Lep.: Geometridae). *Entomologist's Gazette* **15**: 25–28. [Notes on the Wye Valley colony with a discussion of its foodplant.]

HAMMOND, H. E., 1952. A list of previously unrecorded foodplants of lepidopterous larvae with additional notes on preferences etc. *Entomologist's Gazette* **3**: 59–68.

—— & SMITH, K. G. V., 1957. On some parasitic Diptera and Hymenoptera bred from Lepidopterous hosts. *Entomologist's Gazette* **8**: 181–189. [List of hymenopterous parasites on eupitheciine larvae.]

HANNEMANN, H., 1916. Sitzung am 17.Okt., 1916. *Internationale Entomologische Zeitschrift* **11**: 62. [Original description of *E. centaureata* ab. *punctata*.]

HARDING, P. [T.], 1992. Biological recording of changes in British Wildlife. Proceedings of the ITE Symposium no. 26 (13.iii.1990). HMSO, London.

HARRIS, M., 1766. *The Aurelian: or, natural history of English Insects; namely moths and butterflies; together with the plants on which they feed.* London. [First recorded pug moth in the British literature (*E. centaureata*) with description, illustration and notes on life history.]

——, 1775. *The English Lepidoptera or, the Aurelian's Pocket Companion.* J. Robson, London. (Reprinted, 1969, by Classey, Middlesex).

HARRISON, J. W. HESLOP, 1931. The abundance of certain species of *Eupithecia* in Northumberland and Durham. *Entomologist* **64**: 69. [Original description of *E. tenuiata* f. *johnsoni*.]

HATCHER, P. E., 1989. Host-plants and nutrition in conifer-feeding Lepidoptera. Ph.D. Thesis, Oxford Polytechnic.

[HAWORTH, A.H.], 1802. *Prodromus Lepidopterorum Britannicorum.* Holt. (By a Fellow of the Linnean Society. AHH). [Four *Eupithecia* species are listed without full descriptions.]

——, 1809. *Lepidoptera Britannica* **2**. John Murray, London. [Original descriptions of *E. plumbeolata*, *E. vulgata*, *E. subfuscata*, *E. simpliciata*, *C. v-ata* and *G. rufifasciata*; and of *E. icterata* f. *subfulvata*.]

HEINRICH, R., 1916. Beitrag zur Feststellung der Veränderungen der Berliner Grosschmetterlingsfauna in neuester Zeit. *Deutsche Entomologische Zeitschrift* **1916**: 528pp., Pl. 4, Fig.3. [Includes original description of *E. centaureata* ab. *albidior*.]

HELLINS, J., 1861. Note on *Eupithecia succenturiata* and *E. subfulvata*. *Zoologist* **19**: 7797. [Comparative notes on *E. succenturiata* and *E. icterata* larvae.]

——, 1867. Cannibalism of the larva of *Eupithecia minutata*. *Entomologist's monthly Magazine* **3**: 191. [Cannibalism in captive-reared *E. absinthiata* f. *goossensiata*.]

———, 1872. On the habits of *Eupithecia togata*. *Entomologist's monthly Magazine* **9**: 113–114. [Account of larvae feeding inside spruce branches in captivity.]

HENSLOW, J., 1852. Food of microlepidoptera. *Zoologist* **10**: 3358. [First British record of *E. pimpinellata*.]

HENWOOD, B. P., 1996. A wild first-brood larva of *Eupithecia virgaureata* Doubleday (Lepidoptera: Geometridae). *Entomologist's Gazette* **47**: 252. [Record of *E. virgaureata* feeding on *Salix cinerea*.]

HERRICH-SCHÄFFER, G. A. W., 1843–56. *Systematische Bearbeitung der Schmetterlinge von Europa*. Regensburg. [Original descriptions of *E. egenaria*, *E. trisignaria*, *E. distinctaria*, *E. pusillata* subsp. *anglicata*, *E. pusillata* ab. *expressaria*, *E. plumbeolata* ab. *singularia*, *E. subumbrata* f. [in text] *obrutaria* and *G. rufifasciata* ab. *parvularia*; *E. tripunctaria*.]

———, 1861. Die Arten der Spannergattung *Eupithecia*. *Correspondenz-Blatt für Sammler von Insekten* **17**: 131. [Unnamed description of *E. pusillata* subsp. *anglicata*. See Herrich-Schäffer, 1863.]

———, 1863. Zur Spannergattung *Eupithecia*. *Corespondenz-Blatt des zoologisch-mineralogischen vereins in Regensburg*. **17**: 21–24. [First use of the name *anglicata* for the subspecies of *E. pusillata*, although the original description appeared in Herrich-Schäffer, 1861.]

HODGKINSON, J. B., 1885. *Eupithecia curzoni*. *Entomologist* **18**: 76. [Suggests that *curzoni* is a form of *E. satyrata*.]

HOFFMANN, E., 1893. *Die Raupen der Gross-Schmetterlinge Europas*. Stuttgart. [Good colour illustrations of the larvae of most European species. Text in German.]

HOFFMEYER, S., 1952. *De Danske Maalere*. Aarhus. [Colour illustrations of larvae and adults; useful diagnostic notes and figures. Text in Danish.]

———, 1966. *De Danske Målere*. Aarhus. [Good colour illustrations of larvae; black and white colour photographs of adults; useful diagnostic text figures. Text in Danish.]

HOPLEY, E., 1864. *Eupithecia lariciata*. *Entomologist's monthly Magazine* **1**: 50. [First British record.]

HÜBNER, J., 1790. *Beiträge zur Geschichte der Schmetterlinge* **2**. Augsburg. [Original description of *E. insigniata*.]

———, 1796. *Sammlung europäischer Schmetterlinge. Fünfte Horde. Die Spanner*. Pl. 47, fig.246. Augsburg. [First illustration of *E. succenturiata* f. *disparata*.]

———, 1799. *Sammlung europäischer Schmetterlinge*. Augsburg. [Original description of *E. pygmaeata*.]

———, 1813. *Sammlung europäischer Schmetterlinge*. Augsburg. [Original descriptions of *E. tenuiata*, *E. inturbata*, *E. irriguata*, *E. exiguata*, *E. valerianata*, *E. satyrata*, *E. denotata* subsp. *denotata*, *E. indigata*, *E. pimpinellata*, *E. nanata* and *Pasiphila rectangulata* ab. *subaerata*.]

———, 1817. *Sammlung europäischer Schmetterlinge*. Augsburg. [Original description of *P. debiliata*.]

HUFNAGEL, A., 1767. *Phalaena innotata*. *Berlinisches Magazin* **4**: 616. [Original description of *E. innotata*.]

HUGGINS, H. C., 1962a. A new subspecies of *Eupithecia venosata* Fabr. *Entomologist's Record & Journal of Variation* **74**: 171–172. [First description of *E. venosata* subsp. *plumbea*.]

———, 1962b. A new subspecies of *Eupithecia vulgata* Haw. (Lep.: Hydriomenidae). *Entomologist* **95**: 45–46. [Original description of *E. vulgata* f. *clarensis*.]

———, 1968. A new aberration of *Eupithecia venosata* Fabr. *Entomologist's Record & Journal of Variation* **80**: 157. [Original description of *E. venosata* ab. *sepiata*.]

JACKSON, B. C., 1979. A note on breeding the pinion-spotted pug, *Eupithecia insigniata* (Hb.). Lep., Geometridae. *Proceedings of the British Entomological and Natural History Society* **12**: 76–77.

JAGER, J., 1887. Cannibalism among *Eupithecia* larvae. *Entomologist* **20**: 326–327. [Account of *C. v-ata* larvae eating a noctuid larva in captivity.]

JOHNSON, A. C., 2002. Key to pugs. *Atropos* **17**: 29–32.

JOHNSON, W., 1875. *Eupithecia knautiata* of Gregson. *Entomologist* **8**: 22–23. [Remarks on Gregson's (1874b; 1875a,b) claims on the specific status of *E. knautiata* (=*E. absinthiata*) plus an editorial comment by E. Newman.]

JUUL, K., 1948. *Nordens Eupithecier*. Aarhus. [Good colour paintings of larvae and some pupae; useful colour photographs of adults; genitalia drawings rather crude. Text in Danish with brief English summaries.]

KAABER, S., 1980. Om artsretten for *Eupithecia goossensiata* Mab. og *Eup. fraxinata* Crewe. *Lepidoptera* (N.S.) 3(10): 307–311. [Discussion on the specific status of *E. absinthiata* f. *goossensiata*. Text in Danish. English summary in Kaaber, 1981.]

———, 1981. II. *Eupithecia innotata* (Hfn.) og dens danske former. *Lepidoptera* (N.S.) 4: 41–50. [Discussion on the specific status of *E. innotata*. Text in Danish with English summary of this paper and Kaaber, 1980.]

KANE, W. F. de V., 1901. *A Catalogue of the Lepidoptera of Ireland, with supplement.* (Previously published in parts in *Entomologist*: 26–33 (1893–1901).)

KETTLEWELL, B., 1973. *The Evolution of Melanism.* Clarendon Press, Oxford.

KING, H., 1950. Freyer's pug, *Eupithecia arceuthata*, in a London garden. *Entomologist* 83: 267–268. [First capture of a larva on Cypress.]

———, 1953. *Eupithecia plumbeolata* in Dorset. *Entomologist's Record & Journal of Variation* 65: 292. [Note on afternoon flight of *E. plumbeolata*.]

KITT, M., 1925. Neue Lepidopteren-Formen. *Zeitschrift des Österreichen Entomologen-Vereines* 10: 28 + fig.5. [Original description of *E. venosata* ab. *circumflexa*.]

KLOET, G. S. & HINCKS, W. D., 1972. A Checklist of British Insects: Lepidoptera. *Handbooks for the Identification of British Insects.* 11(2) (Edn 2). Royal Entomological Society, London.

KNAGGS, H. G., 1865. Notes on British Lepidoptera for 1864. *Entomologist's Annual* 11: 106–108. [Account of first British record of *E. lariciata* and *E. denotata* subsp. *denotata*. Latter figured in frontispiece.]

KNILL-JONES, S. A., 1997. The Channel Islands Pug *Eupithecia ultimaria* (Boisd.) (Lep.: Geometridae) new to the Isle of Wight. *Entomologist's Record & Journal of Variation* 109: 286. [Short account of discovery of *E. ultimaria* in the Isle of Wight.]

KNUDSEN, K., 1981. *Eupithecia virgaureata* Dbl. Klaekket fra aeg. *Lepidoptera* (N.S.) 4: 51. [Notes on *E. virgaureata* including list of Danish foodplants. Text in Danish.]

LANGMAID, J. R., 1996. The Channel Islands Pug (*Eupithecia ultimaria* Boisduval) (Lepidoptera: Geometridae) resident in England. *Entomologist's Gazette* 47: 239–240.

LEECH, M. J., 1966. *Eupithecia phoeniceata* Rambur in the Channel Islands (Lep.: Geometridae). *Entomologist's Gazette* 17: 147. [First record in the British Isles.]

LEMPKE, B. J., 1947. Catalogus der Nederlandse Macrolepidoptera. *Tijdschrift voor Entomologie* 90: 146–147. [Key to forms of variation referred to again in Lempke (1951). Text in Dutch.]

———, 1951. Catalogus der Nederlandse Macrolepidoptera. *Tijdschrift voor Entomologie* 94: 227–255. [Original records of several forms and discussion of status of *E. absinthiata/goossensiata*. Text in Dutch with some English summaries. *E. absinthiata/goossensiata* discussion in English.]

LEVERTON, R., 1998. *Eupithecia indigata* Hb. (Lep.: Geometridae) larvae eating aphids. *Entomologist's Record & Journal of Variation* 110: 80–81.

LHOMME, L., 1923–1935. *Catalogue des Lépidoptères de France et de Belgique* 1: Macrolépidoptères (*Eupithecia–Gymnoscelis*, pp. 525–557). Le Carriol. [Larval foodplants.]

LINNAEUS, C., 1758. *Systema Naturae* (Edn 10). Stockholm. [Original description of *E. succenturiata* and *P. rectangulata*.]

LONG, R., 1965. Notes on recent additions to the macrolepidoptera of Great Britain and the Channel Islands. *Entomologist's Gazette* 16: 17–19. [Erroneous record of *E. oxycedrata* in Guernsey.]

LUNAK, R., 1936. Die Biologie von *Eupithecia egenaria* H.-S. *Zeitschrift des Österreichen Entomologen-Vereines* 21: 15–17. [Text in German.]

MABILLE, P., 1869a. Lépidoptères des environs de Paris, sur quatre eupithecies nouvelles pour la faune Parisienne. *Annales de la Société Entomologique de France* 38: 78–79. [Original description of *E. absinthiata* f. *goossensiata*.]

———, 1869b. Correspondence. *Petites Nouvelles Entomologiques* 1: 96. [Original description of *P. chloerata*.]

MACKAY, M. R., 1951. Species of *Eupithecia*

reared in the Forest Insect Survey in British Columbia (Lep.: Geometridae). *Canadian Entomologist* **83**(4): 77–91. [Includes a list of foodplants for *E. subfuscata*.]

MACKWORTH-PRAED, C. W., 1962. The macrolepidoptera of Caithness. *Entomologist's Gazette* **13**: 16. [Account of finding *E. fraxinata* pupae under moss on Ash trees.]

MCARTHUR, H., 1884. *Eupithecia nanata* var. *curzoni*. *Entomologist* **17**: 276–277. [Notes on the identity of *E. satyrata* f. *curzoni*.]

MCCORMICK, R., 1995. *Eupithecia abietaria* in Devon. *Entomologist's Record & Journal of Variation* **107**: 275.

MCDUNNOUGH, J. H., 1949. Revision of the North American species of the genus *Eupithecia* (Lep.: Geom.). *Bulletin of the American Museum of Natural History* **93**: 533–728 + figs. 1–20 and pls. 26–32. [Includes many British species, with excellent genitalia drawings.]

MELVILLE, J. COSMO, 1875. *Eupithecia knautiata*. *Entomologist* **8**: 133–134. [Remarks on Gregson's (1874b; 1885a,b) claims on the specific status of *E. knautiata* (=*E. absinthiata*).]

MEARNS, R. & MEARNS, B., 1997. *Chloroclystis debiliata* Hb.) (Lep.: Geometridae): first recent record of Bilberry Pug in Scotland. *Entomologist's Record & Journal of Variation* **109**: 108. [=*P. debiliata*.]

MERE, R. M., 1962. *Eupithecia egenaria* H.-S. (Fletcher's Pug) (Lep.: Geometridae) in the British Isles. *Entomologist's Gazette* **13**: 155–158. [Account of the discovery of *E. egenaria* in the Wye Valley.]

———, 1963. *Eupithecia egenaria* H.-S. (Lep.) in the British Isles. *Entomologist's Gazette* **14**: 23.

———, 1964. A new subspecies of *Eupithecia intricata* Zett. (Lep.: Geometridae). *Entomologist's Gazette* **15**: 73 + plate. [Original description of *E. intricata* subsp. *hibernica* with colour plate.]

METSCHL, C. & SALZL, S., 1935. Die Schmetterlinge der Regensburger Umgebung. *Iris* **49**: 93. [Original description of *E. pulchellata* ab. *defasciata*.]

MEYRICK, E., 1928. *A revised Handbook of British Lepidoptera*. Watkins & Doncaster, London. (Republished, 1968, by E. W. Classey, Middlesex). [Dichotomous key to *Eupithecia* (pp. 219–221); brief text; few diagnostic notes; no illustrations.]

MIKKOLA, K., 1992a. Function of the ventral coupling organs in the genus *Eupithecia* (Geometridae), with a description of a new organ. *Abstracts of the European Congress of Lepidopterology*, Helsinki, 19–23.iv.1992.

———, 1992b. Evidence of a 'lock-and-key' mechanism in the internal genitalia of the *Apamea* moths (Lep.: Noct.). *Systematic Entomology* **17**: 145–153.

———, JALAS, I. & PELTONEN, O., 1989. *Suomen Perhoset Mittarit* **2**. Recallmed, Hangon Kirjapaino. (Good diagnostic figures; excellent colour photographs; many British species covered; poor genitalia figures. Text in Finnish.]

MILLIÈRE, P., 1865. *Iconographie et description de chenilles et Lépidoptères inédits* **2**: 145–147, 184 (pl. 67, figs 1,2). [Original description of '*E. massiliata* Dard. et Mill. (Species nova)', in fifteenth part of this volume, submitted to the *Société Linnéene de Lyon* on 11 December 1865.]

———, 1869. *Iconographie et description de chenilles et Lépidoptères inédits* **3**: 102, pl. 110. Paris. [Original description of *E. intricata* subsp. *millieraria* Wnuk. under the name *anglicata*. See also Wnukowski (1929).]

———, 1875. *Catalogue Raisonné des Lépidoptères du département des Alpes-Maritimes* **3**: 410. Cannes. [Original description of *G. rufifasciata* ab. *incertana*.]

MIRONOV, V., 2003. *The Geometrid moths of Europe* **4**. Larentiinae II. Apollo Books, Stenstrup. [Excellent text and illustrations of 133 European Eupitheciines.]

MORLEY, C. & RAIT-SMITH, W., 1933. The hymenopterous parasites of the British Lepidoptera. *Transactions of the Royal Entomological Society of London* **81**: 133–183.

MORRIS, F. O., 1861. *A natural history of British Moths* **1**. Nimmo, London.

MUTCH, J. P., 1905. *Eupithecia stevensata*. *Entomologist* **38**: 161. [Note on the capture of *E. pusillata* subsp. *anglicata* at Freshwater, Isle of Wight.]

MYERS, A. A., 1982. First Irish record of

Eupithecia phoeniceata (Rambur) (Lepidoptera: Geometridae). *Entomologist's Gazette* **33**(1): 47.

NEWMAN, E., 1845. Description of *Eupithecia togata*, a new British moth of the Family Geometridae. *Zoologist* **3**: 1086. [First British description of *E. abietaria*, with line engraving. See also Stevens, 1845.]

——, 1869. *An Illustrated Natural History of British Moths*. Tweedie, London. [First comprehensive work including details of early stages with engravings of adults.]

NEWMAN, L. W. & LEEDS, H. A., 1913. *Textbook of British Butterflies and Moths*. Privately published, St. Albans.

NEWTON, J. & MEREDITH, G. H. J., 1984. *The Macrolepidoptera of Gloucestershire*. Sutton, Gloucester.

O'MEARA, M., 2001. *The Lepidoptera of Waterford City and County*. Fauna of Waterford Series, No. 1. Waterford Wildlife.

OWEN, J., 1991. *The Ecology of a Garden*. Cambridge University Press. [List of foodplants for *E. exiguata* and *E. vulgata*.]

PALMER, R. M., 1975. Lepidoptera of Aberdeenshire and Kincardineshire. *Entomologist's Record & Journal of Variation* **87**: 218–224.

PEET, T., 1988. An introduction to Guernsey Lepidoptera. *Entomologist's Record & Journal of Variation* **100**: 21–24. [Short account of finding larvae of *E. ultimaria* in Guernsey.]

PELHAM-CLINTON, E. C., 1972. *Chloroclystis chloerata* (Mabille, 1870). A Geometrid moth new to the British list breeding in southern England. *Entomologist's Gazette* **23**: 151–152. [=*Pasiphila chloerata*.]

PENNINGTON, M. G., ROGERS, T. D. & BLAND, K. P., 1997. Lepidoptera new to Shetland. *Entomologist's Record & Journal of Variation* **109**: 273. [Discovery of *E. centaureata*, *E. assimilata* and *E. lariciata* in Shetland.]

PERRING, F. H. & WALTERS, S. M., 1962. *Atlas of the British Flora*. Nelson, London.

PETERSEN, W., 1909. Ein Beitrag zur Kenntnis der Gattung *Eupithecia* Curt. *Iris* **22**(4): 203–314. [Excellent genitalia drawings of most European species. Text in German.]

PICKERING, R. R., 1980. Observations on the cypress pug: *Eupithecia phoeniceata* Rambur at Aldwick Bay, West Sussex. *Entomologist's Record & Journal of Variation* **92**: 274. [Records of several years' captures indicating protracted emergence period in this area.]

PIERCE, J. W., 1914. *The Genitalia of the Group Geometridae of the Lepidoptera of the British Isles*. Privately published, London. [Excellent text, drawings of male and female genitalia, and male abdominal plates of most species.]

PLANT, C. W., 1989. A new aberration of *Gymnoscelis rufifasciata* (Haw.) (Lep.: Geom.) The double-striped pug. *Entomologist's Record & Journal of Variation* **101**: 105. [Original description of *G. rufifasciata* ab. *albofasciata*.]

PORRITT, G. T., 1889. *Eupithecia extensaria* near Hunstanton. *Entomologist's monthly Magazine* **25**: 398–399.

PORTER, J., 1997. *Colour Identification Guide to the Caterpillars of the British Isles*. Viking, Harmondsworth.

PRATT, C., 1981. *A History of the Butterflies and Moths of Sussex*. Booth Museum of Natural History, Brighton.

PRIOR, G., 1977. Observations on the larva of *Eupithecia insigniata* Hb. (Pinion-spotted pug). *Proceedings of the British Entomological and Natural History Society* **10**: 103. [Account of finding early-instar larvae on hawthorn.]

——, 1978. Notes on the larva of *Eupithecia tenuiata* Hb. (Slender pug). *Entomologist's Record & Journal of Variation* **90**: 253. [Record of finding larvae on female sallow catkins.]

——, 1980. The Eupitheciini or 'Pugs'. *Proceedings and Transactions of the British Entomological and Natural History Society* **13**(2): 18–23. Part Two of the Presidential Address. [Brief notes on the early stages of 24 spp.]

PROUT, L. B., 1894. City of London Entomological & Natural History Society: exhibits, July 3rd 1894. *Entomologist's Record & Journal of Variation* **5**: 230. [Notes on the forms of *E. assimilata* adults when continually bred in captivity.]

——, 1896. *Eupithecia succenturiata* and *subfulvata*. *Entomologist's Record & Journal of Variation* **7**: 109–110. [Dis-

cussion of the specific status of *E. succenturiata* and *E. icterata*.]
——, 1901. Some new Geometrid varieties and aberrations. *Entomologist's Record & Journal of Variation* **13**: 336. [Description of *E. venosa* ab. *orcadensis* (=subsp. *ochracae* Gregson).]
——, 1904. *Eupithecia innotata* Hufn. in the Isle of Wight and north Devon. *Entomologist's Record & Journal of Variation* **16**: 336. [Account of supposed *E. innotata* larvae found on *Artemisia vulgaris* and *Crepis virens*.]
——, 1907a. Notes on the genus *Eupithecia*. *Entomologist* **40**: 169–175. [Short bibliography and synonymy and a discussion on the specific status of *E. succenturiata* and *E. icterata*.]
——, 1907b. Notes on the genus *Eupithecia*. *Entomologist* **40**: 206–211. [Notes on the *E. innotata/fraxinata* complex and on the biology of *E. virgaureata* and *E. denotata*.]
——, 1907c. Notes on the genus *Eupithecia*. *Entomologist* **40**: 220–222. [Notes on the biology of *E. virgaureata*, *E. tripunctaria*, *E. trisignaria* and *C. v-ata*.]
——, 1908. Supplemental notes on *Eupithecia*. *Entomologist* **41**: 52–54. [Discussion on *E. innotata/E. fraxinata*; discovery of *C. v-ata* on *Crataegus*; discovery of *E. trisignaria* larvae on *Pastinaca*.]
——, 1914. Eupitheciinae. *In* Seitz, A., *Macrolepidoptera of the world* **4**: 274–299. Stuttgart.
——, 1915. Some new melanic *Eupithecia* aberrations. *Entomologist* **48**: 6–7. [Original descriptions of *E. nanata* f. *oliveri*, *E. lariciata* f. *nigra* and *E. fraxinata* f. *unicolor*.]
——, 1938. Eupitheciinae. *In* Seitz, A., *Macrolepidoptera of the World*. Supplement: 182–211. Stuttgart.
RAMBUR, J. P., 1834. Description de plusieurs espèces inédites de Lépidoptères nocturnes du centre et de midi de la France. *Annales de la Société Entomologique de France* **3**: 393 + Pl. viii, fig.6. [Original description of *E. phoeniceata*.]
RENNIE, J., 1832. *A Conspectus of the Butterflies and Moths found in Britain*. Orr, London.
RICHARDSON, A., 1952. New varieties of British Lepidoptera. *Entomologist's Record & Journal of Variation* **64**: 271 + Pl. XI. [Original descriptions of *E. linariata* ab. *praeruptata* and *G. rufifasciata* ab. *obsolescens*.]
—— & HAGGETT, G., 1963. *Eupithecia phoeniciata* in Britain. *Entomologist's Record & Journal of Variation* **75**: 65–68. [Notes on finding and rearing *E. phoeniciata*.]
—— & MERE, R. M., 1958. Some preliminary observations on the Lepidoptera of the Scilly Isles with particular reference to Tresco. *Entomologist's Gazette* **9**: 115–147. [Brief reference to forms of *E. subfuscata* and *E. dodoneata*.]
RILEY, A. M., 1985a. *Eupithecia ultimaria* Boisd. (Lep.: Geometridae): a pug new to the British list. *Entomologist's Gazette* **36**: 259–261.
——, 1985b. Late capture of *Chloroclystis chloerata* Mab. (sloe pug). *Entomologist's Record & Journal of Variation* **97**: 228. [=*Pasiphila chloerata*.]
——, 1986a. Review of the status of *Eupithecia goossensiata* Mabille (ling pug) and *E. absinthiata* Clerck (wormwood pug) (Lep.: Geometridae). *Entomologist's Record & Journal of Variation* **98**: 85–89.
——, 1986b. Suspected second brood of *Eupithecia lariciata* Freyer (larch pug). *Entomologist's Record & Journal of Variation* **98**: 207–208.
——, 1987. *Eupithecia lariciata* Freyer (Lep.: Geometridae), the larch pug in Dumfries-shire. *Entomologist's Record & Journal of Variation* **99**: 152. [Further evidence of partial bivoltinism.]
——, 1988. *Eupithecia vulgata* Haw. (common pug) ssp. *scotica* Cock (Lep.: Geom.) in Cumbria. *Entomologist's Record & Journal of Variation* **100**: 104. [Discussion on f. *scotica* sympatric with the type.]
——, 1989. White metathoracic crests in *Eupithecia lariciata* Freyer and *E. tripunctaria* H.-S. (Lep.: Geom.), the larch and white-spotted pugs. *Entomologist's Record & Journal of Variation* **101**: 72.
——, 1990a. Delayed emergence in *Chloroclystis v-ata* Haw. *Entomologist's Record & Journal of Variation* **102**: 38–39.
——, 1990b. Recent records of *Eupithecia abietaria* Goeze (Lep.: Geometridae), the cloaked pug, from Rothamsted Insect Survey light traps. *Entomologist's*

Record & Journal of Variation **102**: 238.

———, 1990c. New Lepidoptera records for Guernsey. Entomologist's Record & Journal of Variation **102**: 294–295.

———, 1990d. Unusual flight times of *Eupithecia tripunctaria* H.-S., *Operophtera brumata* L. and *Colostygia multistrigaria* L. (Lep.: Geometridae) in RIS light traps. Entomologist's Record & Journal of Variation **102**: 303–304. [*E. tripunctaria* in Devon in December 1989.]

———, 1991. *Eupithecia ultimaria* Boisduval (Lepidoptera: Geometridae), a third record for the British Isles and the first mainland capture. Entomologist's Gazette **42**: 289–290.

———, 1998. Further unusual records of Lepidoptera from the Rothamsted Insect Survey national light-trap network. Entomologist's Record & Journal of Variation **110**: 49–53. [First records of *E. inturbata* and *G. rufifasciata* in Northumberland.]

——— & PRIOR, G., 1990. A review of the phenology of *Eupithecia tripunctaria* H.-S., the white-spotted pug (Lep.: Geometridae). Entomologist's Record & Journal of Variation **102**: 49–54.

ROBERTSON, G. S., 1933. The resting position in the genus *Eupithecia*. Entomologist's Record & Journal of Variation **45**: 61. [Interesting discussion on the characteristic resting postures of various *Eupithecia* spp.]

ROBSON, J. E., 1902. Catalogue of the Lepidoptera of Northumberland and Durham, 1899–1905. *Natural History Transactions of Northumberland, Durham and Newcastle-upon-Tyne* **12**(2): 256–276.

——— & GARDNER, J., 1886. A List of British Lepidoptera and Their Named Varieties, p. 41. Kempster, London. [Original description of *E. satyrata* f. *fagicolaria*. In this document, *fagicolaria* is attributed to Henry Harpur Crewe but, as his description was not published, Robson & Gardner stand as the authority.]

RÖSSLER, A., 1866. Verzeichniss der Schmetterlinge des Herzogthums Nassau. Wiesbaden. [Original description of *E. millefoliata*.]

RUTHERFORD, C. I., 1988. *Eupithecia abietaria* (Goeze) (Lep.: Geometridae) breeding in Wales. Entomologist's Record & Journal of Variation **100**: 140–141. [Record of larva feeding on Noble Fir in Clwyd.]

SAWYER, J., 1875. *Eupithecia extensaria*. Entomologist **8**: 132–133. [Discussion of possible origin.]

SCHÜTZE, E., 1954. Eupithecien-Studien III., (Lep.: Geom.) *Eupithecia egenaria* H.-S. *Abhandlungen und Bericht des Vereins für Naturkunde zu Kassel* **59**: 1–9 + figs. [Account of the biology and identification characters of *E. egenaria*. Includes illustrations of adult, larva, pupa and genitalia. Text in German.]

SCHWINGENSCHUSS, L., 1953. Beitrag zur Lepidopterenfauna von Niederösterreich. *Zeitschrift der Wiener Entomologischen Gesellschaft* **38**: 251–255. [Original descriptions of *E. expallidata* f. *pseudoabsinthiata* and ab. *pallida*.]

———, 1954. Einige neue Lepidopterenformen aus Niederösterreich: *Chloroclystis rectangulata* L. *Zeitschrift der Wiener Entomologischen Gesellschaft* **39**: 177. [Original description of *Pasiphila rectangulata* ab. *mediospoliata*.]

SCORER, A. G., 1913. The Entomologist's Log Book. Routledge, London. [Extensive phenological and foodplant lists for each species.]

SHAW, M. R. & FITTON, M. G., 1978. Survey of parasitoids of British butterflies. Entomologist's Record & Journal of Variation **101**: 69–71.

SHAYER, C. J., 1959. *Eupethecia* [sic] *oxycedrata* (Ramb.) in Guernsey. (Lep., Geometridae). Entomologist's Gazette **10**: 134.

SHELDON, W. G., 1896. *Eupithecia succenturiata* and *subfulvata*. Entomologist's Record & Journal of Variation **7**: 197–198. [Discussion of the specific status of *E. succenturiata* and *E. icterata*.]

———, 1899. Variation of *Eupithecia pulchellata* with description of var. *hebudium*, n. var. Entomologist's Record & Journal of Variation **11**: 344. [Notes on adult behaviour with a original description of f. *hebudium*.]

SHOWLER, A. J., 1962. *Eupithecia inturbata* at Chattenden. Entomologist's Record & Journal of Variation, **74**: 246. [Account of *E. inturbata* larvae feeding on small Field Maple trees.]

SILBERNAGEL, A., 1943. Několik Novch

Forem Macrolepidopter Z. Čech. *Casopis Ceskoslovenské Spolecnosti Entomologické. Praha* **40**: 7. [Original description of *E. tantillaria* ab. *mediopallens*.]

SIMSON, E. C. L., 1980. British Pugs. *Entomologist's Record & Journal of Variation* **92**: 261–266. [Notes on collecting pugs.]

———, 1981a. British Pugs. *Entomologist's Record & Journal of Variation* **93**: 7–10. [More notes on collecting pugs.]

———, 1981b. British Pugs. *Entomologist's Record & Journal of Variation* **93**: 30–35. [Further notes on collecting pugs.]

SIRCOM, J., 1851. Captures of macrolepidoptera near Bristol in 1851. *Zoologist* **9**: 3287–3288.

SKINNER, B., 1984. *Colour Identification Guide to the Moths of the British Isles*. Viking, Harmondsworth. (Edn 2, 1998). [Good colour photographs of all the British species and common forms; brief text.]

———, 1990. Protracted emergence of *Eupithecia pusillata* D.& S. (Lep.: Geometridae) juniper pug. *Entomologist's Record & Journal of Variation* **102**: 213.

SKOU, P., 1986. *The Geometroid Moths of Northern Europe*. Scandinavian Science Press, Leiden and Copenhagen. [Good colour plates of adults with some black and white figures. English text.]

SLADE, B. E. and AGASSIZ, D. J. L., 1991. *Eupithecia sinuosaria* Eversmann (Lep.: Geom.) new to the British Isles. *Entomologist's Record & Journal of Variation* **104**: 287–288.

SOUTH, R., 1890. Is *Eupithecia abietaria* Goeze identical with *E. togata* Hübner? *Entomologist* **23**: 205. [Discussion of the specific status of *E. abietaria*.]

———, 1908 (Edn 2, 1920; Edn 3, 1939). *Moths of the British Isles*, Series I. Warne, London. [Standard text; some good colour life-size photographs of adults; a few poor colour drawings of adults; a few monochrome illustrations of larvae.]

———, 1961 (Edn 4). *Moths of the British Isles*, Series II. Warne, London. [Similar text to previous editions. Four colour plates of adults (twice life-size) of limited value; some larvae illustrated in black and white.]

SPÜLER, A., 1910. *Die Schmetterlinge Europas* **2**, **3**. Stuttgart. [Fair life-size colour lithographs of adults. Brief German text.]

STACE, C., 1991. *New Flora of the British Isles*. Cambridge University Press, Cambridge.

STAINTON, H. T., 1855. New British species since 1835. *Entomologist's Annual* **1**: 5–21. [Short accounts of the discovery in Britain of *E. abietaria* (as *togata*), *E. pusillata* and its subsp. *anglicata*, *E. pygmaeata* (as *palustraria*), *E. satyrata* f. *callunaria* (as *E. callunaria*), *E. satyrata* f. *satyrata*, *E. fraxinata* (as *E. innotata*), *E. exiguata* (as *E. lanceolaria*), *E. tenuiata*, *E. indigata* and *E. pimpinellata*.]

———, 1858. New British species in 1857 *Eupithecia helveticaria*. *Entomologist's Annual* **4**: 87–88. [Reviews the first capture in Britain of *E. intricata* subsp. *millieraria*.]

———, 1859a. *A Manual of British Butterflies and Moths* **2**. Van Voorst, London. [Contains good descriptions of 39 species; five engravings of adults.]

———, 1859b. New British species in 1858. *Entomologist's Annual* **5**: 148. [Account of the first British record of *E. valerianata*.]

———, 1862. New British species in 1861. *Entomologist's Annual* **8**: 109–110. [Review of the first British records of *E. trisignaria* and *E. tripunctaria*.]

———, 1866. New British species since 1853. *Entomologist's Annual* **12**: 25–26. [Short review of the first British records of *E. tripunctaria*, *E. trisignaria*, *E. intricata* subsp. *arceuthata*, *E. valerianata*, *E. fraxinata*, *E. lariciata* and *E. denotata*.]

STAUDINGER, O., 1861. *Eupithecia altenaria* nov. spec. *Entomologische Zeitung* **22**: 401–402. [Original description of *E. virgaureata* ab. *altenaria*.]

———, 1871. *Catalog der Lepidopteren des Europaeischen Faunengebiets*. I. *Macrolepidoptera*, p. 197. Dresden. [Original descriptions of *E. satyrata* ab. *subatrata* and *E. subumbrata* f. *obrutaria*.]

———, 1892. *Eupithecia Oblongata* Thunb. var. *Centralisata* Stgr. *Iris* **5**: 250. [Original description of *E. centaureata* ab. *centralisata*.]

———, 1901. *Catalog der Lepidopteren des palaearktischen Fauengebiets*. (Edn 3 of 1871 *Catalog*, p. 315.) *Succenturiata*. v. (et. ab.) *Exalbidata*. [Original description of *E. succenturiata* ab. *exalbidata*.]

STEP, E., 1892. Exhibits. *Proceedings of the South London Entomological and Natural History Society* **1892**: 56. [Discussion on status of *E. pusillata* subsp. *anglicata*.]

STEPHENS, J. F., 1829. *A systematic Catalogue of British Insects*. Baldwin & Cradock, London. [Contains checklist of 24 species.]

——, 1829[–31]. *Illustrations of British Entomology*: ... Haustellata **III**. Baldwin & Craddock, London. [35 species listed, including six new to Britain, some subsequently synonymized.]

STERLING, D. H., 1977. *Eupithecia phoeniceata* ... in the Winchester district. *Entomologist's Record & Journal of Variation* **89**: 315. [Capture of an adult on 6/7.vii.1977, very early for this species.]

——, 1985. *Eupithecia*-delayed emergence. *Entomologist's Record & Journal of Variation* **97**: 93. [Account of *E. insigniata* pupa lying over two winters.]

STEVENS, S., 1845. Capture of *Eupithecia togata* (Hüb.), a new British Moth at Black Park, Bucks. *Zoologist* **3**: 1086. [Account of discovery of *E. abietaria* in Britain.]

——, 1882. *Eupithecia ultimaria* Duponchel. *Entomologist* **15**: 18. [Note on finding *E. pusillata* subsp. *anglicata*.]

STOKOE, W. J., 1958. *Caterpillars of British Moths*. Warne, London. [Pug illustrations of limited use.]

STYLES, J. H., 1961. Additions to the list of foodplants of Lepidoptera larvae not recorded in *Larval Foodplants* by P. B. M. Allan and *The Caterpillars of the British Moths* by W. J. Stokoe. *Entomologist* **94**: 86–88.

SUTTON, S. L. & BEAUMONT, H. E., 1989. *Butterflies and Moths of Yorkshire*. Yorkshire Naturalists' Union, Doncaster.

SYMES, H., 1958. Notes on some Eupithecidae. *Entomologist's Record & Journal of Variation* **70**: 230–233. [Brief extracts from the diaries of Dr H. King.]

TAYLOR, L. R., 1986. Synoptic dynamics, migration and the Rothamsted Insect Survey. Presidential address to the British Ecological Society, Dec. 1984. *Journal of Animal Ecology* **55**: 1–38.

TORRE-BUENO, J. R. DE LA, 1930. *A Glossary of Entomology*. Brooklyn Entomological Society, Brooklyn.

TOWNSEND, M. C. and RILEY, A. M., 1991. *Eupithecia sinuosaria* Eversmann (Lep.: Geom.) in Hertfordshire. The second British record. *Entomologist's Record & Journal of Variation* **104**: 323.

TREITSCHKE, F., 1828. *Die Schmetterlinge von Europa* **6** (2): 114. Leipzig. [Original description of *E. icterata* f. *oxydata*.]

TUGWELL, W. H., 1892. The Paisley 'Pug' (*Eupithecia castigata* var.). *Entomologist* **25**: 41–42. [Note on the specific status of the 'Paisley Pug' (=*E. subfuscata*).]

TUTT, J. W., 1885. On the variation of *Eupithecia nanata*. *Entomologist* **18**: 75–76. [Suggestion that *E. satyrata* f. *curzoni* is a form of *E. nanata*.]

——, 1894. Obituary. William Machin. *Entomologist's Record & Journal of Variation* **5**: 209–210.

——, 1896. *British Moths*. Routledge, London. [Short notes on all the species.]

——, 1906a. A puzzling group of Eupitheciids. *Entomologist's Record & Journal of Variation* **18**: 157–158. [Discussion of the specific status of *E. innotata*, *E. fraxinata* and *E. tamarisciata*.]

——, 1906b. Practical hints relating to the Eupitheciids. *Entomologist's Record & Journal of Variation* **18**: 179–182. [Notes on how to collect pugs.]

——, 1906c. Practical hints relating to the Eupitheciids. *Entomologist's Record & Journal of Variation* **18**: 201–204. [More notes on how to collect pugs.]

——, 1906d. Practical hints relating to the Eupitheciids. *Entomologist's Record & Journal of Variation* **18**: 218–222. [Further notes on how to collect pugs.]

——, 1906e. A puzzling group of Eupitheciids. *Entomologist's Record & Journal of Variation* **18**: 260. [Continued discussion of the specific status of *E. innotata*.]

——, 1906f. Another puzzling group of Eupitheciids. *Entomologist's Record & Journal of Variation* **18**: 261–263. [Discussion of the specific status of *E. succenturiata* and *E. icterata*.]

—— (Ed.), 1906g. Obituary. Mrs Emma Sarah Hutchinson. *Entomologist's Record & Journal of Variation* **18**: 56. [Details of Mrs Hutchinson's breeding of *E. insigniata*.]

——, 1908. *Eupithecia tamarisciata* as a British insect. *Entomologist's Record & Journal of Variation* **20**: 102–104.

VILLERS, C. J. DE, 1789. *Caroli Linnaei entomologia, faunae suecicae descriptionibus aucta* ... Lyons. [Original description of *E. icterata*.]

VOJNITS, A. M., 1987. Falces and clavulus, hitherto disregarded parts of the male genitalia in *Eupithecia* species (Lep.: Geometridae). *Annales Historico-Naturales Musei Nationalis Hungarici* **79**: 179–184. [Very brief discussion with figures of 27 spp.]

WAKELEY, S., 1957. *Eupithecia egenaria* H.-S. *Entomologist's Record & Journal of Variation* **69**: 199–200. [The article which prompted the successful search for *E. egenaria* in Britain.]

WARING, P., 1997. The Pauper Pug *Eupithecia egenaria* H.-S. (Lep.: Geometridae) discovered in Lincolnshire, and other interesting moth records from a survey of the Bardney Limewoods Site of Special Scientific Interest in 1995. *Entomologist's Record & Journal of Variation* **109**: 1–9. [Contains a list of all known localities for *E. egenaria*.]

———, 2002. Wildlife reports – Moths. *British Wildlife* **14**(1): 58. [*E. egenaria* records from Lincolnshire and Worcestershire.]

———, 2003. Wildlife reports – Moths. *British Wildlife* **14**(5): 361. [*E. egenaria* records from Somerset, Suffolk and Oxfordshire.]

——— & TOWNSEND, M. C., 2003. *Field guide to the moths of Great Britain and Ireland*. Illustrations by Richard Lewington. British Wildlife Publishing, Hook. [Illustrates moths in natural resting postures.]

WEBB, S., 1881. *Eupithecia ultimaria* Duponchel. *Entomologist* **14**: 300. [Note on finding *E. pusillata* subsp. *anglicata*.]

WEIGT, H.-J., 1976. Die Blütenspanner Westfalens, Teil 1: Die Imagines und ihre Verbreitung. *Dortmunder Beiträge zur Landeskunde* **10**: 66–152. [Account of Westphalian species, with photographs of adults and typical habitat for each, including many found in Britain. Text in German.]

———, 1977. Die Blütenspanner Westfalens (Lep.: Geometridae). Teil 2. Die Raupen und ihre Futterpflanzen. *Dortmunder Beiträge zur Landeskunde* **11**: 41–98. [Thorough illustrated account of larvae, including many British species, with their foodplants. Text in German.]

———, 1978. Die Blütenspanner Westfalens. *Dortmunder Beiträge zur Landeskunde* **12**: 9–77. [Line drawings for some difficult species; some larvae illustrations; photographs of antennae, heads, female genitalia, male aedeagi and some adults; drawings of the male valves and abdominal plates. Text in German.]

———, 1979. Blütenspanner-Beobachtungen 1. *Eupithecia actaeata* (Lep.: Geom.). *Entomologische Zeitschrift* **89**(3): 17–23. [Full account of the biology with photographs of adult, larva and pupa. Text in German.]

———, 1980. Blütenspanner-Beobachtungen 2. Bemerkungen zur *Eupithecia absinthiata* Gruppe (Lep.: Geometridae). *Abhandlungen aus dem Landesmuseum für Naturkunde zu Münster* **42**: 31–50. [Discussion of the taxonomy of *E. absinthiata*, (+ f. *goossensiata* treated as a distinct species), *E. expallidata* and *E. assimilata*. Illustrations of adults, larvae, pupae and genitalia. Text in German.]

———, 1981. Blutenspanner-Beobachtungen 5. *Dortmunder Beiträge zur Landeskunde* **15**: 59–66. [Excellent descriptions, photographs and genitalia drawings of *C. v-ata*, *C.* (=*P.*) *chloerata* and *G. rufifasciata*. Text in German.]

———, 1985. Blütenspanner-Beobachtungen 8. Vorkommen und Lebensweise von *Eupithecia pygmaeata* Hb. (Lep.: Geometridae). *Dortmunder Beiträge zur Landeskunde* **19**: 9–18. [Excellent account of the life history of *E. pygmaeata* with beautiful colour photographs of all stages. Text in German.]

———, 1987. Die Blütenspanner Mitteleuropas (Lep., Geom: Eupitheciini) Teil I. Biologie der Blütenspanner. *Dortmunder Beiträge zur Landeskunde* **21**: 5–57. [Excellent details on the biology of the pugs with a useful synonymy. Text in German.]

———, 1988. Die Blütenspanner Mitteleuropas. (Lep., Geom.: Eupitheciini) Teil 2. *Gymnoscellis rufifasciata* bis *E. insigniata*. *Dortmunder Beiträge zur Landeskunde* **22**: 5–81. [Account of many British species including excellent colour photographs of adults, larvae and ova, with sketches of genitalia. Text in German.]

———, 1990. Die Blütenspanner Mitteleuropas (Lep.: Geometridae: Eupitheciini) Teil 3. *Eupithecia sinuosaria* bis *pernotata*. *Dortmunder Beiträge zur Landeskunde* **24**: 5–100. [Superb colour illustrations of many British species. Text in German.]

———, 1991. Die Blütenspanner Mitteleuropas (Lep.: Geometridae: Eupitheciini) Teil 4. *Eupithecia satyrata* bis *indigata*. *Dortmunder Beiträge zur Landeskunde* **25**: 5–106. [Superb colour photographs of many British species. Text in German.]

WEST, B. K., 1989. The evidence for bivoltinism in *Eupithecia tripunctaria* H.-S. (Lep.: Geometridae) in S.E. England. *Entomologist's Record & Journal of Variation* **101**: 57–59.

———, 1994. *Eupithecia icterata* Vill. (Lepidoptera: Geometridae): Larval foodplants. *Entomologist's Record & Journal of Variation* **106**: 71. [Lists Feverfew as a foodplant.]

WESTWOOD, J. O., 1851. Exhibits. *Proceedings of the Entomological Society*, **1851** (Sept.): 108. [First record of *E. pusillata* subsp. *anglicata*.]

WHITE, F. BUCHANAN, 1891. Structure of the terminal abdominal segments in the males of the genus *Eupithecia*. *Entomologist* **24**: 129–130 + 1 plate. [First discovery of male abdominal plates as a diagnostic feature. Some of the species figured are incorrect.]

WHITTAKER, O. and PROUT, L. B., 1904. *Eupithecia innotata* Hufn. in the Isle of Wight and north Devon. *Entomologist's Record & Journal of Variation* **16**: 336. [Notes on the foodplants of *E. fraxinata*.]

WILSON, O. S., 1880. *The Larvae of the British Lepidoptera and their Foodplants*. Lovell Reeve, London. [Excellent descriptive text and colour figures.]

WINTER, T. G., 1983. *A catalogue of phytophagous insects and mites on trees in Great Britain*. Forestry Commission Booklet 53. Edinburgh.

———, 1990. An annotated checklist of British conifer-feeding macrolepidoptera and their foodplants. *Entomologist's Gazette* **41**: 177–196.

WNUKOWSKY, W., 1929. Einige Nomenklatur – Notizen über die paläarktischen Lepidopteren. *Zoologische Anzeiger* **83**: 221–224. [New name proposed for *E. intricata* subsp. *anglicata* Millière (*E. intricata* subsp. *millieraria*) as the name *anglicata* was pre-occupied. See Millière (1869) for original description.]

WOLFF, N. L., 1929. Lepidoptera. *In* Jensen, S, Lundbeck, W. & Mortensen, Th. (Eds) *Zoology of the Faroes* **2**(1): 13–14. Copenhagen. [Description of *E. satyrata* f. *curzoni* ab. *trifasciata* nov. and *E. nanata* subsp. *zebrata* nov.]

WOOD, W., 1839. (Edn 2, 1854) *Index Entomologicus, or, a Complete Illustrated Catalogue of the Lepidopterous Insects of Great Britain*. Ward, London. [Twenty-four species listed; good descriptions; hand-coloured illustrations of adults.]

ZELLER, P. C., 1847. Bemerkungen über die auf einer Reise nach Italien und Sicilien gesammelten Schmetterlingsarten. *Isis* **7**: 502–503. [Original description of *G. rufifasciata* ab. *tempestivata*.]

ZETTERSTEDT, J. W., 1840. *Insecta Lapponica descripta*. Leipzig. [Original description of *E. intricata*.]

Synonymic Index

Principal entries are given in **bold type**. Plate references are shown as (Pl.3: 13). The index includes references to figures in the text, as 65 (text fig. 6g).

ab. = aberration   f. = form   subsp. = subspecies

abbreviaria = *E. abbreviata* 24, 117
abbreviata, Eupithecia 20, 24, 27, 65 (text fig. 6g), 69 (text fig. 8r), **117** (Pl.3: 13, 14; Pl.6: 9; Pl.8: 2; Pl.D: 1) (text fig. 9), 119, 120, 146 (text fig. 12j), 152 (text fig. 18a), 158 (text fig. 24b), 163 (text fig. 29a), 168 (text fig. 33b), 225, 228, 231, 232
abietaria, Eupithecia 21, 27, 30, **43** (Pl.1: 7, 8; Pl.6: 16, 17; Pl.A: 5) (text fig. 2a), 147 (text fig. 13f), 154 (text fig. 20f), 161 (text fig. 27e), 165 (text fig. 30e), 224, 226, 230, 231, 232
absinthiata, Eupithecia 19, 21, 22, 25, 27, 67 (text fig. 7a), 69 (text fig. 8c), 75, **76** (Pl.1: 32, 33); Pl.7: 15, 16; Pl.B: 6, 7), 78, 79, 81, 83, 84, 89, 107, 116, 141, 142, 149 (text fig. 15f(ii)), 156 (text fig. 22b), 161 (text fig. 27t), 166 (text figs 31h,i), 224, 226, 227, 228, 229, 230, 231, 232, 233, 234, 235
achilleata = *E. millefoliata* 23, 99
achromata, E. pusillata ab. 123
actaeata, Eupithecia 145
aequistrigata, E. subumbrata f. 98
albescens Cock., *P. debiliata* f. 137
albescens Cock., *E. extensaria* f. 111
albescens Lempke, *G. rufifasciata* ab. 139
albidior, E. centaureata ab. 63
albipunctata = *E. tripunctaria* 23, 86
albofasciata, E. exiguata ab. 51
Aleiodes (Braconidae) 31
alliaria = *E. pygmaeata* 141
altenaria, E. virgaureata f. 114
analoga, Eupithecia 145
angelicata Barrett, *E. tripunctaria* f. 19, 86, 87
angelicata Prout, *E. trisignaria* f. 18, 64, 65, 66, 84, 127
anglicata H.-S., *E. pusillata* subsp. 20, 22, 24, 28, 122, 142, 143

anglicata Mill. = *E. intricata* subsp. *millieraria* 67, 68
angusta, E. nanata subsp. 20, 23, 108
angustata = *E. nanata* subsp. *angusta* 23, 108
anthrax, P. rectangulata f. 66, 85, 87, 93, 116, 135
approximata Lempke, *E. dodoneata* ab. 119
approximata Lempke, *E. linariata* ab. 46
approximata Lempke, *E. pulchellata* ab. 47, 48
arceuthata Freyer, *E. intricata* subsp. 18, 22, 28, 67, 68, 71, 73, 144
assimilaria = *E. assimilata* 22, 81
assimilata, Eupithecia 21, 22, 28, 30, **67** (text fig. 7g), 69 (text fig. 80), 77, 79, **81** (Pl.2: 7, 8; Pl.7: 18, 20; Pl.B: 9), 84, 85, 89, 107, 149 (text fig. 15d), 156 (text fig. 22a), 162 (text fig. 28b), 166 (text fig. 31j), 224, 230, 233
athalia, Mellicta (Nymphalidae) 42
atropicta, E. vulgata f. 17, 19, 64, 83, 85, 87, 128
austeraria = *E. vulgata* 23, 83
austerata = *E. vulgata* 23, 83

bandanae, E. venosata fumosae ab. 58
basifasciata, E. lariciata ab. 127
basinigrata, E. venosata ab. 58
begrandaria = *E. plumbeolata* 21, 41
bicolor, E. nanata ab. 109
bifasciata = *basifasciata, E. lariciata* ab. 127
bimaculata, Lomographa (Geometridae) 53
bistrigata Lempke, *C. v-ata* ab. 131
bistrigata Lempke, *E. nanata* ab. 109
bistrigata Dietze, *E. satyrata* ab. 72
bistrigata Dietze, *E. subumbrata* ab. 98
bistrigata Lempke, *E. succenturiata* ab. 96
bistrigata (Haw.) = *G. rufifasciata* 24, 138
bistrigata Dietze, *P. rectangulata* ab. 135
blancheata = *E. subfuscata* 23, 91

257

brunnea, *E. simpliciata* ab. 101
brunneata, *P. rectangulata* ab. 135
bucovinata, *G. rufifasciata* ab. 139

caeca, *E. satyrata* ab. 73
callunaria, *E. satyrata* f. 19, 28, 70, 72, 73, 74, 75, 84
campanulata = *E. denotata* subsp. *denotata* 23, 88
Casinaria (Ichneumonidae) 32
castigaria = *E. subfuscata* 23, 91
castigata = *E. subfuscata* 23, 75, 91, 144
cauchiata, *Eupithecia* 22, 28, 33, 72, 73, **75** (Pl.2: 5; Pl.6: 15), 149 (text fig. 15b), 156 (text fig. 22e), 161 (text fig. 27r)
centaureata, *Eupithecia* 22, 25, 27, **62** (Pl.1: 25; Pl.4: 22; Pl.B: 2), 146 (text fig. 12e), 148 (text fig. 14e), 155 (text fig. 21f), 161 (text fig. 27o), 166 (text fig. 31c), 224, 226, 227, 228, 229, 230, 231, 232, 233, 234, 235
centaurearia = *E. centaureata* 22, 62
centralisata, *E. centaureata* ab. 63
chloerata, *Pasiphila* 16, 24, 28, **132** (Pl.3: 29; Pl.5: 3; Pl.D: 11), 133 (text fig. 11a), 136, 137, 153 (text fig. 19d), 160 (text fig. 26c), 163 (text fig. 29h), 168 (text fig. 33j), 225, 232
CHLOROCLYSTIS 11, **131** *et seq.*
cinerae, *E. tenuiata* f. 18, 33
circumflexa, *E. venosata* ab. 58
clarensis, *E. vulgata* f. 19, 28, 83
clusterata = *E. vulgata* 23, 83
coaequata, *E. tenuiata* ab. 33
cognata, *E. icterata* f. 19, 20, 27, 94, 96
concolor, *E. satyrata* ab. 73
confluens Dietze, *E. pusillata* f. 123
confluens Dietze, *E. venosata* ab. 58
conjuncta = *E. pusillata* ab. *confluens* 123
consignaria = *E. insigniata* 21, 52
consignata = *E. insigniata* 21, 52
constrictaria = *E. distinctaria* subsp. *constrictata* 23, 103
constrictata Guen., *E. distinctaria* subsp. 23, 28, 103
constrictata Prout, *E. abietaria* ab. 44
conterminata, *Eupithecia* 145
contrastata Dannehl, *E. satyrata* ab. 73
contrastata Lempke, *G. rufifasciata* f. 139
Copidosoma (Encyrtidae) 32
coronaria = *C. v-ata* 24, 131
coronata = *C. v-ata* 24, 131
Cotesia (Braconidae) 31
curzoni, *E. satyrata* f. 19, 22, 28, 70, 72, 73, 74, 75, 109, 142

cydoniata, *P. rectangulata* f. 135

debiliaria = *P. debiliata* 24, 137
debiliata, *Pasiphila* 24, 27, 133 (text figs 11b,c), 134, 136, **137** (Pl.3: 30; Pl.5: 4; Pl.D: 13), 153 (text fig. 19f), 160 (text fig. 26d), 163 (text fig. 29j), 168 (text fig. 33l), 225, 235
decussata = *E. venosata* 22, 57
defasciata, *E. pulchellata* ab. 47
denotata, *Eupithecia* 19, 23, 28, 61 (text fig. 5i), 64, 65 (text fig. 6e), 67 (text fig. 7h), 69 (text fig. 8f), 70, 78, 81, 84, 87, **88** (Pl.2: 18, 19; Pl.7: 21; Pl.8: 1; Pl.B: 13, 14), 107, 115, 119, 146 (text fig. 12g), 150 (text fig. 16a), 156 (text fig. 22c), 162 (text fig. 28e), 167 (text fig. 32a), 224, 227, 228, 230
denotata sensu Doubl. = *E. pimpinellata* 23, 106
dietzii, *E. icterata* ab. 94
disparata, *E. succenturiata* f. 20, 23, 96
distinctaria, *Eupithecia* 23, 30, 34, 35 (text figs 1: h,j), **103** (Pl.2: 32; Pl.7: 1; Pl.C: 7), 105, 126, 146 (text fig. 12h), 151 (text fig. 17a), 157 (text fig. 23e), 162 (text fig. 28m), 167 (text fig. 32h), 225, 231, 235
dodonearia = *E. dodoneata* 24, 119
dodoneata, *Eupithecia* 24, 28, 34, 35 (text fig. 1m), 38, 69 (text fig. 8s), 90, 92, 104, 105, 115, 117 (text fig. 9b), 118, **119** (Pl.3: 15, 16; Pl.6: 13; Pl.7: 4; Pl.8: 14; Pl.D: 2), 121, 152 (text fig. 18c), 158 (text fig. 24i), 163 (text fig. 29b), 225, 228, 232
Dusona (Ichneumonidae) 32

effusa, *P. rectangulata* ab. 135
egenaria, *Eupithecia* 22, 28, **59** (Pl.1: 24; Pl.8: 8; Pl.B1), 61 (text fig. 5a), 69 (text fig. 8l), 70, 73, 86, 89, 92, 114, 127, 146 (text fig. 12d), 148 (text fig. 14f), 155 (text fig. 21d), 161 (text fig. 27n), 166 (text fig. 31b), 168 (text fig. 33c), 224, 235
elongaria = *E. absinthiata* 22, 76
elongata = *E. absinthiata* 22, 76, 77
Ennominae 44
enuncleata, *E. plumbeolata* ab. 41
Epitriptus cowini 141
EUPITHECIA **33** *et seq.*
Eupitheciini 11, 16
exalbidata, *E. succenturiata* f. 96
excelsa, *E. icterata* ab. 94
exiguaria = *exiguata* 21, 50

exiguata, Eupithecia 18, 21, 27, **50** (Pl.1: 13, 14; Pl.4: 9, 10; Pl.A: 10), 146 (text fig. 12c), 147 (text fig. 13i), 155 (text fig. 21a), 161 (text fig. 27i), 165 (text fig. 30i), 224, 226, 227, 228, 229, 230, 231, 232, 233, 234
expallidaria = *E. expallidata* 22, 79
expallidata, Eupithecia 21, 22, 28, 67 (text fig. 7f), 77, **79** (Pl.2: 6; Pl.7: 19; Pl.B: 8), 81, 107, 116, 149 (text fig. 15f(i)), 156 (text fig. 22b), 162 (text fig. 28a), 166 (text fig. 31k), 224, 227, 234
explicata, *E. plumbeolata* ab. 41
expressaria, *E. pusillata* f. 123
extensaria, Eupithecia 23, 28, **110** (Pl.3: 7; Pl.4: 14; Pl.C: 12), 151 (text fig. 17f), 158 (text fig. 24f), 162 (text fig. 28r), 167 (text fig. 32n), 225, 227, 234
extrema, *E. succenturiata* f. 96

fagicolaria, *E. satyrata* f. 72
famelica, *E. distinctaria* f. 103
ferrearia, *E. lariciata* ab. 127
flaveolata, *E. plumbeolata* ab. 41
flavescens, *E. icterata* ab. 95
flavofasciata, *E. linariata* ab. 46
franconica, *E. irriguata* ab. 49
fraxinata, Eupithecia 17, 20, 23, 25, 27, 28, **112** (Pl.3: 8–10; Pl.4: 20, 21; Pl.C: 13, 14), 143, 144, 151 (text fig. 17g), 158 (text fig. 24g), 162 (text fig. 28s), 168 (text fig. 33a), 225, 227, 229, 230, 232, 233, 235
fumosae, *E. venosata* subsp. 17, 18, 22, 28, 57, 58
fuscosparsata, *E. tenuiata* ab. 33

gelidata, Eupithecia 145
Geometridae 44, 53
globulariata = *G. rufifasciata* 138
goodsoni, *E. icterata* f. 94
goossensiata, *E. absinthiata* f. 17, 19, 22, 25, 27, 76, 78, 79, 110, 142
grabei, *E. pygmaeata* ab. 55
griseata, *E. satyrata* ab. 73
grisescens Dietze, *E. assimilata* ab. 81
grisescens Petersen, *E. fraxinata* ab. 112
grisescens Lempke, *E. icterata* f. 20, 94
grisescens Dietze, *P. debiliata* ab. 137
grisescens Lempke, *P. rectangulata* ab. 135
guttata, *E. pulchellata* ab. 47
GYMNOSCELIS 11, **138** et seq.

hadenata, *P. chloerata* ab. 133
haworthiaria = *E. haworthiata* 21, 39

haworthiata, Eupithecia 21, 28, 33, 34, 35 (text fig. 1e), 37, **39** (Pl.1: 5; Pl.7: 11; Pl.A: 3), 41, 54, 56, 147 (text fig. 13c), 154 (text fig. 20c), 161 (text fig. 27c), 165 (text fig. 30c), 224, 228
hebridensis, *E. venosata* f. 18, 28, 57, 58
hebudium, *E. pulchellata* f. 18, 28, 47, 48
helveticaria = *E. intricata* subsp. *millieraria* 2, 67, 68
helveticata, Eupithecia 68
hibernica, *E. intricata* subsp. 18, 22, 28, 67, 68, 70, 71, 73
hirschkei, *E. abbreviata* f. 20, 117
hyperboreata, *E. gelidata* subsp. 145

iberica, *E. pulchellata* ab. 48
icterata, Eupithecia 19, 23, **94** (Pl.2: 23–25; Pl.4: 6–8; Pl.C: 1), 96, 150 (text figs 16c, 1(ii)), 157 (text fig. 23a(ii)), 162 (text fig. 28g), 167 (text fig. 32c), 225, 226, 227, 229, 234, 235
immundata, Eupithecia 145
impuncta Lempke, *E. icterata* ab. 95
impuncta Lempke, *E. pusillata* f. 123
impuncta Lempke, *E. subumbrata* f. 98
impuncta Lempke, *E. vulgata* ab. 83
incertata, *G. rufifasciata* f. 139
indigaria = *E. indigata* 23, 105
indigata, Eupithecia 23, 28, 33, 35 ((text figs 1f,i), 37, 39, 41, 54, 69 (text fig. 8m), 84, 104, **105** (Pl.3: 1; Pl.7: 3; Pl.C: 8), 126, 151 (text fig. 17c), 158 (text fig. 24b), 162 (text fig.28o), 167 (text fig. 32i), 225, 229, 230, 232
innotata Hufn., *E.* 25, 27, 28, 143
innotata Wood = *E. absinthiata* 22, 76
innotata sensu Stephens = *E. fraxinata* 112
insigniata, Eupithecia 21, 27, **52** (Pl.1: 15; Pl.4: 23; Pl.A: 11), 146 (text fig. 12b), 148 (text fig. 14a), 155 (text fig. 21b), 161 (text fig. 27j), 165 (text figs 30k,m), 224, 228, 231, 232, 233
intermedia Dietze, *E. icterata* f. 94
intermedia Lempke, *E. tripunctaria* ab. 86
intricata, Eupithecia 18, 61(text fig. 5d), **67** (Pl.1: 29–31; Pl.6: 18; Pl.8: 6, 16; Pl.B: 4), 69 (text fig. 8b), 73, 89, 92, 100, 115, 128, 149 (text fig. 15a), 155 (text fig. 21i), 161 (text fig. 27q), 166 (text fig. 31g), 224, 228, 229, 230, 232, 235
inturbata, Eupithecia 21, 28, 30, 34, 35 (text fig.1a), **37** (Pl.1: 4; Pl.7: 10; Pl.A: 2), 39, 41, 54, 56, 147 (text fig. 13b), 154 (text fig. 20e), 161 (text fig. 27b), 165 (text fig. 30b), 224, 226

irriguaria = *E. irriguata* 21, 49
irriguata, Eupithecia 21, 27, **49** (Pl.1: 12; Pl.2: 10; Pl.A: 9) (text fig. 4c), 98, 130, 147 (text fig. 13d), 154 (text fig. 20d), 161 (text fig. 27h), 165 (text fig. 30g), 224, 227, 233
isogrammaria = *E. haworthiata* 21, 39

joannista, P. rectangulata f. 135
jasioneata, E. denotata subsp. 16, 19, 23, 28, 60, 73, 84, 88, 89, 90, 91, 128
johnsoni, E. tenuiata f. 18, 34
jubata, Alcis (Geometridae) 43 (text fig. 2b), 44

knautiata = *E. absinthiata* 22, 76, 141, 142

laevigata = *E. pusillata* 24, 122
lanceata, Eupithecia 145
lanceolaria Wood = *E. exiguata* 21, 50
lariciata, Eupithecia 20, 24, 28, 60, 61 (text fig. 5k), 69 (text fig. 8d), 70, 74, 87, 90, 93, 115, 118, 123, **127** (Pl.3: 23, 24; Pl.6: 8; Pl.8: 4; Pl.D: 6), 152 (text fig. 18d), 159 (text fig. 25c), 163 (text fig. 29e), 168 (text fig. 33g), 225, 230, 231
limbofasciaria, E. indigata ab. 105
limbopunctata, E. satyrata ab. 73
limbosignata, E. pimpinellata ab. 106
Lime-Speck, The – *see* pug, lime-speck 25
linaria = *E. linariata* 21, 45
linariaria = *E. linariata* 21, 45
linariata, Eupithecia 21, 27, **45** (Pl.1: 9; Pl.5: 9; Pl.A: 6) (text fig. 3a), 47, 48, 147 (text fig.13g), 154 (text fig. 20g), 165 (text fig. 30f), 224, 226, 231
lividata, E. plumbeolata ab. 41
luneburgensis, E. pusillata f. 123
luxuriosa, E. lariciata ab. 127

malaisei, E. succenturiata ab. 96
massiliata, Eupithecia 24, 28, 33, **120** (Pl.3: 25), 232, 233
mediofaciata, E. nanata f. 109
mediofasciata Lempke, *E. absinthiata* f. *goossensiata* ab. 77
mediofasciata Dietze, *E. intricata* subsp. *arceuthata* ab. 68
mediofasciata Dietze, *E. lariciata* ab. 127
mediofasciata Dietze, *P. debiliata* ab. 137
medionotata, E. satyrata ab. 72
mediopallens Dietze, *E. lariciata* ab. 127
mediopallens Silbern., *E. tantillaria* ab. 130
mediopallens Plant, *G. rufifasciata* ab. 139
mediospolinata, P. rectangulata ab. 135

melaena, E. icterata ab. 95
millefoliata, Eupithecia 23, 28, 61 (text fig. 5h), 67 (text fig. 7b), 70, 74, **99** (Pl.2: 22; Pl.8: 19; Pl.C: 4), 107, 150 (text fig. 16e), 157 (text fig. 23b), 162 (text fig. 28k), 167 (text fig.32f), 225, 226
millieraria, E. intricata subsp. 18, 22, 28, 67, 70, 71, 73, 84
minutaria = *E. absinthiata* 22, 76
minutata = *E. absinthiata* 22, 76, 141
muricolor, E. exiguata f. 18, 28, 51

nanata, Eupithecia 20, 27, 30, 61 (text fig. 5j), 72, 74, **108** (Pl.3: 5, 6; Pl.4: 19; Pl.6: 16; Pl.C: 10, 11), 113, 142, 151 (text fig.17e), 158 (text fig. 24d), 162 (text fig. 28q), 167 (text fig. 32k), 225, 227, 229, 230, 231, 232, 235
nebulata = *E. abbreviata* 24, 117
nigra Cock., *E. abbreviata* f. 20, 65, 85, 87, 92, 115, 117, 118
nigra Prout, *E. lariciata* f. 20, 65, 85, 127, 128
nigra Cock., *E. satyrata* ab. 73
nigra Lempke, *E. virgaureata* f. 20, 84, 114
nigricata, E. tantillaria ab. 130
nigrofasciata Dietze, *E. linariata* ab. 46
nigrofasciata Dietze, *E. nanata* f. 109
nigrofasciata Dietze, *E. pusillata* ab. 123
nigrofasciata Dietze, *E. satyrata* ab. 72
nigrofasciata Dietze, *E. virgaureata* ab. 114
nigrofasciata Dietze, *E. vulgata* ab. 83
nigrofasciata Dietze, *G. rufifasciata* ab. 17, 139
nigrofasciata Dietze, *P. chloerata* ab. 133
nigronotata, E. virgaureata ab. 114
nigropunctata, P. debiliata f. 24, 137
nigrosericeata, P. rectangulata f. 20, 24, 135, 144
nigrostriata, G. rufifasciata f. 139
niveipicta, E. tenuiata ab. 33
notata Dietze, *E. virgaureata* ab. 114
notata Stephens = *E. absinthiata* 22, 76, 77
Nymphalidae 42

oblongata = *E. centaureata* 22, 62
obrutaria, E. subumbrata f. 98
obscura Cock., *E. absinthiata* ab. 77
obscura Dietze, *E. absinthiata* f. *goossensiata* ab. 77
obscura Dietze, *E. centaureata* f. 63
obscura Dietze, *E. subfuscata* f. 19, 92
obscurata Lempke, *E. subumbrata* f. 98
obscurata Lempke, *E. succenturiata* f. 20, 96

obscurevirescens, P. debiliata ab. 137
obscurissima, E. subfuscata f. 19, 84, 91, 92, 93, 144
obsolescens, G. rufifasciata ab. 139
occidua, E. extensaria subsp. 23, 28, 110
ochracae, E. venosata subsp. 18, 22, 28, 57, 58
ochraceata, E. denotata denotata ab. 89
ochrea, P. rectangulata f. 135
ochreata = E. exiguata 21, 50, 51
oliveri, E. nanata angusta f. 20, 109
oxycedrata, Eupithecia 124, 141
oxydata = E. icterata f. *cognata* 20, 94

pallescens, E. linariata ab. 46
pallida Lempke, *E. tantillaria* ab. 130
pallida Schwin., *E. expallidata* ab. 79
palustraria = E. pygmaeata 22, 55, 141
parvularia, G. rufifasciata f. 139
PASIPHILA 11, **132** *et seq.*
paupera, E. fraxinata ab. 112
pauxillaria = E. nanata f. 109
pernotaria = E. cauchiata 22, 75
pernotata = E. cauchiata 22, 75
peyerimhoffata = E. massiliata 24, 120
phoeniceata, Eupithecia 24, 28, **124** (Pl.3: 17; Pl.4: 13; Pl.D: 4), 141, 146 (text fig. 12k), 152 (text fig. 18b), 159 (text fig. 25b), 163 (text fig. 29d), 168 (text fig. 33f), 225, 228, 229, 230
piceata = E. tantillaria 24, 129, 130
pilcheri, P. rectangulata ab. 135
pimpinellaria = E. pimpinellata 23, 106
pimpinellata Hübn., *Eupithecia* 23, 28, 61 (text fig. 5e), 67 (text fig. 7d), 69 (text fig. 8a), 78, 79, 81, 90, 100, 101, **106** (Pl.3: 2–4; Pl.7: 24; Pl.8: 7, 20; Pl.C: 9), 151 (text fig. 17d), 158 (text fig. 24c, 162 (text fig. 28p), 167 (text fig. 32j), 225, 226, 227, 232, 234
pimpinellata sensu Doubl. = *E. virgaureata* 24, 114
pini = E. abietaria 21, 43
piperaria = E. subumbrata 23, 98
piperata = E. subumbrata 23, 98
piperitata = E. subumbrata 23, 98
Platylabus (Ichneumonidae) 32
plumbalbeolata, E. plumbeolata ab. 41
plumbea, E. venosata subsp. 18, 22, 28, 58
plumbeolaria = E. plumbeolata 21, 41
plumbeolata, Eupithecia 21, 27, 30, 33, 34, 35 (text fig. 1d), 37, 39, **41** (Pl.1: 6; Pl.7: 12; Pl.A: 4), 54, 56, 147 (text fig. 13e), 154 (text fig. 20b), 161 (text fig. 27d) 165 (text fig. 30d), 224, 228, 229, 231, 233

praeruptata, E. linariata ab. 46
Pretty Widow, The – *see* pug, netted 25, 57
privata, E. tripunctaria ab. 86
prolongata sensu Dietze = *E. extensaria* subsp. *occidua* 23, 110
pseudoabsinthiata, E. expallidata f. 79
pseudozibellinata, E. pygmaeata f. 56
pug, angelica = white-spotted pug 86
pug, angle-barred 143
pug, apple = green pug 135
pug, ash 20, 23, 27, 28, **112**, 143
pug, ash-tree = ash pug 112
pug, barberry = mottled pug 50
pug, beautiful = toadflax pug 45
pug, beech = satyr pug 72
pug, bell-flower = campanula pug 88
pug, bilberry 24, 27, **137**
pug, black-streaked = cypress pug
pug, black silk 135, 144
pug, bleached 22, 28, **79**
pug, bordered 20, 23, 27, **96**
pug, bordered lime-speck = bordered pug 96
pug, brindled (1) 20, 24, 27, **117**
pug, brindled (2) = grey pug 91
pug, brown grey = grey pug 91
pug, burnet = pimpinel pug 106
pug, campanula 19, 24, 28, **88**
pug, chalkhill = Freyer's pug 67
pug, Channel Islands 24, 28, **126**, 143
pug, chickweed = marsh pug 55
pug, clematis = Haworth's pug 39
pug, cloaked 21, 27, **43**
pug, coast = angle-barred pug 143
pug, common 19, 23, 27, 28, **83**
pug, Cornish tamarisk = tamarisk pug 143
pug, coronet = V-pug 131
pug, cow-wheat = lead-coloured pug 41
pug, currant 22, 28, **81**
pug, Curzon's 142
pug, cypress (1) 24, 28, **124**, 141
pug, cypress (2) = Freyer's pug
pug, dentated 11
pug, Devon = bilberry pug 137
pug, double-striped 24, 27, **138**
pug, Doubleday's 22, 28, **75**
pug, dwarf 24, 28, **129**
pug, Edinburgh 18, 22, 28, **67**
pug, Epping 24, 28, **120**
pug, Fletcher's = pauper pug 22, **59**
pug, foxglove 18, 21, 27, 28, **47**
pug, Freyer's 18, 21, 28, **67**, 144
pug, garden = double-striped pug 138
pug, golden-rod 20, 24, 28, **114**

pug, goosefoot (1) 23, 28, **102**
pug, goosefoot (2) = plain pug 101
pug, green 20, 24, 27, **135**
pug, grey 19, 23, 27, **91**
pug, Guenée's = Doubleday's pug 75
pug, Guernsey = Channel Islands pug 126
pug, hawkbit = shaded pug 98
pug, Haworth's 21, 28, **39**
pug, hawthorn = mottled pug 50
pug, Hunstanton = scarce pug 110
pug, Isle of Man 141
pug, jasione 19, 23, 28, **88**, 89
pug, juniper 20, 24, 28, **122**, 142
pug, Kentish tamarisk 20, 24, 28, **122**, 142
pug, larch 20, 24, 28, **127**
pug, large = cloaked pug 43
pug, large oak = brindled pug 117
pug, lead-coloured 21, 27, **41**
pug, lime-speck 22, 27, **62**
pug, ling 19, 22, 27, **76**, 142
pug, little = bilberry pug 137
pug, long-winged (1) = angle-barred pug 143
pug, long-winged (2) = goldenrod pug 114
pug, long-winged (3) = wormwood pug 77
pug, maple 21, 28, **37**
pug, marbled 21, 27, **49**
pug, marsh 22, 28, **55**, 141
pug, Mere's 18, 22, 28, **67**
pug, milfoil = yarrow pug 99
pug, moorland = satyr pug 72
pug, mottled 18, 27, 28, **50**
pug, narrow-winged 20, 23, 27, **108**
pug, netted 18, 22, 25, 27, **57**
pug, northern juniper = Edinburgh pug 67
pug, oak-tree 24, 28, **119**
pug, ochreous 23, 28, **105**
pug, Paisley = grey pug 92, 144
pug, pauper 22, **59**
pug, Pelham-Clinton's = sloe pug 132
pug, pimpinel 23, 28, **106**
pug, pine = ochreous pug 105
pug, pinion-spotted 21, 27, **52**
pug, plain 23, 27, **101**
pug, pretty = foxglove pug 47
pug, red-barred = double-striped pug 138
pug, ribbed = thyme pug 103
pug, sallow = slender pug 33
pug, saltern = scarce pug 110
pug, satin 135, 144
pug, satyr 19, 22, 28, **72**
pug, scabious 22, 141
pug, scarce 23, 28, **110**
pug, shaded 23, 27, **98**

pug, short = brindled pug 117
pug, slender 18, 21, 28, **33**
pug, sloe 24, 28, **132**
pug, small brindled = oak-tree pug 119
pug, small grey = dwarf pug 129
pug, small oak = oak-tree pug 119
pug, speckled (1) = shaded pug 86
pug, speckled (2) = white-spotted pug 53
pug, spotted = pinion-spotted pug 52
pug, spruce = dwarf pug 129
pug, tamarisk 143
pug, tawny = tawny-speckled pug 94
pug, tawny-speck = tawny-speckled pug 94
pug, tawny-speckled 19, 23, 27, **94**
pug, thyme 28, **103**
pug, toadflax 21, 27, **45**
pug, triple-spotted 18, 21, 28, **64**
pug, unspotted = ash pug 112
pug, V- 24, 27, **131**
pug, valerian 22, 28, **53**
pug, Welsh = marbled pug 49
pug, white-spotted 19, 23, 28, **86**
pug, wild-thyme = thyme pug 103
pug, wormwood 19, 22, 27, **76**
pug, yarrow 23, 28, **99**
pulchellaria = *E. pulchellata* 21, 47
pulchellata, Eupithecia 18, 21, 27, 30, 45 (text fig. 3b), 46, **47** (Pl.1: 10, 11; Pl.5: 6–8; Pl.A: 7, 8), 147 (text fig. 13h), 154 (text fig. 20h), 161 (text fig. 27g), 165 (text fig. 30h), 224, 229
pumilata = *G. rufifasciata* 24, 138
puncta, G. rufifasciata ab. 139
punctata Cock., *E. linariata* ab. 46
punctata Hann., *E. centaureata* ab. 63
pusillata, Eupithecia 20, 23, 27, 28, 31, 65 (text fig. 6f), 69 (text fig. 8p), 85, 87, 92, **122** (Pl.3: 19–22; Pl.5: 14, 15; Pl.6: 4; Pl.D: 3), (text fig. 10), 128, 129, 143, 146 (text fig. 12i), 152 (text fig.18e), 159 (text fig. 25d), 163 (text fig. 29c), 168 (text figs 33d,e), 225, 229, 230, 235
pusillata sensu auctt. = *E. tantillaria* 24, 129
pygmaearia = *E. pygmaeata* 22, 25
pygmaeata, Eupithecia 22, 28, 34, 35 (text fig. 1g), 37, 39, 41, 54, **55** (Pl.1: 17, 18; Pl.7: 13, 14); Pl.A: 13), 141, 148 (text fig. 14c), 155 (text fig. 21c), 161 (text fig. 27l), 165 (text fig. 30l), 224, 228, 234

recictaria = *G. rufifasciata* 138
rectangularia = *P. rectangulata* 24, 135
rectangulata, Pasiphila 16, 20, 24, 27, 29,

65 (text fig. 6h), 69 (text fig. 8q), 132, 133 (text figs 11d–f), **135** (Pl.3: 31–34; Pl.5: 1, 2; Pl.6: 10, 11; Pl.D: 11, 12), 137, 144, 153 (text fig. 19e), 160 (text fig. 26a), 163 (text fig. 29i), 168 (text fig. 33k), 225, 229, 231, 232
reducta Bastelb., *E. pulchellata* ab. 47
reducta Folkin, *E. linariata* ab. 46
rittichi, *E. pusillata* ab. 123
rotundata, *E. fraxinata* ab. 112
rufifasciata, *Gymnoscelis* 24, 27, 29, **138** (Pl.3: 35, 36; Pl.4: 16, 17; Pl. D: 14, 15), 146 (text fig. 12n), 153 (text fig. 19b), 160 (text fig. 26d), 163 (text fig. 29k), 168 (text fig. 33m), 225, 226, 227, 228, 229, 230, 231, 232, 233, 234, 235

satyraria = *E. satyrata* 22, 72
satyrata, *Eupithecia* 19, 22, 28, 60, 61 (text fig. 5g), 68, 69 (text fig. 8j), **72** (Pl.2: 1–4; Pl.6: 14, 17; Pl.8: 9, 18; Pl.B: 5), 75, 76, 87, 89, 92, 100, 115, 116, 128, 144, 149 (text fig. 15c), 155 (text fig. 21h), 161 (text fig. 27s), 166 (text figs 31e,f), 224, 226, 227, 228, 229, 230, 231, 232, 233, 234, 235
scabiosata = *E. subumbrata* 23, 98
scotica Cock., *E. vulgata* f. 17, 19, 28, 83
scotica Dietze, *E. pusillata* f. 20, 24, 122
selinata, *Eupithecia* 145
sepiata Curtis, *E. tripunctaria* ab. 86
sepiata Huggins, *E. venosata* ab. 58
sericeata = *P. rectangulata* f. *anthrax* 24, 135, 144
simpliciata, *Eupithecia* 23, 27, 29, 61 (text fig.5c), 67 (text fig. 7c), **101** (Pl.2: 30; Pl.8: 21, Pl.C: 5), 107, 150 (text fig. 16f), 157 (text fig. 23c), 162 (text fig. 28l), 167 (text fig. 32g), 225, 227, 228
singularia, *E. plumbeolata* ab. 41
singulariata = *E. subfuscata* 23, 91
sinuosaria, *Eupithecia* 23, 28, 33, **102** (Pl.2: 31; Pl.4: 18; Pl.C: 6), 145, 150, 157 (text fig. 23d), 162 (text fig. 28j), 167
sobrinaria = *E. pusillata* 24, 122
sobrinata = *E. pusillata* 24, 122
solidaginis, *E. denotata* subsp. *denotata* ab. 89
sparsata, *Anticollix* (Geometridae) 11
spissilineata, *E.* 41
stevensata = *E. pusillata* subsp. *anglicata* 24, 122, 142, 143
strandi, *E. satyrata* f. *callunaria* ab. 73
striata, *E. abbreviata* ab. 117
strigata, *E. lariciata* ab. 127

strobilata sensu Stephens = *G. rufifasciata* 138
strobilata Borkh. = *E. abietaria* 21, 43
subaerata = *P. rectangulata* 24, 135
subatrata = *E. satyrata* ab. 73, 75
subciliaria = *E. inturbata* 21, 37
subciliata = *E. inturbata* 21, 37
subfasciata = *E. abbreviata* 24, 117
subfulvata, *E. icterata* f. 19, 20, 27, 94
subfuscata Haw., *Eupithecia* 19, 23, 27, 60, 61 (text fig. 5l), 65 (text fig. 6d), 69 (text fig. 8g), 70, 73, 75, 84, 87, 90, **91** (Pl.2: 20, 21; Pl.6: 2; Pl.8: 3; Pl.B: 15), 115 (text fig. b), 118, 119, 123, 128, 150 (text fig. 16b), 156 (text fig. 22f), 162 (text fig. 28f), 167 (text fig. 32b), 224, 226, 227, 228, 229, 230, 231, 232, 233, 234, 235
subnotaria = *E. simpliciata* 23, 101
subnotata = *E. simpliciata* 23, 101
subumbraria = *E. subumbrata* 23, 98
subumbrata D. & S., *Eupithecia* 23, 27, 47 (text fig. 4a), **98** (Pl.2: 28, 29; Pl.5: 13, Pl.8: 13; Pl.C: 3), 150 (text fig.16d), 156 (text fig. 22n), 162 (text fig. 28i), 167 (text fig. 32e), 225, 227, 228, 229, 230, 231, 232, 233, 234
subumbrata sensu Hübn. = *E. tantillaria* 24
succenturiaria = *E. succenturiata* 96
succentaurearia = *E. succenturiata* 23
succenturiata, *Eupithecia* 20, 23, 27, 94, 95, **96** (Pl.2: 26, 27; Pl.4: 11, 12; Pl.C: 2), 150 (text fig. 16c(i)) 157 (text fig. 23a(i)) 162 (text fig. 28h), 167 (text fig. 32d), 225, 226, 227, 228, 229, 234, 235
suffusa, *E. intricata* subsp. *millieraria* ab. 68
suspectata, *E. fraxinata* ab. 112

tamarisciata = *E. fraxinata* 25, 112, 113, 143, 144
tantillaria, *Eupithecia* 24, 28, 49 (text fig. 4b), **129** (Pl.3: 26; Pl.5: 11; Pl.D: 7), 146 (text fig. 12l), 153 (text fig. 19a), 159 (text fig. 25a), 163 (text fig. 29f), 168 (text fig. 33h), 225, 226, 228, 230, 231, 232, 235
tempestivata, *G. rufifasciata* f. 139
tenebrata, *G. rufifasciata* ab. 139
tenuiaria = *E. tenuiata* 21, 33
tenuiata, *Eupithecia* 18, 21, 28, 29, **33** (Pl.1: 1–3; Pl.7: 5–7; Pl.A: 1), 35 (text fig. 1b), 37, 39, 41, 54, 56, 119, 126, 147 (text fig. 13a), 154 (text fig. 20a), 161 (text fig. 27a), 165 (text fig. 30a), 224, 233

tenuita = *E. tenuiata* 21
togaria = *E. abietaria* 21, 43
togata = *E. abietaria* 43
transversa, *E. satyrata* ab. 72
trifasciata, *E. satyrata curzoni* ab. 73
trilineata, *E. satyrata* ab. 73
trimaculata = *E. exiguata* 21, 50
tripunctaria, *Eupithecia* 19, 23, 28, 29, 64, 65 (text fig. 6c), 66, 67 (text fig. 7i), 69 (text fig. 8h), 70, 74, 75, 81, 84, **86** (Pl.2: 16, 17; Pl.6: 7; Pl.8: 5; Pl.B 12), 89, 92, 114, 123, 128. 144, 149 (text fig. 15g), 156 (text fig. 22g), 166 (text figs 31m,n), 224, 226, 228, 229, 230, 231, 233, 234
trisignaria, *Eupithecia* 18, 22, 28, **64** (Pl.1: 26–28; Pl.6: 1; Pl. 7: 17; Pl.8: 12; Pl.B: 3), 65 (text fig. 6a), 67 (text fig. 7j), 69 (text fig. 8k), 77, 81, 84, 86, 89, 92, 114, 118, 123, 144, 146 (text fig. 12f), 148 (text fig.14g), 155 (text fig. 21g), 161 (text fig. 27p), 162 (text fig. 28d), 166 (text fig. 31d), 224, 226, 230, 231, 232
trisignata = *E. trisignaria* 64
tristrigata, *E. indigata* ab. 105

ultimaria Boisd., *Eupithecia* 24, 28, 34, 35 (text fig. 1l), 38, 104, 105, **126** (Pl.3: 18; Pl.7: 2; Pl.D: 5), 142, 151 (text fig. 17b), 158 (text fig. 24a), 162 (text fig. 28n)
ultimaria Dup. = *E. pusillata* subsp. *anglicata* 24, 122
undosata = *E. egenaria* 22, 59
unedonata, *Eupithecia* 145
unicolor Lempke, *E. vulgata* f. 19, 83
unicolor Prout, *E. fraxinata* f. 20, 112
uniformis Dietze, *E. lariciata* ab. 127
uniformis Dietze, *E. millefoliata* ab. 100
uralensis, *E. plumbeolata* ab. 41

v–ata, *Chloroclystis* 24, 27, **131** (Pl.3: 27, 28; Pl.4: 15; Pl.5: 5; Pl.D: 8, 9), 146 (text fig. 12m), 153 (text fig. 19c), 159 (text fig. 25e), 163 (text fig. 29g), 225, 226, 227, 228, 229, 230, 231, 232, 233, 234, 235
valerianata, *Eupithecia* 22, 28, 34, 35 (text fig. 1c), 37, 39, 41, **53** (Pl.1: 16; Pl.7: 8; Pl.A: 12), 56, 69 (text fig. 8i), 84, 148 (text fig. 14b), 154 (text fig. 20i), 161 (text fig. 27k), 165 (text fig. 30j), 224, 228, 235
variegata = *E. irriguata* 21, 49
venosaria = *E. venosata* 22, 57
venosata, *Eupithecia* 18, 22, 25, 27, **57** (Pl.1: 19–23; Pl.4: 1–5; Pl.A: 14, 15), 148 (text fig. 14d), 155 (text fig. 21e), 161 (text fig. 27m), 166 (text fig. 31a), 224, 234
viminaria = *E. valerianata* 22, 53
viminata = *E. valerianata* 22, 53, 54
virgata, *E. lariciata* ab. 127
virgaurearia = *E. virgaureata* 24, 114
virgaureata, *Eupithecia* 20, 24, 28, 60, 61 (text fig. 5l), 65 (text fig. 6d), 69 (text fig. 8g), 70, 73, 84, 87, 90, 92, **114** (Pl.3: 11, 12; Pl.6: 3; Pl.8: 2; Pl.C: 15) 114, 115 (text fig. a), 118, 119, 123, 128, 144, 151 (text fig. 17h), 158 (text fig. 24e), 162 (text fig. 28l), 167 (text figs 32l,m), 225, 226, 227, 228, 229, 230, 231, 232, 233, 234
viridulata = *P. rectangulata* 24, 135
vulgaria = *E. vulgata* 23, 83
vulgata, *Eupithecia* 17, 19, 23, 27, 29, 35 (text fig. 1k), 54, 61 (text fig. 5b), 67 (text fig. 7e), 69 (text fig. 8e), 70, 73, 77, 81, **83** (Pl.2: 9–15; Pl.5: 12; Pl.6: 5; Pl.7: 9, 22, 23; Pl.8: 10, 11, 17; Pl.B: 10, 11), 87, 89, 92, 114, 118, 123, 128, 149 (text fig. 15e), 156 (text fig. 22d), 162 (text fig. 28c), 166 (text fig. 31l), 224, 226, 228, 230, 232, 233, 234, 235

Xanthia (Noctuidae) 36

Zele (Braconidae) 31
zibellinata = *E. pygmaeata* f. 56